Synthesis Series in Digital Circuits and Systems

Volume 1

Microcontrollers Fundamentals for Engineers and Scientists
Steven F. Barrett, Daniel J. Pack
2006

High-Speed Digital System Design
Justin Davis
2006

Introduction to Logic Synthesis using Verilog HDL
Robert B. Reese, Mitchell A. Thornton
2006

Finite State Machine Datapath Design, Optimization and Implementation
Justin Davis, Robert B. Reese
2008

Volume 2

Pragmatic Circuits : DC and Time Domain
William J. Eccles
2006

Pragmatic Circuits: Frequency Domain
William J. Eccles
2006

Pragmatic Circuits: Signals and Filters
William J. Eccles
2006

Pragmatic Power
William J. Eccles
2008

Volume 3

Pragmatic Logic

William J. Eccles

2007

PSpice® for Circuit Theory and Electronic Devices

Paul Tobin

2007

PSpice® for Digital Communications Engineering

Paul Tobin

2007

Volume 4

PSpice® for Analog Communications Engineering

Paul Tobin

2007

PSpice® for Filters and Transmission Lines

Paul Tobin

2007

PSpice® for Digital Signal Processing

Paul Tobin

2007

Volume 5

Atmel AVR Microcontroller Primer: Programming and Interfacing

Steven F. Barrett, Daniel J. Pack

2008

MultipleValued Logic: Concepts and Representations

D. Michael Miller, Mitchell A. Thornton

2008

An Introduction to Logic Circuit Testing

Parag K. Lala

2009

Volume 6
Asynchronous Sequential Machine Design and Analysis
Richard F. Tinder
2009
Mismatch and Noise in Modern IC Processes
Andrew Marshall
2009
Developing Embedded Software using DaVinci and OMAP Technology
B.I. (Raj) Pawate
2009

Volume 7
Embedded Systems Interfacing for Engineers using the Freescale HCS08 Microcontroller: Part I Assembly Language Programming
Douglas Summerville
2009
Embedded Systems Interfacing for Engineers using the Freescale HCS08 Microcontroller: Part II Digital and Analog Hardware Interfacing
Douglas Summerville
2009
Designing Asynchronous Circuits using NULL Convention Logic (NCL)
Scott C. Smith, Jia Di
2009

Synthesis Series in Digital Circuits and Systems 1

Microcontrollers Fundamentals for Engineers and Scientists
Steven F. Barrett, Daniel J. Pack

High-Speed Digital System Design
Justin Davis

Introduction to Logic Synthesis using Verilog HDL
Robert B. Reese, Mitchell A. Thornton

Finite State Machine Datapath Design, Optimization and Implementation
Justin Davis, Robert B. Reese

ISBN: 9781608453108

www.morganclaypool.com

A Publication in the Morgan & Claypool Publishers

SYNTHESIS SERIES IN DIGITAL CIRCUITS AND SYSTEMS

Series Editor: Mitchell A. Thornton

Synthesis Series in Digital Circuits and Systems 1

This volume includes:

Microcontrollers Fundamentals for Engineers and Scientists

Steven F. Barrett, Daniel J. Pack

High-Speed Digital System Design

Justin Davis

Introduction to Logic Synthesis using Verilog HDL

Robert B. Reese, Mitchell A. Thornton

Finite State Machine Datapath Design, Optimization and Implementation

Justin Davis, Robert B. Reese

Microcontrollers Fundamentals for Engineers and Scientists

Steven F. Barrett
Department of Electrical and Computer Engineering,
University of Wyoming, Laramie, Wyoming USA

Daniel J. Pack
Department of Electrical Engineering,
United State Air Force Academy, Colorado Springs, Colorado, USA

MORGAN & CLAYPOOL PUBLISHERS

ABSTRACT

This book provides practicing scientists and engineers a tutorial on the fundamental concepts and use of microcontrollers. Today, microcontrollers, or single integrated circuit (chip) computers, play critical roles in almost all instrumentation and control systems. Most existing books are written for undergraduate and graduate students taking an electrical and/or computer engineering course. Furthermore, these texts have been written with a particular model of microcontroller as the target discussion. These textbooks also require a requisite knowledge of digital design fundamentals. This textbook presents the fundamental concepts common to all microcontrollers. Our goals are to present the over-arching theory of microcontroller operation and to provide a detailed discussion on constituent subsystems available in most microcontrollers. With such goals, we envision that the theory discussed in this book can be readily applied to a wide variety of microcontroller technologies, allowing practicing scientists and engineers to become acquainted with basic concepts prior to beginning a design involving a specific microcontroller. We have found that the fundamental principles of a given microcontroller are easily transferred to other controllers. Although this is a relatively small book, it is packed with useful information for quickly coming up to speed on microcontroller concepts.

KEYWORDS

Microcontrollers, embedded systems design, systems-on-chip technology, control, digital design, computer engineering, digital design

Contents

Preface

The purpose of this text, "Microcontrollers Fundamentals for Engineers and Scientists," is to provide practicing scientists and engineers a tutorial on the fundamental concepts and use of microcontrollers. Today, microcontrollers, or single integrated circuit (chip) computers, play critical roles in almost all instrumentation and control systems. There are a number of books that explore the fascinating world of microcontroller theory and applications. However, most of these are geared toward undergraduate and graduate students taking an electrical and/or computer engineering course. Furthermore, these texts have been written with a particular model of microcontroller as the target discussion. These textbooks also require a requisite knowledge of digital design fundamentals.

In this textbook we present the fundamental concepts common to all microcontrollers. Our goals for writing this book are to present the over-arching theory of microcontroller operation and to provide a detailed discussion on constituent subsystems available in most microcontrollers. With such goals, we envision that the theory discussed in this book can be readily applied to a wide variety of microcontroller technologies, allowing practicing scientists and engineers to become acquainted with basic concepts prior to beginning a design involving a specific microcontroller. Each of us have used a wide variety of microcontrollers from various manufacturers. We have found that the fundamental principles of a given microcontroller are easily transferred to other controllers. Although this is a relatively small textbook, it is packed with useful information in quickly coming up to speed on microcontroller concepts.

FLOW OF THE BOOK

In Chapter 1 we begin with a basic review of digital design fundamentals. Chapter 2 provides a brief review of the design and development process and discusses some of the common tools and procedures used to implement a working design. Chapter 3 provides an overview of a microcontroller and its constituent subsystems. We also provide practical advice about how to select a specific microcontroller for a specific application. In Chapters 4 and 5 we provide additional information on the timing and analog-to-digital subsystems. Chapter 6 describes the concepts required to link multiple interacting microcontrollers together in a network. This is commonplace in the entertainment and automotive industries. Chapter 7 discusses some of the real world practicalities of interfacing microcontrollers to external components.

ACKNOWLEDGMENTS

Space does not permit us to thank everyone who has provided encouragement along the way. We would like to thank Joel Claypool and John Enderle for inviting us to participate in their efforts to develop a series of short tutorial textbooks on select engineering topics. Most of all we would like to thank our families. We acknowledge our parents. Thank you moms, Eleanore and Jackie, and thank you dad, Frank, for always believing in me (sb). Thank you moms, Young Shin and Rana, and thank you dads, Sung Bock and Chong Kon, for your encouragement and unfailing support (dp). Finally, our work could not have come to fruition without the sacrifices of our family members: Cindy, Heidi, Heather, Jon R., Christine, Jon B., Andrew, and Graham. Without you none of this would matter. We love you!

Steve Barrett and Daniel Pack
Laramie and Colorado Springs,
February 2006

CHAPTER 1

Digital Design Fundamentals

Objectives: After reading this chapter, the reader should be able to

- Define different methods of specifying binary logic levels.
- Perform binary addition and subtraction using the two's complement system.
- Define the basic combinational operations of digital logic.
- Specify the difference between combinational and sequential logic.
- Describe the basic operation of flip-flops using waveform diagrams.
- List and explain the operation of medium scale integration (MSI) combinational and sequential circuits.
- Summarize digital design technology approaches available for implementing digital circuits.

In this chapter, we briefly describe the prerequisite knowledge required for designing with microcontrollers. Typically, this information is provided in a sophomore level digital design fundamentals course. For a more in-depth coverage on a specific topic, the interested reader is referred to a partial list of excellent textbooks provided at the end of the chapter.

1.1 INTRODUCTION

At its most fundamental level, digital design is the orderly manipulation of digital signals by hardware components. The most fundamental piece of information in digital design is a binary digit or bit. A bit can hold a single piece of information. The bit can be set to a logic one or zero level. It will remain at this logic level until changed by later processing. In active-high convention, the logic one level represents a logic true condition. Similarly, a logic zero level represents a logic false condition.

Logic representations in digital design are processed by hardware devices. Logic levels within a hardware device are represented by different direct current voltage (VDC) levels. A logic one is typically represented by a 5 VDC signal while a logic zero condition is represented by a 0 VDC signal. Newer digital hardware technologies represent logic signals with 3.3 VDC for logic one and 0 VDC for logic zero.

1.2 BINARY MATH

Digital designs must have the capability to process and manipulate mathematical data. This includes methods to represent positive and negative integers and real numbers and how to perform basic mathematical operations such as addition and subtraction. Also, nonmathematical data such as alphanumeric characters must be represented as a series of logic signals compatible for hardware storage and manipulation.

1.2.1 Representation of Integers and Floating Point Variables

In the decimal (base 10) number system, a quantity is expressed by using allowable digits (0 through 9) weighted by their positions in the number to represent the quantity. The weighting scheme for each position is determined by taking the number system's base and applying an exponent specific for that position. Starting from the decimal point (or more generally the radix point) and moving to the left we are familiar with the one's place (10^0), the ten's place (10^1), the hundred's place (10^2), etc. Moving to the right of the decimal point we encounter the tenth's place (10^{-1}), the hundreth's place (10^{-2}), etc.

Exactly the same rules are applied in the binary (base 2) number system. Again we have a set of allowable values (0 and 1) weighted by the position at a particular location. We use the identical weighting system as we did for the base 10 system except that we change the base in our calculations to two. Starting from the radix point and moving to the left we encounter the one's place (2^0), the two's place (2^1), the four's place (2^2), etc. Moving to the right of the decimal point we encounter the half's place (2^{-1}), the fourth's place (2^{-2}), etc. For example, the decimal number 31.75 may be represented in the binary number system as 11111.11. Do you agree?

To be precise in our notation we need to bracket the number in parenthesis and provide the number's base as a subscript outside the parenthesis. For example, we can summarize our earlier calculation with $(31.75)_{10} = (11111.11)_2$.

1.2.2 Two's Complement

In the decimal number system we indicate a number value less than zero by simply preceding the numerical quantity with a minus sign. Things are a bit more complicated in the binary number system, because all representations in a digital-based system must be expressed as a logic value one or zero. Various methods have been used to represent negative numbers in the binary number system including sign-magnitude, one's complement, and two's complement notation. The two's complement notation is the one used by the vast majority of digital-based systems.

When representing a number in the two's complement notation, it is important to indicate the number of bits that will be used to represent the number. To represent a positive number in the two's complement system, we simply represent it as we did in the previous section. For example, representing $(31)_{10}$ in two's complement notation using eight bits results in $(00011111)_2$.

To represent a negative number in two's complement notation requires a three-step process:

1. Represent the number's magnitude with the specified number of bits.
2. Perform a bit-by-bit inversion.
3. Add one (increment).

For example, to represent $(-31)_{10}$ in two's complement using eight binary bits, we would perform the three-step conversion process.

1. Represent the number's magnitude with the specified number of bits: $(00011111)_2$
2. Perform a bit-by-bit inversion: $(11100000)_2$
3. Add one (increment): $(11100001)_2$

1.2.3 Floating Point Notation

To store very large or very small numbers in a digital-based system requires considerable storage space. For example, we can calculate the largest unsigned integer that we can store with a given number of bits by applying the expression $2^{bits} - 1 =$ *greatest unsigned integer*. For example, the largest unsigned integer we can store with eight bits is 255. For signed numbers, approximately half of our storage allocation must be used for positive numbers and half for negative numbers.

To allow storage of very large or very small numbers floating point notation is used. This provides an efficient method that does not require considerable storage space. In floating point notation a real number is stored in three separate parts: the sign bit, an exponent, and a fractional portion representing the significant digits of the real number. There are many different formats used to represent floating point numbers. The most common is the IEEE (Institute for Electrical and Electronic Engineers) ANSI/IEEE Standard 754-1985. With this standard, floating point numbers may be represented in either a single precision (32-bit) or double (64-bit) precision format. In the single precision format a single bit is allocated for the sign bit (s), eight bits are allocated for the exponent (e), and 23 bits are allocated for the fraction (f). The actual real number (r) may be obtained using the following formula:

$$r = (-1)^s \times 1.ff \cdots ff \times 2^{e - 127}$$

Note in the equation that the binary representation of the significant portion of the number is normalized such that the leading bit is one. Also, the exponent is stored as the desired exponent plus 127. For more information on various floating point format standards, the interested reader is referred to Horowitz and Hill (1990).

Example: Two numbers have been stored using the 32-bit single precision format. The stored values are $(4060\_0000)_{16}$ and $(C0F0\_0000)_{16}$. The two equivalent decimal numbers that were stored are $(3.5)_{10}$ and $(-7.5)_{10}$. Do you agree?

1.2.4 Basic Math Operations: Addition and Subtraction

In the binary number system addition is performed as follows:

- $0 + 0 = 0$
- $0 + 1 = 1$
- $1 + 0 = 1$
- $1 + 1 = 10$

The last equation simply means that when one is added to one, the result is zero with a carry to the next higher place holder (2^1). For example, adding $(00100111)_2$ to $(00000110)_2$ results in $(00101101)_2$. Do you agree?

In the two's complement system, subtraction is performed as addition. Instead of performing $(a - b)$, we change the operation to $(a + (-b))$. In other words, we represent "b" as a negative number and add it to "a." For example, using four-bit two's complement notation, $7 - 3 = 7 + (-3)$ becomes $(0111)_2 + (1101)_2 = (0100)_2$. One of the motivations to use the two's complement notation in digital systems is to reduce hardware complexities. Changing a subtraction operation into an addition operation with the help of two's complement notation removes the need to develop a separate hardware component that substracts.

1.3 CODES

There is considerable non-numerical information that must be stored in a digital-based system. Various codes allow the storage of such data.

1.3.1 Unicode and ASCII

The American Standard Code for Information Interchange or ASCII is a standardized, seven-bit method of encoding alphanumeric data. It has been in use for many decades, so some of the characters and actions listed in the ASCII table are not in common use today. However, ASCII is still the most common method of encoding alphanumeric data. The ASCII code is provided in Figure 1.1. For example, the capital letter "G" is encoded in ASCII as 0x47. The "0x" symbol

	Most significant digit							
Least significant digit	0x0_	0x1_	0x2_	0x3_	0x4_	0x5_	0x6_	0x7_
0x_0	NUL	DLE	SP	0	@	P	`	p
0x_1	SOH	DC1	!	1	A	Q	a	q
0x_2	STX	DC2	“	2	B	R	b	r
0x_3	ETX	DC3	#	3	C	S	c	s
0x_4	EOT	DC4	$	4	D	T	d	t
0x_5	ENQ	NAK	%	5	E	U	e	u
0x_6	ACK	SYN	&	6	F	V	f	v
0x_7	BEL	ETB	‘	7	G	W	g	w
0x_8	BS	CAN	(	8	H	X	h	x
0x_9	HT	EM	)	9	I	Y	i	y
0x_A	LF	SUB	*	:	J	Z	j	z
0x_B	VT	ESC	+	;	K	[	k	{
0x_C	FF	FS	‘	<	L	\	l	|
0x_D	CR	GS	–	=	M	]	m	}
0x_E	SO	RS	.	>	N	^	n	~
0x_F	SI	US	/	?	O	_	o	DEL

FIGURE 1.1: ASCII code. The ASCII code is used to encode alphanumeric characters. The “0x” indicates hexadecimal notation in the C programming language

indicates the hexadecimal number representation. Unicode is the international counterpart of ASCII. It provides standardized 16-bit encoding format for the written languages of the world. ASCII is a subset of Unicode. The interested reader is referred to the Unicode home page website *www.unicode.org* for additional information on this standardized encoding format.

1.3.2 Gray Code

Gray code is a coding standard used to encode different occurrences of an event. The power of the Gray code is that neighboring events will have a code representation that only varies by a single bit. For example, the Gray code is commonly used to encode the angular displacement of a rotating wheel or disk as shown in Figure 1.2. As the wheel rotates its angular displacement from the reference position is encoded with four bits. As the wheel transitions from one sector to another the Gray code indicating its rotational position also changes by a single bit. If a binary number encoding scheme was used instead of a Gray code, incorrect values would result near the boundaries of adjacent sectors since all bits would not change simultaneously as shown in Figure 1.2. Horowitz and Hill (1990) describe a simple rule to generate Gray code states: begin with a state of all zeros. To obtain the next state, change the single least significant bit that results in a new state. A four-bit Gray code generated by this scheme is provided in Figure 1.2. Also, note how the first eight Gray code values are a mirror image about the centerline of the last eight Gray code values.

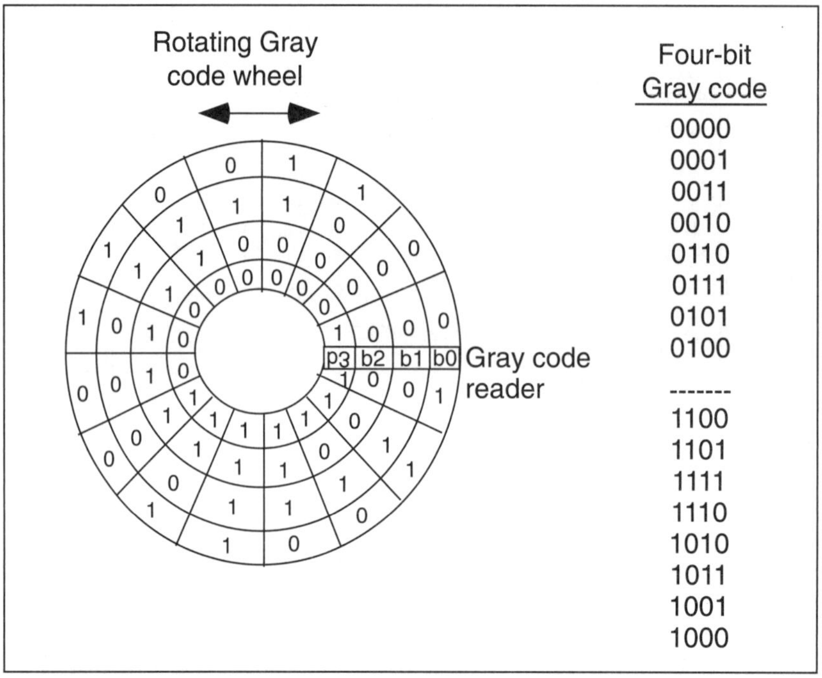

FIGURE 1.2: Gray code. The Gray code is only varied by a single bit for neighboring events

1.4 COMBINATIONAL AND SEQUENTIAL CIRCUITS

In hardware digital design, circuits may be classified into combinational or sequential circuits. An ideal combinational circuit (one without propagation delays) immediately provides a change in its output when a change in its input(s) occurs. In other words, the inputs are combined by the circuit's function to render the correct output. A block diagram of a combinational circuit is provided in Figure 1.3(a). A classic example of a combinational circuit is the binary full adder. Here the current inputs (two addend bits and the carry in bit) are combined to form the correct outputs (sum and carry out).

A sequential circuit provides an output based on its current input(s) and current state. A block diagram of a sequential circuit is provided in Figure 1.3(b). As you can see a sequential circuit combines the input with the current state stored in memory (flip-flops) to render the correct circuit output. An elevator controller is a classic example of a sequential circuit application. Here the input (button depressed by the user) is combined with current state information (floor the elevator is currently at) to render the correct output (move up or down a given number of floors). In the next several sections we review the basic building blocks of combinational and sequential circuits.

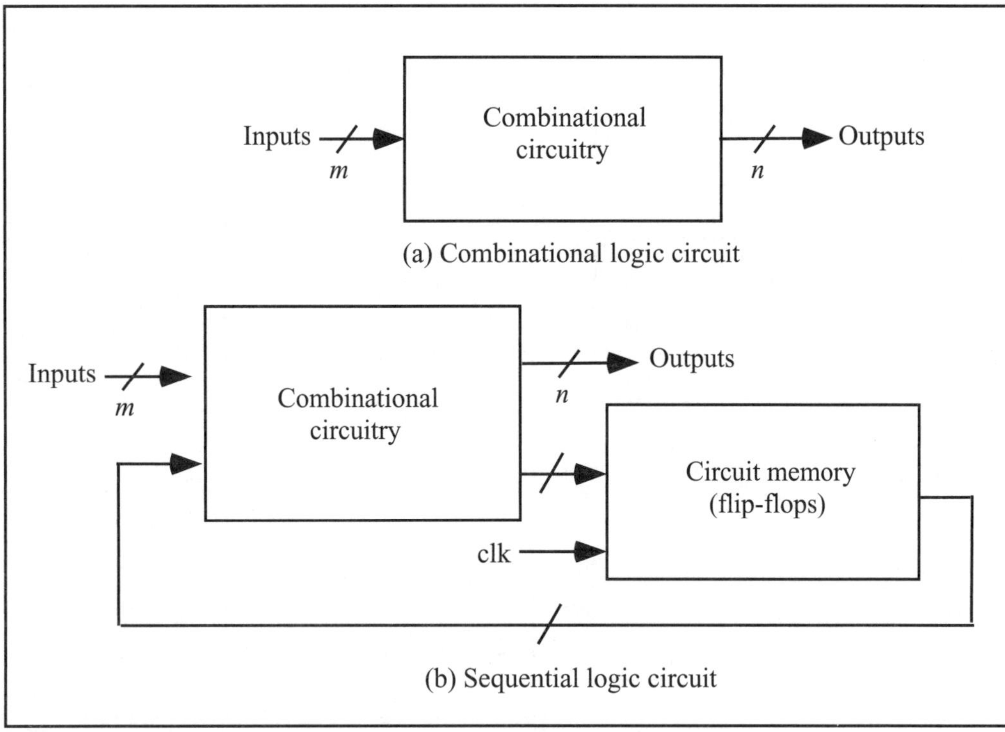

FIGURE 1.3: Combinational versus sequential circuits

1.4.1 Digital Design Building Blocks

As a child you may have played with building block sets. The blocks came in a wide variety of shapes, colors, and sizes. Each block with its own unique function. With the basic building blocks you could design and construct virtually anything your imagination would allow. In the next two sections we will review the basic building blocks of digital design combinational and sequential circuits. As with the building blocks of your past, you can design and construct virtually anything your imagination will allow.

Basic Combinational Circuit Functions

The basic combinational building block functions are summarized in Figure 1.4. For each circuit we have provided a circuit diagram, its corresponding truth table providing the relationship between inputs and outputs, and the characteristic equation. With the exception of the inverter and the buffer, each of the basic functions can be extended to multiple inputs. These components are commonly referred to as small scale integration (SSI) components since they contain tens of switching transistors.

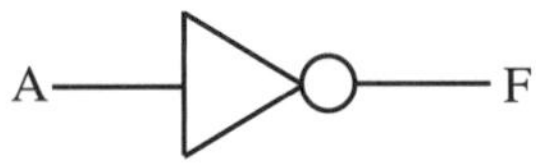

(a) Inverter: F = A′

Input	Output
A	F
0	1
1	0

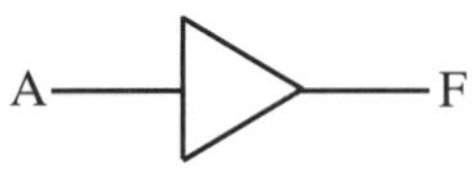

(b) Buffer: F = A

Input	Output
A	F
0	0
1	1

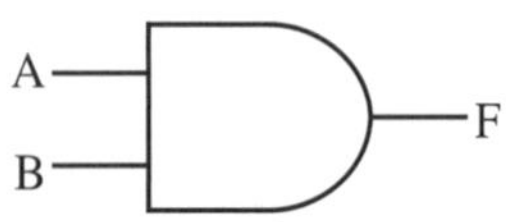

(c) AND: F = AB

Inputs		Output
A	B	F
0	0	0
0	1	0
1	0	0
1	1	1

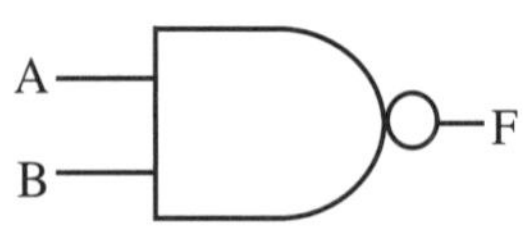

(d) NAND: F = (AB)′

Inputs		Output
A	B	F
0	0	1
0	1	1
1	0	1
1	1	0

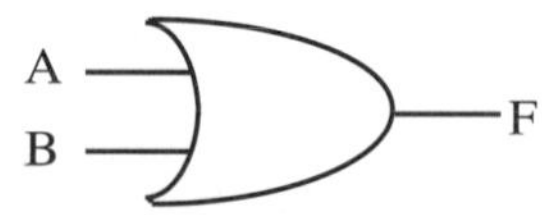

(e) OR: F = A+B

Inputs		Output
A	B	F
0	0	0
0	1	1
1	0	1
1	1	1

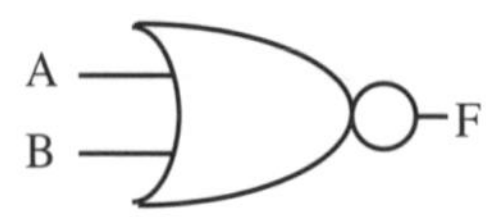

(f) NOR: F = (A+B)′

Inputs		Output
A	B	F
0	0	1
0	1	0
1	0	0
1	1	0

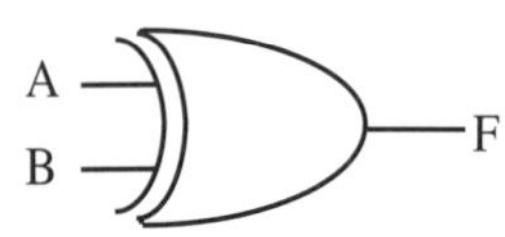

(g) EXOR: F = A⊕B

Inputs		Output
A	B	F
0	0	0
0	1	1
1	0	1
1	1	0

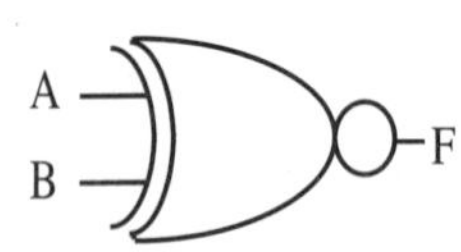

(h) EXNOR: F = (A⊕B)′
= A⊙B

Inputs		Output
A	B	F
0	0	1
0	1	0
1	0	0
1	1	1

FIGURE 1.4: Combinational functions

MSI Combinational Circuits

The next level of component integration beyond SSI is medium scale integration (MSI) devices. MSI components contain hundreds of switching transistors per device. Some of the common MSI combinational components are illustrated in Figure 1.5. An MSI device usually has a specific advanced function. Within the MSI device are the basic combinational building blocks discussed in the previous section. The functions of the combinational MSI devices provided in Figure 1.5 are

- *Full adder:* Adds to single bit addends A and B with the Carry In value. The resulting sum is reflected as a Sum and a Carry Out value.

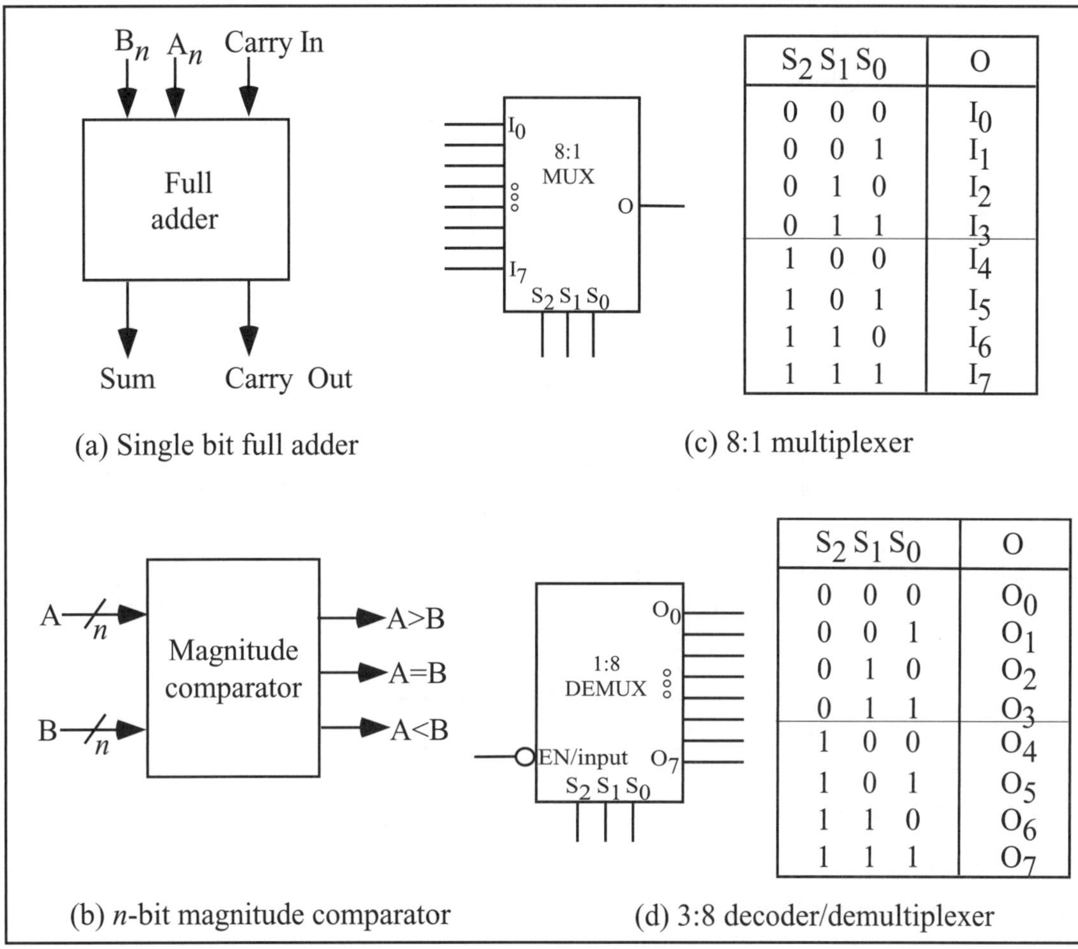

FIGURE 1.5: MSI combinational circuits

- *Magnitude comparator:* Compares the magnitude of two n-bit inputs (A and B) and provides a single active one-bit output indicating the relationship between A and B ($A < B$, $A = B$, $A > B$).
- *Multiplexer:* A multiplexer is a multiposition switch. It has multiple inputs that may be connected one at a time to a single output. The binary value applied to the select lines determine which input is connected to the output.
- *Decoder/demultiplexer:* It is used to connect a single input to one of many outputs. The select lines are used to determine which specific output is connected to the input.

In the next section we investigate the sequential building blocks of digital design, flip-flops.

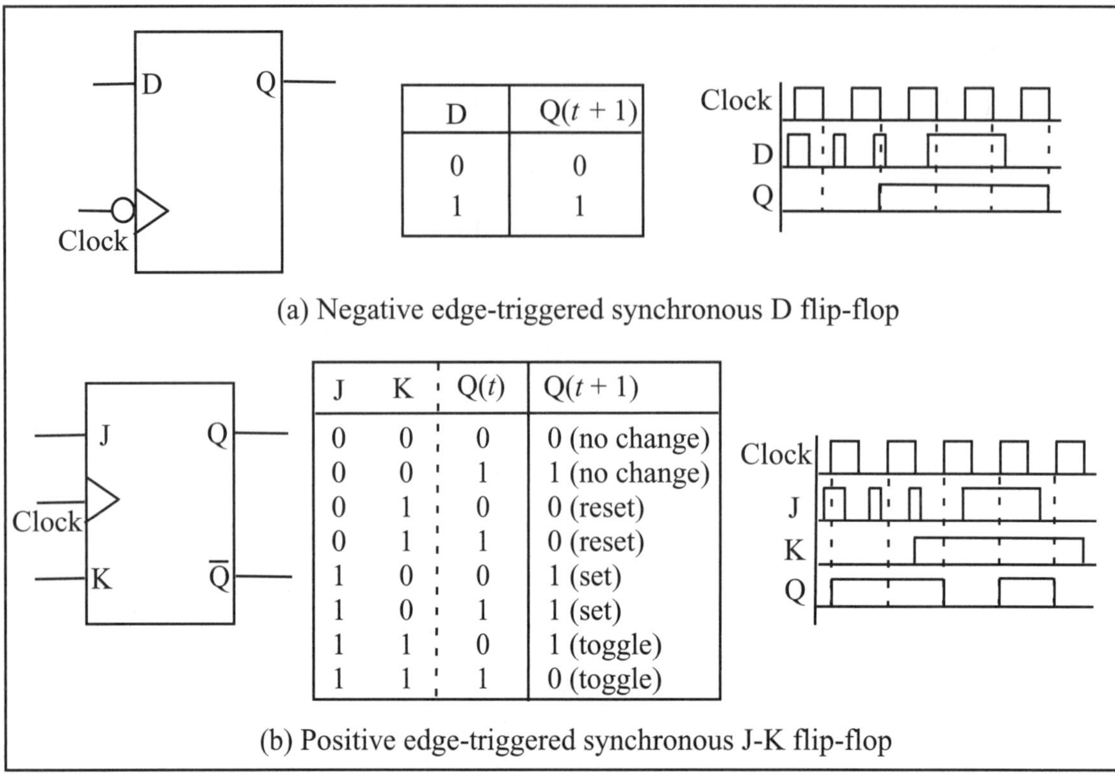

D	Q(t + 1)
0	0
1	1

J	K	Q(t)	Q(t + 1)
0	0	0	0 (no change)
0	0	1	1 (no change)
0	1	0	0 (reset)
0	1	1	0 (reset)
1	0	0	1 (set)
1	0	1	1 (set)
1	1	0	1 (toggle)
1	1	1	0 (toggle)

FIGURE 1.6: Flip-flops

Basic Sequential Circuit Functions

The basic building block for sequential circuits is a flip-flop. A single flip-flop can store a single bit of information either a logic one or a logic zero. Data is stored within the flip-flop by configuring its input(s) to a given value and then providing a clock edge to the flip-flop. There are two basic types of flip-flops: D and J-K. The action of each of these flip-flops is summarized in Figure 1.6. It must be emphasized that these are synchronous devices. That is, action is initiated by clock edges. The flip-flop inputs are shown in the left column(s) of the table. The Q(t) and the Q($t + 1$) columns provide information on flip-flop outputs. The Q(t) column represents the flip-flop output before the clock edge. The Q($t + 1$) column represents the flip-flop output after the clock edge. For example, if a J-K flip-flop currently has a zero on output Q and a logic one applied to both of its J and K inputs, the flip-flop will toggle to a logic one at the next positive clock edge.

MSI sequential circuits

A number of MSI sequential circuits may be configured from the basic flip-flop circuits. A representative sample of MSI sequential circuits is provided in Figure 1.7.

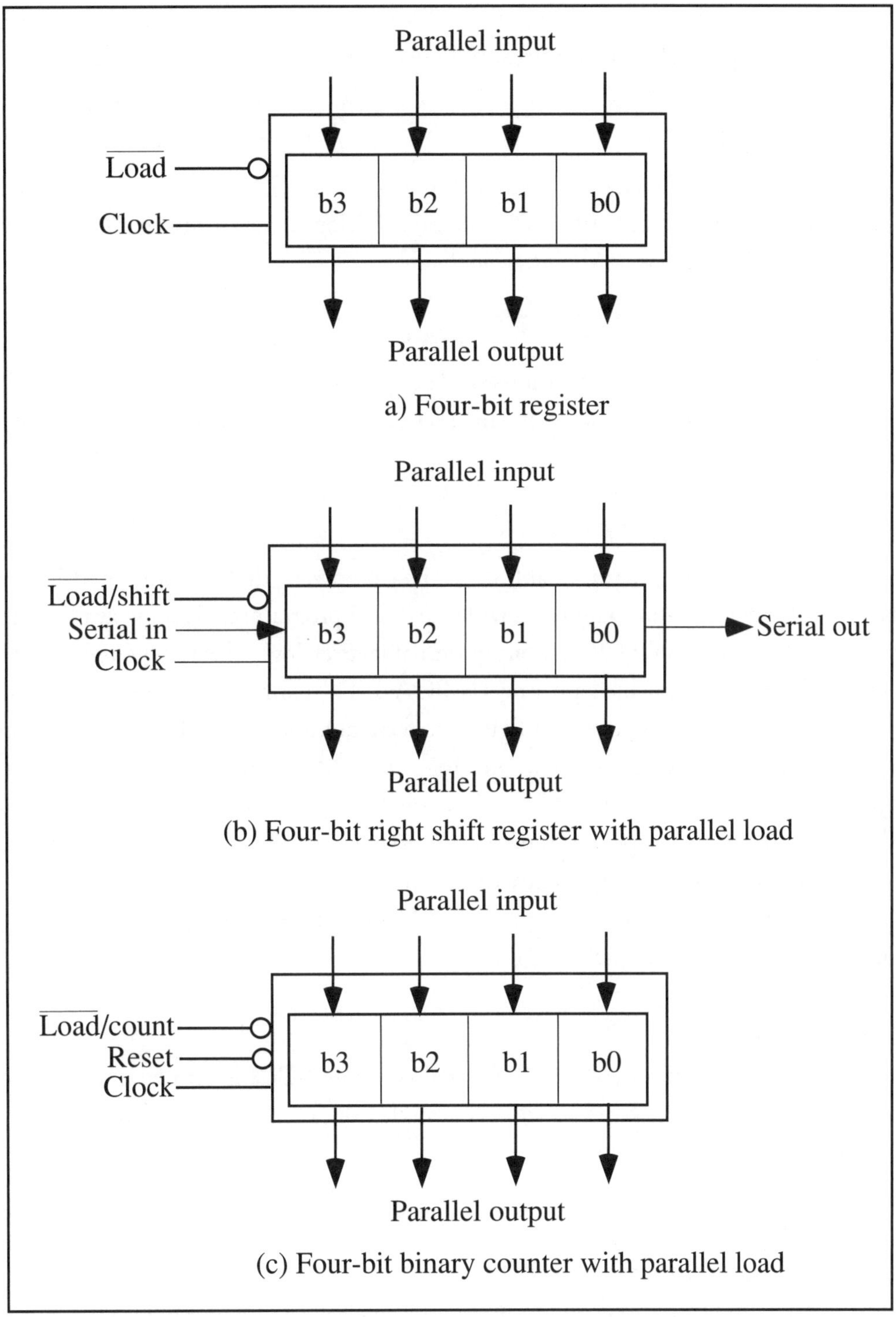

FIGURE 1.7: MSI sequential circuits

- *Register:* A register is a collection of flip-flops that may be simultaneously loaded (written) in parallel or read from simultaneously.
- *Shift register:* A shift register is a storage register that has the capability to shift its contents one bit to the right (or left) with each clock pulse.
- *Binary counter:* A binary counter increments by one for each successive clock pulse. When the counter reaches its maximum value (all logic ones) it automatically increments back to all logic zeros on the next clock edge.

1.5 DIGITAL DESIGN SOLUTIONS

One of the most important design decisions while implementing a digital design is choosing the most appropriate solution technology. In this section we provide a brief overview of available alternatives.

1.5.1 Programmable Gate Arrays

The basic concept behind programmable logic is a hardware that is configured into digital systems via software programs. There is a wide variety of programmable logic available including programmable logic arrays (PLAs), programmable array logic (PALs), generic logic arrays (GALs), and field programmable gate arrays (FPGAs). The configurable hardware may consist of multiple input AND gates that are configured to feed multiple input OR gates. This allows for the implementation of complex combinational logic in standard product of sums representation. Many of these programmable devices also allow for the implementation of sequential circuits since they also contain flip-flops. For more advanced designs, FPGAs may be used to program extremely complex algorithms. Typically, these complex digital systems are programmed using Verilog Hardware Descriptive Language or VHDL. These techniques allow for the rapid implementation of complex digital designs.

1.5.2 Microprocessors

The main integrated circuit or chip of a personal computer (PC) is termed a microprocessor. Rarely is a dedicated PC required to solve a complex digital design. Usually the design can be rendered in a standalone technology. A PC is usually employed to solve a wide variety of different challenges.

1.5.3 Digital Signal Processors (DSPs)

A DSP, as its name implies, is used when considerable analysis of signals is required. For example, a DSP would be a sound choice for processing and conditioning of cellular phone signals. A DSP is extremely efficient in processing floating point data.

1.5.4 Microcontrollers

A microcontroller is a self-contained single chip processor with all constituent subsystems of a larger computer system. Within the confines of a single integrated circuit it contains input/output capability, a time base, a timing system, memory, an Arithmetic and Logic Unit (ALU) providing the capability to perform arithmetic and logic processes, and also the capability to generate output control signals. A microcontroller is usually employed when a moderate amount of local intelligence is required within a given application. It is best suited for applications involving integer-based processing although floating point calculations are possible.

1.5.5 Mixed Mode Processing Microcontroller with FPGA

As mentioned previously, it is important to match the appropriate technology to the design challenge at hand. However, a single technology may not be the best choice. The current trend in digital design is to employ a mixed mode technology such as an FPGA coupled with a microcontroller.

1.6 SUMMARY

We began this chapter by reviewing different methods of specifying binary logic levels followed by a brief review of binary addition and subtraction techniques using the two's complement system. We also reviewed coding techniques to store non-numeric data within a hardware design including ASCII and Gray code. We then reviewed the fundamentals of combinational and sequential circuit design. The chapter concluded with a review of digital design technology approaches available for implementing a specific application.

BIBLIOGRAPHY

M Mano and C Kime, Logic and Computer Design Fundamentals, 3rd ed., Prentice Hall, Upper Saddle River, NJ, 2003.

J Wakerly, Digital Design Principles and Practices, 4th ed., Prentice Hall, Upper Saddle River, NJ, 2006.

P Horowitz and W Hill, The Art of Electronics, 2nd ed., Cambridge University Press, Cambridge, U.K., 1990.

CHAPTER PROBLEMS

- **Fundamental**
  1. *Question*: What is ASCII used for?
  2. *Question*: What is the advantage of using Gray code encoding in certain applications?

3. *Question*: What is the primary difference between combinational and sequential circuits?

- **Advanced**
  1. *Question*: Represent $(-110)_{10}$ in eight-bit two's complement notation.
  2. *Question*: Provide a five-bit Gray code.
- **Challenging**
  1. *Question*: How do you tell if binary data stored in a computer is a two's complement number, a Gray code representation of the angular displacement of a wheel attached to a motor, or the ASCII representation of an alphanumeric character?
  2. *Question*: Prepare a four-panel chart that summarizes the advantages and disadvantages of using programmable hardware, DSP processors, microcontrollers, and microprocessors.
  3. *Question*: List and describe five sequential circuits you have used today.

CHAPTER 2

The Design and Development Process

Objectives: After reading this chapter, the reader should be able to

- Design an electrical and computer system from conceptual design through final testing.
- Describe how to transition from a system description to detailed system requirements.
- Apply top-down design, bottom-up implementation techniques to partition a system description into manageable pieces.
- Apply system design tools such as structure charts and UML activity diagrams.
- Transform a UML activity diagram to a code prototype.
- Develop a detailed test plan based on system requirements.
- Apply techniques to document system design.

We believe you will find this chapter extremely interesting and well worth your time. In this chapter we take you through the systematic, logical process of transforming a system description into a fully functional, operational system that meets the defined requirements. Furthermore, we discuss how to properly document the operation and testing of a system.

2.1 THE DESIGN PROCESS

The design process is illustrated in Figure 2.1. This figure is based on our experience developing multiple and diverse embedded control systems from stereo amplifier controllers, to autonomous robots, to industrial gate controllers. We employ a running scenario to illustrate the design process.

Scenario: You have been hired as a consultant by a small engineering firm that designs and manufactures industrial gate controllers such as those found at parking lots to control access. The firm has been successfully manufacturing the controllers for a number of years based on analog Application Specific Integrated Circuit (ASIC) technology. ASIC technology is based on custom designed integrated circuits. Fabrication of an ASIC device usually requires a lead time of 6 months and a large set up cost to turn out the first custom IC. From then the cost of the IC is amortized over the production run of the IC.

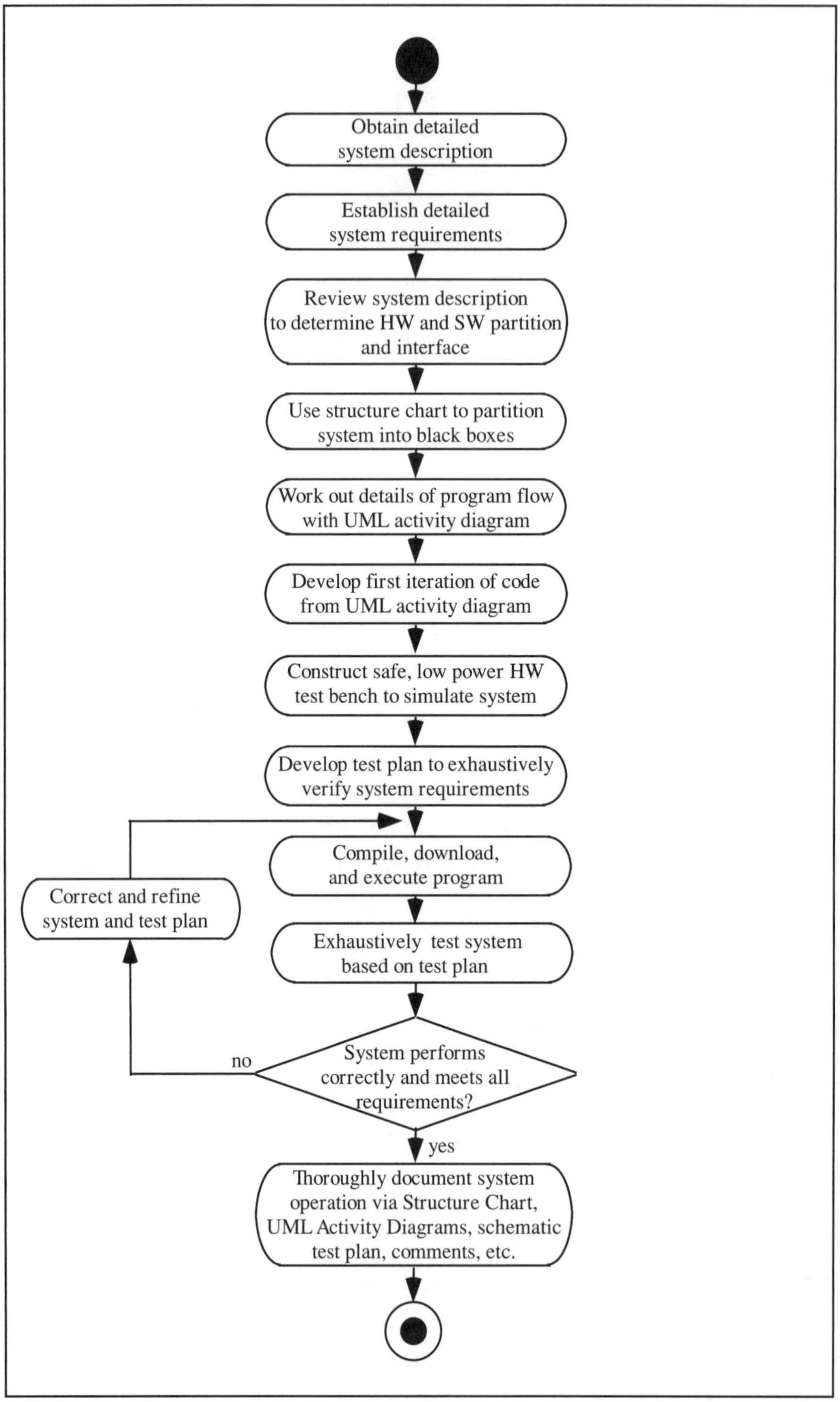

FIGURE 2.1: The design process

The president of the company, a gifted analog design engineer, has contacted you to convert the company's proprietary control algorithms from an analog (operational amplifier) implementation to a microcontroller-based system. The president is intrigued by the short lead time to develop a microcontroller-based project and the ability to customize each algorithm design if need be. You have had several meetings to discuss the overall picture of how the gate controller will operate. You have received a ten page text document describing what the controller is supposed to do. The motor used to open and close the gate in this specific application is a large, high power direct current motor. You will be using pulse width modulation (PWM) techniques to activate the motor. You are responsible for implementing the control algorithm. All of the details concerning interfacing the low power microcontroller to the gate control hardware are being handled by another engineer.

Theory: Where do you begin on such a difficult, complex project? A natural reaction might be to panic. Instead, constructively use your creative energy to logically and methodically overcome the design challenge by the design process provided in Figure 2.1.

For starters you should gather all of the available information on the desired system. You already have a lengthy project description to read and absorb. As you read through the documentation make a list of questions where answers are not covered in the description. The answers should be obtained from the president and the other engineer who is working the project interface issues. You may end up iterating multiple times with the company representatives to develop the overall system description. This would also be a good time to do some homework and review concepts that are unfamiliar to you. For example, if you are not familiar with PWM motor control techniques, now would be a good time to review the concepts. Today, considerable technical background information may be obtained from the Internet.

Once you have absorbed all of the information, you should develop a detailed list of system requirements. In other words, what is the system supposed to do. For a microcontroller-based system, it is often helpful to take a scenario-based approach. That is, you provide a list of required system activities and what the controller is supposed to do for each of these activities.

Scenario: In the case of the gate control system, the gate will be activated by a number of pushbutton switches including open, close, emergency open, system lock, and emergency shutdown. For each of these inputs, you should develop a detailed list of required actions. Your requirements list should also include how the system will react to other circumstances such as a malfunctioning gate, an incorrect input sequence (pushing the gate close button when the gate is already closed), and also an incorrect sequence of inputs (two pushbuttons depressed at the same time).

Theory: Once the requirements list is completed (you will probably add additional requirements as the project progresses), you will actually begin trying to figure out how to implement the design. An effective tool to start partitioning the design is based on the techniques of top-down design, bottom-up implementation. In this approach, you start with the overall system and begin to partition it into systems. At this point of the design, you are not concerned with how the design will be accomplished but how the different pieces of the project will fit together. A handy tool to use at this design stage is the structure chart. The structure chart shows the hierarchy of how system hardware and software components will interact and interface with one another. You should continue partitioning system activity until each subsystem in the structure chart has a single definable function.

Scenario: A structure chart for the gate controller is provided in Figure 2.2. As you can see the main functions are partitioned into subsystems. These functions are further subdivided into their constituent parts.

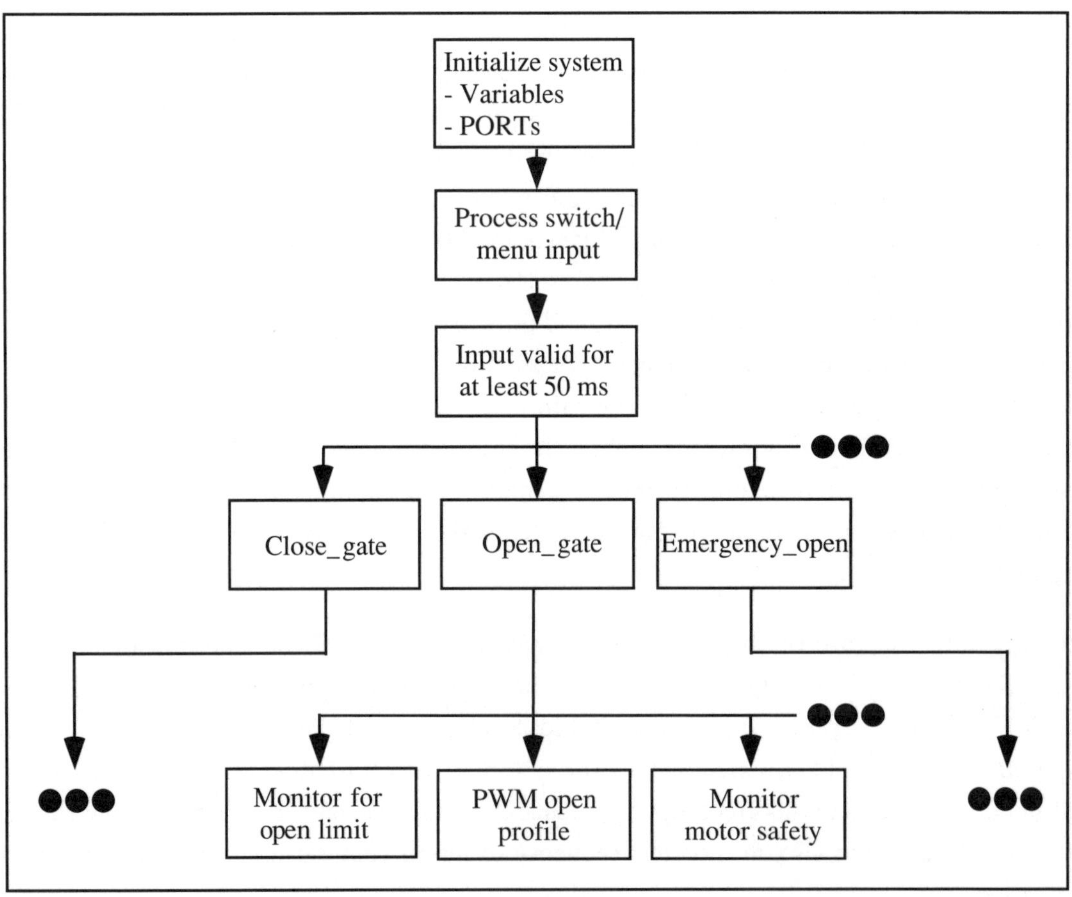

FIGURE 2.2: The structure chart for a gate controller

Theory: In a microcontroller-based system, many tasks can be accomplished in hardware or software or a combination of both. At this step in the design you need to decide how different portions of the design will be implemented. In general a hardware-based solution is typically faster; however, it is inflexible. On the other hand, a software-based solution is typically slower, but is easily adapted for specific applications.

Scenario: The gate control design is a good example where we let the microcontroller do the majority of the processing. Since the microcontroller will be controlling a slow mechanical system (which is often the case), speed is not a primary concern. Therefore, we will shift as much of the control algorithm to the microcontroller as possible.

Theory: The next step in the design process is to start working out the details of the operation of each subsystem we previously identified. Rather than beginning to code each subsystem as a function, we will work out the information and control flow of each subsystem using another design tool: the Unified Modeling Language (UML) activity diagram. The activity diagram is simply a UML compliant flow chart. UML is a standardized method of documenting systems. The activity diagram is one of the many tools available from UML to document system design and operation. The basic symbols used in a UML activity diagram for a microcontroller-based system are provided in Figure 2.3.

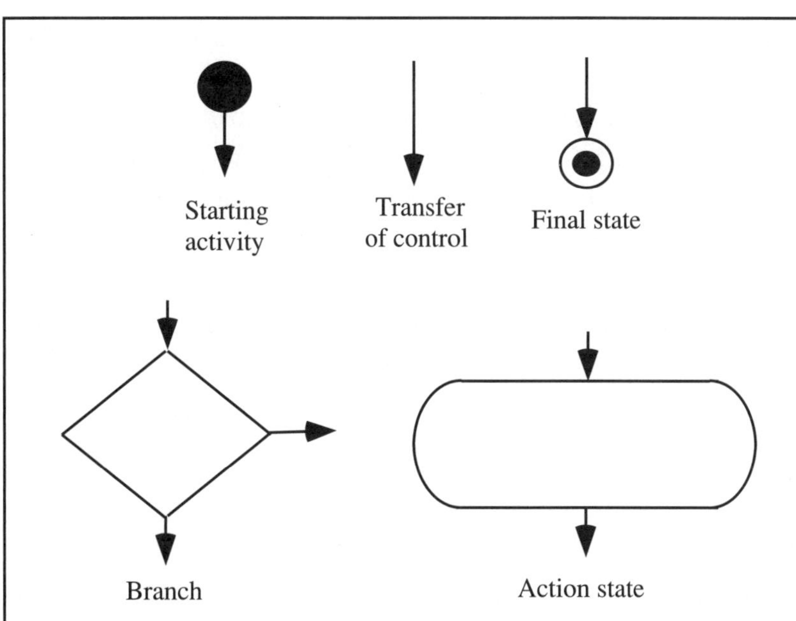

FIGURE 2.3: UML activity diagram symbols

As we begin developing the UML activity diagrams for the system we can use a top-down, bottom-up, or a , hybrid approach. In the top-down approach, we would begin by modeling the overall flow of the algorithm from a high level. If we choose to use the bottom-up approach, we would begin at the bottom of the structure chart and choose a subsystem for flow modeling. The specific course of the action chosen depends on the specific project. Often, a combination of both techniques, a hybrid approach, is used. You should work out all algorithm details at the UML activity diagram level prior to coding any software. If you cannot explain system operation at this higher level first, you have no business being down in the detail of the code. Therefore, the UML activity diagram should be of sufficient detail so that you can code the algorithm directly from it.

Scenario: In our gate controller design we will employ a hybrid approach. We will first model the flow of how the system will respond to the pushbutton switches and develop a menu-based flow to incorporate different pushbutton activity. This is top-down design. We will also work on some of the more complicated subsystems to insure that some of the more challenging technical issues are resolved early in the design process. For example, we will model the flow of the PWM motor control algorithm. A sample of both is provided in Figure 2.4. The details of the specific application illustrated in the figure are not important. What you need to take from these diagrams is the level of detail provided.

Theory: Once the UML activity diagram is completed for the entire project, then coding may commence. Here the detailed algorithms are committed to a microcontroller compatible language (typically C or assembly language.) The C language has the advantage of portability between processors, reusability, readability, and coding efficiency. Assembly language is not portable and is not easy to read; however, applications originally coded in assembly language typically execute faster than those originally coded in C. We recommend C for all but the most time sensitive applications. Again a top-down, bottom-up, or hybrid approach should be followed.

Scenario: For the gate control project, a hybrid approach would be a wise choice. The menuing software could be first written and tested. This would provide a framework for activating the remaining low-level functions in an orderly manner. The lower level functions should be written and individually tested and incrementally brought on line.

Theory: How do you test a microcontroller-based system that will be controlling high power and/or expensive hardware? It is a bit of a dilemma. You need to test the code in actual hardware to insure it operates correctly and the interface between the hardware and software is correct However, you might be reluctant to put untested software in an actual hardware control

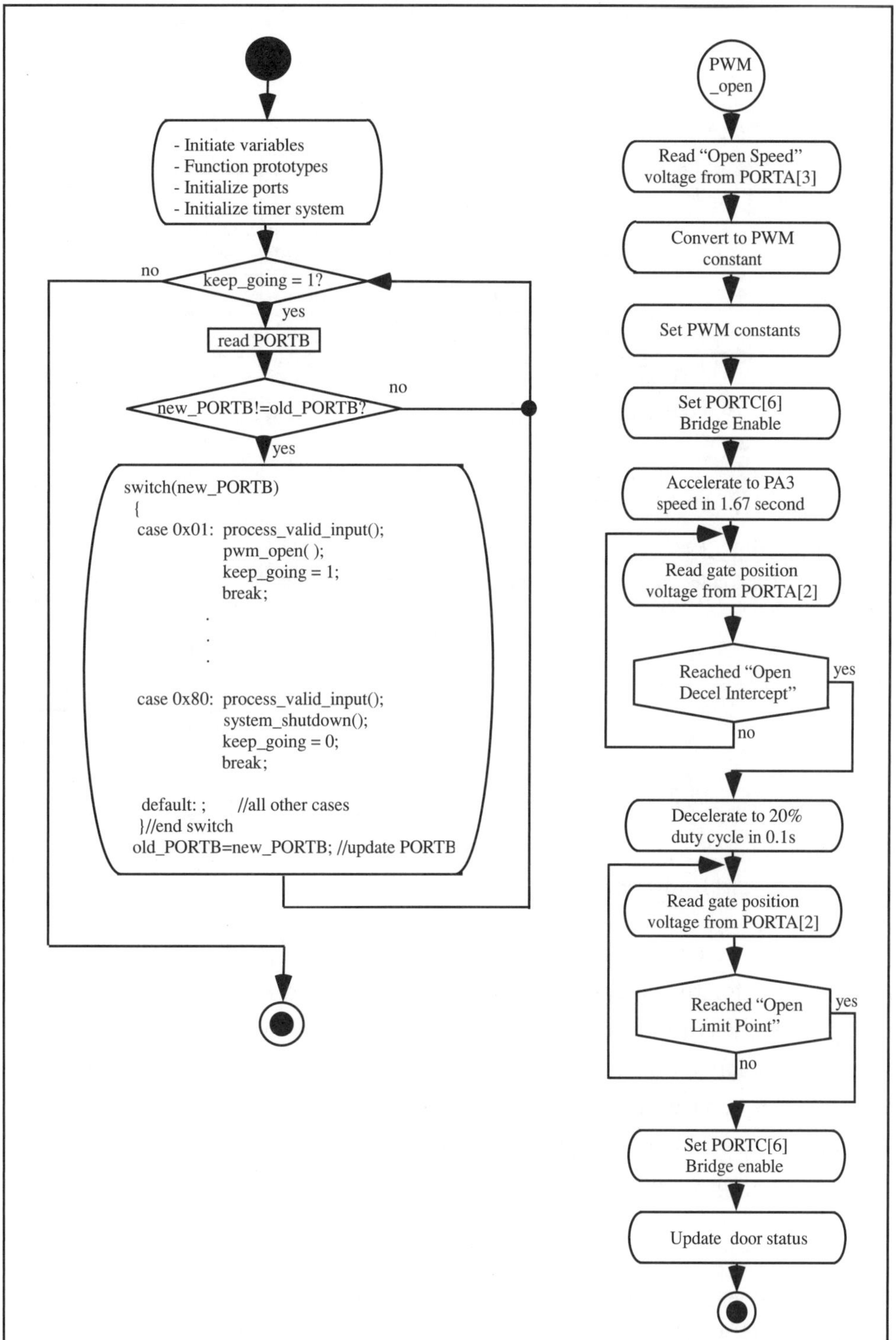

FIGURE 2.4: A hybrid approach to gate design. The UML activity diagram is provided for the menuing algorithm and also the low level PWM motor control algorithm. The details of the specific application illustrated in the figure are not important. What you need to take from these diagrams is the level of detail provided

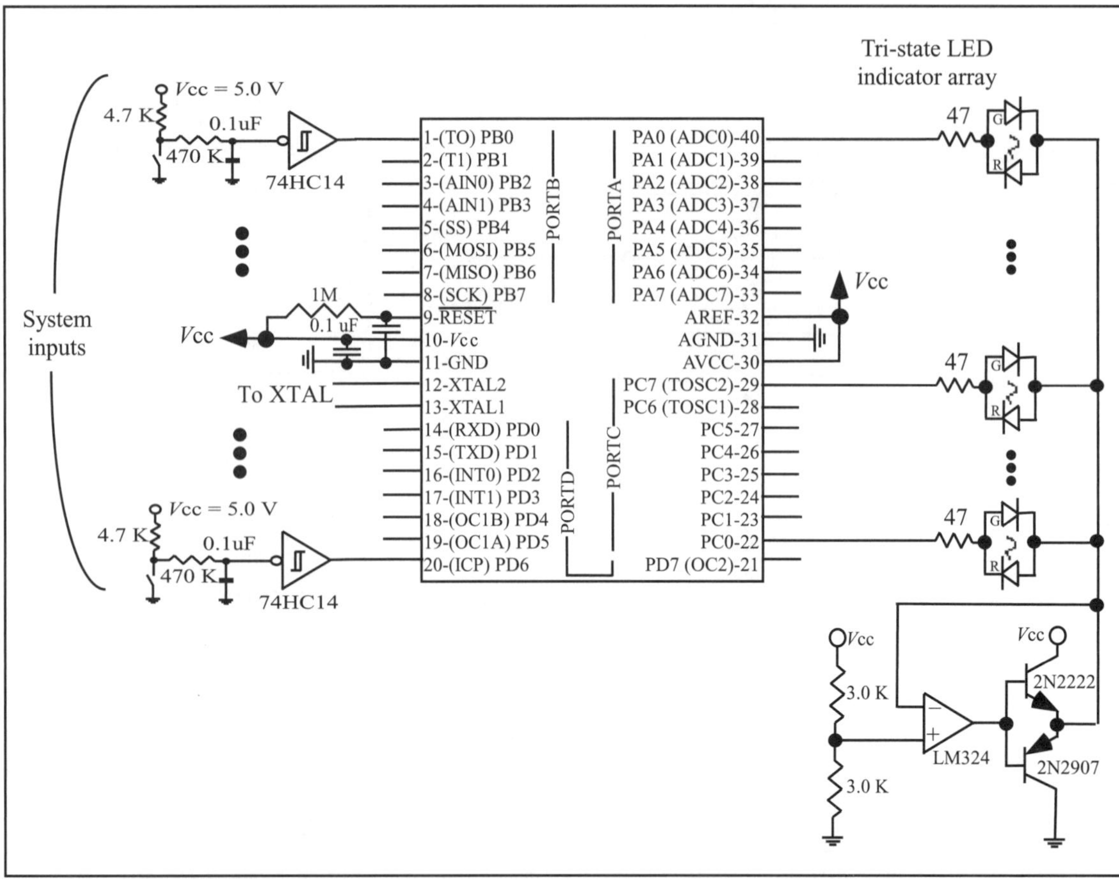

FIGURE 2.5: Low cost hardware simulator

system until you are sure it operates correctly. What to do? A low cost hardware simulator will serve as a stand-in for the actual expensive hardware system until you are sure the software is operating correctly. An inexpensive hardware simulator is illustrated in Figure 2.5. System inputs can be simulated with low cost tact or dual inline package (DIP) switches. Output control signals can be viewed by illuminating light-emitting diodes (LEDs). Also, in certain applications you can monitor the output signals with an oscilloscope or a multichannel logic analyzer. Also, sending status "prints" to a host PC or a liquid crystal display (LCD) from the microcontroller is an effective method of displaying ongoing status during algorithm testing.

Scenario: For the gate control algorithm, system inputs can be simulated with low cost tact switch inputs. System output control signals may be sent to LEDs to indicate system status during program execution. Feedback signals from the actual gate hardware can be simulated with potentiometers for analog signals and switches for digital signals. The PWM signals may be verified with an oscilloscope or logic analyzer.

Theory: Once a low cost simulator has been developed to test the system, a detailed test plan must be developed. Tests should be developed to verify that the system meets all of its requirements and also intended system performance in an operational environment. The test plan should also include scenarios in which the system is used in an unintended manner. As before a top-down, bottom-up, or hybrid approach can be used to test the system.

Once the test plan is completed, actual testing may commence. The results of each test should be carefully documented. As you go through the test plan you will probably uncover a number of run time errors in your algorithm. After you correct a run time error, the entire test plan must be performed again. This insures that the new fix does not have an unintended affect on another part of the system. Also, as you process through the test plan, you will probably think of other tests that were not included in the original test document. These tests should be added to the test plan. As you go through testing, realize that your final system is only as good as the test plan that supports it!

Once testing is completed, you might try another level of testing where you intentionally try to "jam up" the system. In other words, try to get your system to fail by trying combinations of inputs that were not part of the original design. A robust system should be intolerant to this type of testing. It is imperative that you design robustness into your system. When testing on a low cost simulator is complete, the entire test plan should be performed again with the actual system hardware. Once this is completed you should have a system that meets its requirements!

Scenario: For the gate controller testing a formal, exhaustive test plan was developed to document controller operation under approximately 50 different operational scenarios. The controller response to each scenario was carefully documented. When a run time error was discovered, it was corrected, and the test plan reaccomplished from the beginning. The scenarios covered anticipated operational scenarios as well as potential misuse of the system. This provided for a robust design. When the testing was completed, a prototype control unit was sent to the company for "jam up" testing. The goal was to test the system in an anticipated real world environment and try to get it to fail. Once this stage of testing was completed, the prototype controller was installed in an operational field unit and exhaustively tested in an actual operating environment.

Theory: With testing completed, the system design should be thoroughly documented. Much of the documentation will have already been accomplished during system development. Documentation will include the system description, system requirements, the structure chart, the UML activity diagrams documenting program flow, the test plan, results of the test plan, system schematics, and properly documented code. To properly document code you should carefully comment all functions describing their operation, their inputs, and outputs. Also, comments should be included within the body of the function describing key portions of the code. Enough detail should be provided such that code operation is obvious. It is also extremely helpful to provide variables and functions within your code names that describe their intended use.

You might think that completed system documentation is not worth the time or effort to complete it. Complete documentation pays rich dividends when it is time to modify or repair an existing system. Also, well-documented code may be often reused in other projects. This provides for the efficient and timely development of new systems.

Scenario: As mentioned earlier, good documentation pays rich dividends during the maintenance phase of a deployed system. When a system update is required, it is easy to readily identify the portion of the system requiring update. Also, a properly documented system is readily adapted to new, similar designs. You just received an e-mail from the company president. He was so pleased with your first controller project that he has hired you to design three more similar systems.

2.2 IMPLEMENTATION AND TESTING TOOLS

To aid in the design and development process, there are a number of support tools commercially available to make the process easier. We provide a brief definition of some of these design tools.

- *Assembler:* An assembler converts a microcontroller control algorithm written in assembly language to processor specific machine code.
- *Compiler:* A compiler translates a high-level language such as C first into assembly language and then into processor specific machine code.
- *Emulator:* An emulator is a software program that simulates the operation (emulates) of a hardware microcontroller. The emulator is used frequently during system development and testing.
- *Programmer:* Usually microcontroller algorithms are programmed in a host personal computer (PC) using an assembler or compiler. The resulting machine code is downloaded from the host PC to the microcontroller via a hardware programmer. For some microcontrollers this is simply a serial RS-232 compatible cable. For other microcontrollers it consists of a programming pod which converts the machine code to appropriate programming signals to program the memory resident within the microcontroller.
- *Logic analyzer:* A logic analyzer is a piece of test equipment that allows the simultaneous viewing of multiple channels of signals from a microcontroller. It is especially useful for examining the timing relationships between related microcontroller signals.

- *Oscilloscope:* An oscilloscope normally can display two to four channels of data simultaneously. It is especially useful for examining analog signals.
- *In system programming (ISP):* Some microcontrollers are equipped with ISP capability. This means that a program can be downloaded into the microcontroller while the microcontroller is resident within its systems.

2.2.1 Software Development Process

The software development process is illustrated in Figure 2.6. This is a general overview description of the code development process. It provides a generic description of the end-to-end process from C source file development to downloading machine code in the target processor. Details for a specific software suite coupled with a specific target controller may have slight differences.

As the system designer, you will write the C source files that implement the required control algorithm. The source files consist of C source code and also header files. Header files may include functions you have written, functions provided with your compiler, and also personality data for the specific processor you are using as the target system.

The C source files are compiled using a compiler for the specific target processor. The output from the compiler will be assembly code which is automatically routed to the assembler portion of the software development suite. The output from the assembler is object code which

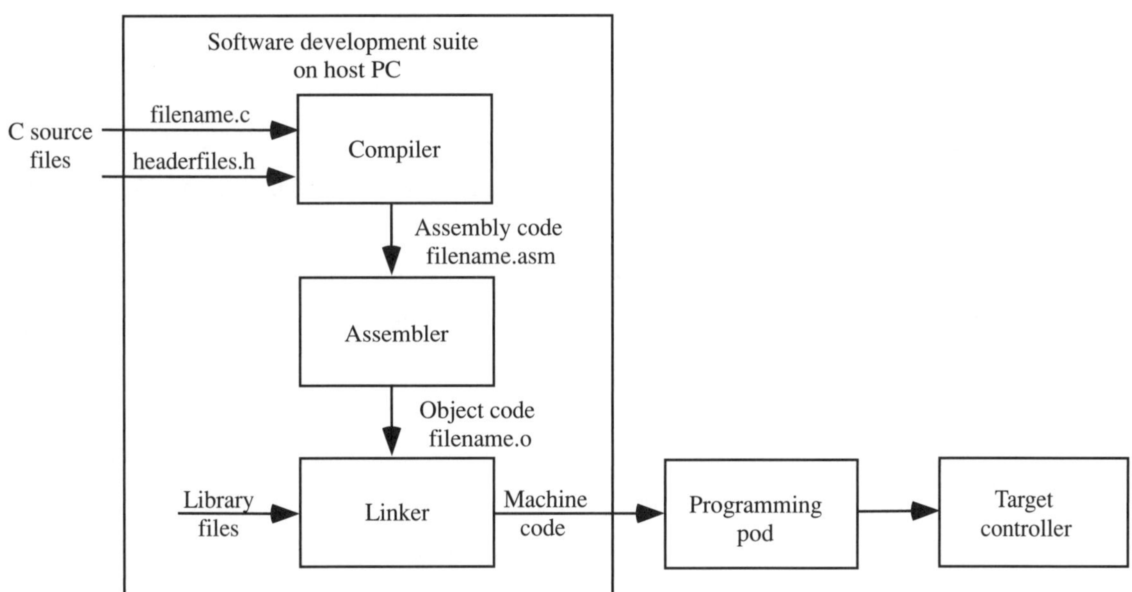

FIGURE 2.6: Software development process

is linked with library files provided with the compiler. The overall output will be machine code for download into the target processor. The software development suite may also provide some auxiliary files. These files provide information on memory usage and code placement within memory. These files are handy if system troubleshooting is required. The machine code is downloaded into the target controller via a programming cable or pod.

2.3 SUMMARY

In this chapter we discussed logical, methodical processes to transform a conceptual design into a fully functional, well-documented system that meets system requirements. We also investigated some associated design tools including structure charts and UML activity diagrams. These processes are well suited for a wide variety of design activities.

BIBLIOGRAPHY

P- J Meilir, The Practical Guide to Structured Systems Design, 2nd ed., Yourdon Press, Upper Saddle River, NJ, 1988.

K Chris, UML 2001, Communications of the ACM, October 1999, Volume 42, Number 10, pp. 29–37.

F Martin and S Kendall, UML Distilled – A Brief Guide to the Standard Object Modeling Language, 2nd ed., Addison-Wesley, Reading, MA, 2000.

B P Douglass, Real-Time UML – Developing Efficient Objects for Embedded Systems, 2nd ed., Addison-Wesley, Reading, MA, 2000.

CHAPTER PROBLEMS

- **Fundamental**
  1. *Question*: What is the difference between a structure chart and a UML activity diagram. What is each used for?
  2. *Question*: What is top-down design/bottom-up implementation?
  3. *Question*: Why is it necessary to break an overall system into multiple subsystems during the design process?
- **Advanced**
  1. *Question*: Provide an example where an oscilloscope might be preferred over a logic analyzer? What is the difference in the information provided by each?
  2. *Question*: Describe the tools available to the system designer.
  3. *Question*: When should coding begin for a microcontroller-based system?

- **Challenging**
  1. *Question*: When is a microcontroller-based control system ready for production?
  2. *Question*: It was stated that a system is only as good as the test plan that supports it. Provide a paragraph to support this bold statement.
  3. *Question*: What is meant by a fully documented system? What is the expected payoff for a well-documented system?

CHAPTER 3
Microcontroller

Objectives: After reading this chapter, the reader should be able to

- Identify and describe the main subsystems available in a typical microcontroller.
- Describe the relationship between the width of the data and address buses as it relates to memory capacity.
- Describe the differences among common microcontroller architectures.
- Describe the primary function(s) of each subsystem aboard a microcontroller.
- Sketch a functional block diagram of a microcontroller.
- Discuss the operating parameters (voltage, current, and frequency) of a microcontroller and their impact on application design.

3.1 SO WHAT EXACTLY IS A MICROCONTROLLER

A microcontroller in its most fundamental form is an entire computer system contained within a single integrated circuit. One of the primary challenges in a microcontroller-based design is choosing the best controller for a specific design. The goal is to choose the most economical microcontroller that has the desired parameters and features for the application at hand. Microcontrollers range from small four-bit processors with limited features to full featured, high speed 32-bit processors. In this chapter we will review the subsystems found in most microcontrollers and also discuss their electrical and timing characteristics. Our overall goal is to help readers effectively choose a microcontroller for a given application. We will not discuss a specific microcontroller or manufacturer for general coverage of microcontroller technologies. However, we provide a list of additional reference material at the end of the chapter on specific processors.

3.1.1 Microcontroller Overview

A basic block diagram of a generic microcontroller is provided in Figure 3.1. In the next several sections we briefly describe the systems commonly found aboard a typical microcontroller. This is merely an overview chapter to provide insights into the power and flexibility provided by microcontrollers. Follow on chapters will provide additional information on selected subsystems.

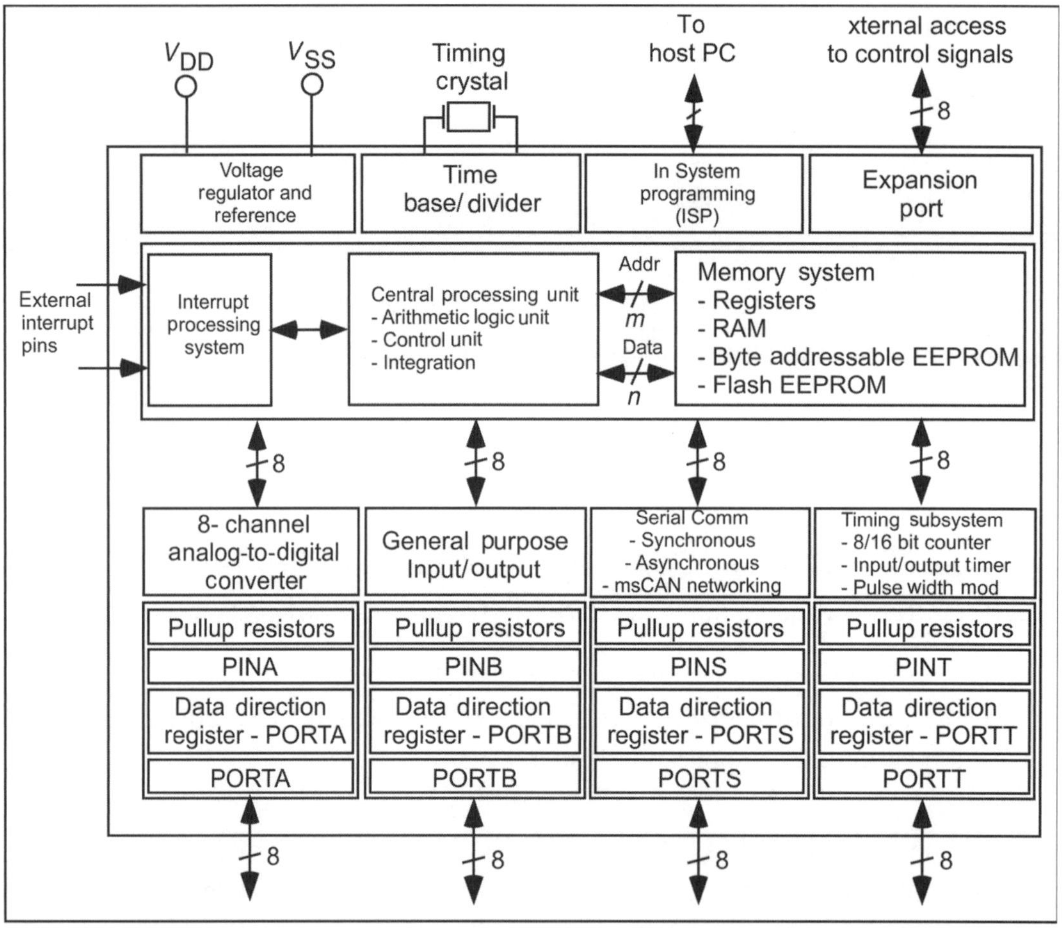

FIGURE 3.1: Microcontroller block diagram. Most microcontrollers are equipped with the subsystems shown in the figure. The ports are used to provide access to the microcontroller to the outside world. Typically, ports are bidirectional and also have alternate functions such as analog-to-digital conversion, serial communications, and a flexible timing system. Microcontrollers are also equipped with a complement of different memory components. Normal microcontroller operation can be interrupted by an external event using the external interrupt pins. This allows the microcontroller to respond to high priority events. The microcontroller is programmed via In System Programming (ISP) features using a host personal computer. The time base for the microcontroller is provided by an external crystal oscillator or resonator

3.1.2 Basic Architecture

The central processing unit (CPU) within a microcontroller is a complex sequential circuit whose primary function is to execute programs that are stored within its Flash EEPROM (Electrically Erasable Programmable Read Only Memory). A program is simply a series of instructions to perform a specific task. Programs are developed by microcontroller system designer (you) using program development tools. The program development tools are usually hosted on a personal

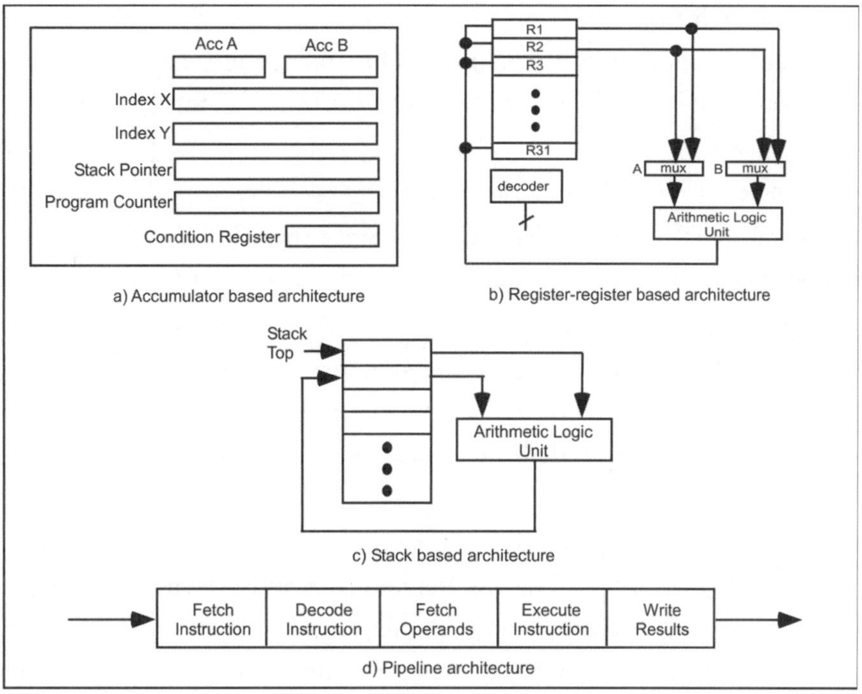

FIGURE 3.2: CPU architectures

computer (PC). Once program development is complete, the program is downloaded into the microcontroller, and the microcontroller becomes a stand-alone processing system.

Once reset, the microcontroller's CPU will sequentially fetch a program instruction from memory, decode the instruction's content (what it is supposed to do), and execute (perform) the instruction. The central processing unit (CPU) is the main control center for the entire microcontroller. While responding to different program instructions, the CPU will call upon its resident subsystems to perform their tasks.

The basic architecture of a CPU can usually be placed into one of the several general categories, as illustrated in Figure 3.2. It should be emphasized that a given architecture is not better than the other. Each one has its inherent advantages and disadvantages.

- *Accumulator-based architecture:* In an accumulator-based architecture, illustrated in Figure 3.2(a), instructions begin and end in specially designated registers called accumulators (A and B). Typically, an operation is performed in which one operand is found in an accumulator and the other is fetched from memory. The result is then placed in the accumulator. This architecture tends to run slower than the other two configurations since operands must be continually fetched from memory. Typically, the memory runs at a slower speed than the main processor so the processor must slow down to accommodate an operand fetch from memory. An accumulator-based architecture has the ability to execute fairly complicated instructions. The architecture may also be modified such that one operand is located in a register and the other is found in memory.

- *Register-based architecture:* In a register-based architecture, both operands are stored in registers that are typically collocated with the central processing unit. The result of a given operation is also stored in a register. Since the CPU and the registers operate at the same speed, the processor does not have to slow down to read or write operands. Register contents are read from and written to memory using a background operation.
- *Stack-based architecture:* In a stack-based architecture, both operands and the operation to be performed are stored on the stack. The result is then placed back on the stack. The stack may be based in dedicated registers or may be a special portion of random access memory.
- *Pipeline architecture:* A pipeline-based microcontroller architecture has the general form illustrated in Figure 3.2(d). The architecture consists of separate hardware subsystems called stages to fetch an instruction from memory, decode the instruction, fetch instruction operands from memory or registers, execute the instruction, and then write the results back to memory. Each stage is simultaneously processing a different instruction such that the overall result is that an instruction completes execution on every clock cycle. For example, in a five-stage pipeline, five instructions are simultaneously being processed through the pipeline each at a different stage. Typically, instructions in a pipeline processing system are simple instructions easily implemented within a single stage. More complex instructions are built up from these small instruction building blocks.

3.1.3 RISC versus CISC Instruction Set

Closely related to the hardware architecture of the microcontroller is the architecture of the instruction set. There are two basic types of instruction set architectures: Reduced Instruction Set Computer (RISC) and Complex Instruction Set Computer (CISC) Architecture. A RISC processor as its name implies has a complement of simple building block instructions. More complex instructions are built up from the basic instructions in the RISC processor. RISC-based instruction architectures lend themselves to systems with less complex CPU architectures. A CISC-based architecture has a complement of fuller feature, more complex instructions than the RISC-based architecture. It is difficult to predict whether a given program will be more efficiently coded with a RISC- or CISC-based instruction set. It largely depends on how well the specific algorithm matches the feature set of a given processor.

As a system designer you need to be intimately familiar with the hardware and software architecture of a given microcontroller, particularly if you will be coding the system using an assembly language. However, if you will be programming using a high-level language such as C, knowledge of some of the lower level architectural details are not required as long as you have a thorough understanding of the microcontroller subsystems at the register level. We will assume this approach throughout the remaining portions of the text. In the next several sections we briefly describe the hardware subsystems common to most microcontrollers.

3.2 REGISTER SET

Most microcontrollers have a complement of registers designated as the register set. The register set is the interface between you the user and the different subsystems aboard the microcontroller. Each register consists of a collection of flip-flops. Each flip-flop can either be set to a logic one or logic zero. Each flip-flop can be viewed as a software configurable switch. Sending a logic one to the flip-flop asserts or turns on the switch while sending a logic zero to the flip-flop de-asserts or turns the flip-flop off.

The flip-flops are categorized by subsystem as shown in Figure 3.3. Usually all registers associated with a given subsystem are grouped together. To configure a specific subsystem the system designer will determine the appropriate setting for each bit within the register. The

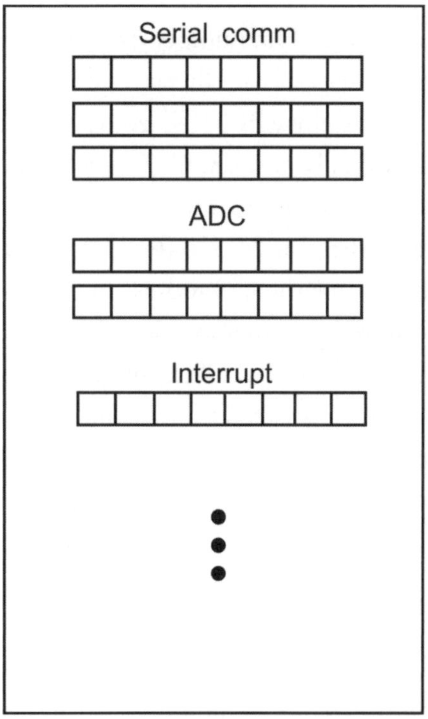

FIGURE 3.3: Register set

function of each register and each register bit are carefully defined in the specific microcontroller's documentation.

All bits within a register are programmed simultaneously by setting the name of the register (as defined in the compiler definitions) to the desired bit pattern. For example, to set a register called SCIBaudRate to the binary value 1010_1110 the following command may be used:

```
SCIBaudRate = 0xAE;          //0x is used to designate a hexadecimal number
```

It is important to note that compilers provide a complement of header files. These header files contain the "personality data" for a given microcontroller. Specifically, they provide a link between the name of a specific register used within a program and its location within the microcontroller.

3.3 BUS STRUCTURE

Different subsystems within a microcontroller are connected with several different buses. A bus is a collection of parallel conductors that have a similar function. Most microcontrollers are equipped with an address bus, a data bus, and a control bus.

3.3.1 Address Bus

The address bus provides a connection between the central processing unit and the memory subsystem aboard the microcontroller. The number of conductors in the address bus sets the upper limit of memory locations that may be linearly addressed by the microcontroller. The first address in the memory subsystem will be all zeroes while the final address will be all logic ones. The number of individually addressable memory addresses may be determined by evaluating $2^{address\ lines} = addressable\ locations$. For example, a microcontroller equipped with a 16-bit address bus is capable of addressing 65 536 (64 kB) separate locations. The first address in this memory space is $(0000)_{16}$ while the last address in this space will be $(FFFF)_{16}$.

To expand the span of addressable memory locations, some microcontrollers employ a paged memory addressing scheme. In paged addressing, a memory system is subdivided into memory pages. Memory page length is usually some smaller block of memory such as a 4 kB page. Many 4 kB pages of memory can be assigned to the same 4,096 addresses in the linearly addressable memory space. To select a specific memory page for access additional addressing or select bits are required.

3.3.2 Data Bus

The data bus as its name implies is used to route parallel data about different subsystems within the microcontroller. Microcontrollers are commonly available with data bus widths of 4, 8, 16, or 32 bit. The width of the data path generally determines the size of a data argument that the microcontroller can process. For example, the largest unsigned integer that may be stored in a microcontroller with an eight-bit data path is 255. It should be emphasized that a 32-bit microcontroller is not a better microcontroller than its four-bit counterpart. Recall that the primary objective of the embedded system designer is to choose the most economical microcontroller that accomplishes the requirements of a specific design.

For example, a four-bit microcontroller may be an ideal choice to host the control algorithm for the irrigation (water sprinkler) system used to water your lawn. However, a 32-bit processor may be required to host the control features required of a cellular phone.

3.3.3 Control Bus

Microcontrollers are equipped with paths to send and receive a collection of control signals designated as the control bus. These signal lines carry control signals to different subsystems throughout the microcontroller. Most of the control signals are internal to the microcontroller integrated circuit (chip). However, they are often conveniently provided at a microcontroller port so that external components may be added to the processor. Control signals are issued by the CPU in response to program instructions to insure the instruction is properly executed.

3.4 MEMORY

As previously mentioned the number of uniquely addressable memory locations in a microcontroller is determined by the width of the address bus. This span of addressable memory usually contains several different types of memory including Static Random Access Memory (SRAM), byte-addressable Electrically Erasable Programmable Read Only Memory (EEPROM), and bulk programmable Flash EEPROM. It should be emphasized that these memory components are contained within the confines of the microcontroller chip.

To keep a track of the memory locations in use and the type of memory present within the system, a visual tool called a memory map is employed. The memory map provides the size in bytes of each memory component and its start and stop address within the memory system. A sample memory map is provided in Figure 3.4. Note that there are portions of the memory map not in use. These open spaces are provided for system expansion.

The following memory components are shown in Figure 3.4 and are commonly available in most microcontrollers:

- *RAM:* RAM memory is volatile. That is, if the microcontroller loses power, the contents of RAM memory are lost. It can be written to and read from during program execution. It is typically used during system development to store a program. Once development is complete, the completed program is stored in nonvolatile memory such as Flash EEPROM. During program execution, RAM is used to store global variables, support

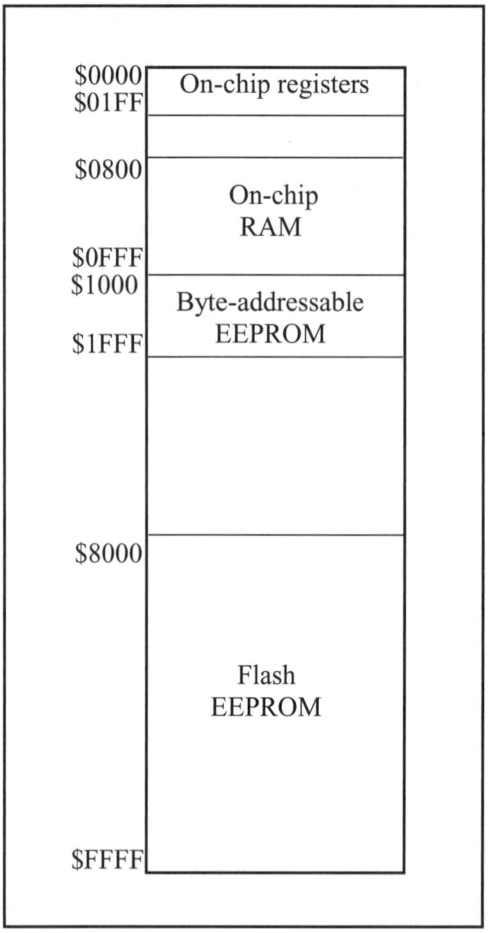

FIGURE 3.4: Memory map. The memory map shows which memory addresses are currently in use and the type of memory present

dynamic memory allocation of variables, and to provide a location for the stack (to be discussed later).

- *Byte-addressable EEPROM:*This type of memory is used to permanently store and recall variables during program execution. It is especially useful for logging system malfunctions and fault data during program execution. It is also useful for storing data that must be retained during a power failure but might need to be changed periodically. Examples where this type of memory is used are found in applications to store system parameters, electronic lock combinations, and automatic garage door electronic unlock sequences.
- *Flash EEPROM:* Bulk programmable Flash EEPROM is used to store programs. It can be erased and programmed as a whole. Some microcontroller systems provide a large complement of both RAM and Flash EEPROM. Therefore, a system program can be developed in RAM and then transferred to Flash EEPROM when complete.

Other microcontrollers provide only a large Flash EEPROM and a smaller RAM component. With this memory configuration, system development takes place in Flash EEPROM. Flash EEPROM is typically programmed using In System Programming (ISP) techniques. That is, a host PC is connected via a cable to a microcontroller while it is resident within its application circuit. The host PC downloads the program to the microcontroller.

3.5 TIME BASE

As previously mentioned a microcontroller is a complex synchronous state machine that responds to program instructions in a predictable fetch-decode-execute sequence. The speed at which a microcontroller sequences through its actions is controlled by an external time base. The time base may be provided by an external quartz crystal, a programmable oscillator, a ceramic resonator, or an internal time base. Some microcontrollers are designed to operate at a set specific frequency such as 8 MHz while others have been flexibly designed to operate at a wide range of frequencies. Currently, microcontrollers are available that operate up to approximately 50 MHz. Frequently a microcontroller is employed to control a slow (relative to the microcontroller) mechanical system. In this type of application a slow time base may be required. A 32.768-kHz time base is frequently used in systems that track 24 h clock time. Some microcontrollers are equipped with an internal time base.

3.5.1 Timing Subsystem

The main clock source is routed throughout the microcontroller to provide synchronicity to the subsystems. Most microcontrollers are also equipped with a timing subsystem. To better understand the features of this subsystem, we review common terminology associated with signal timing illustrated in Figure 3.5.

- *Frequency:* Signal frequency is the number of cycles per second completed by a repetitive signal. It is expressed in units of Hertz (Hz).
- *Period:* The period is the time increment in seconds required for a repetitive signal to complete a single cycle. The period is the reciprocal of the frequency ($T = 1/f$).
- *Duty cycle:* The duty cycle indicates the percentage of time for which the signal is active in a single period.
- *Pulse width modulation (PWM):*PWM signals are frequently used to control motor speed. The digital PWM signal is converted to an effective DC value by the mechanical inertia of the motor as well as the low pass filter characteristics of the inductance inherently present in a motor. Note in Figure 3.5 how various effective DC voltages can be delivered to a load by simply adjusting the duty cycle of the digital PWM signal.

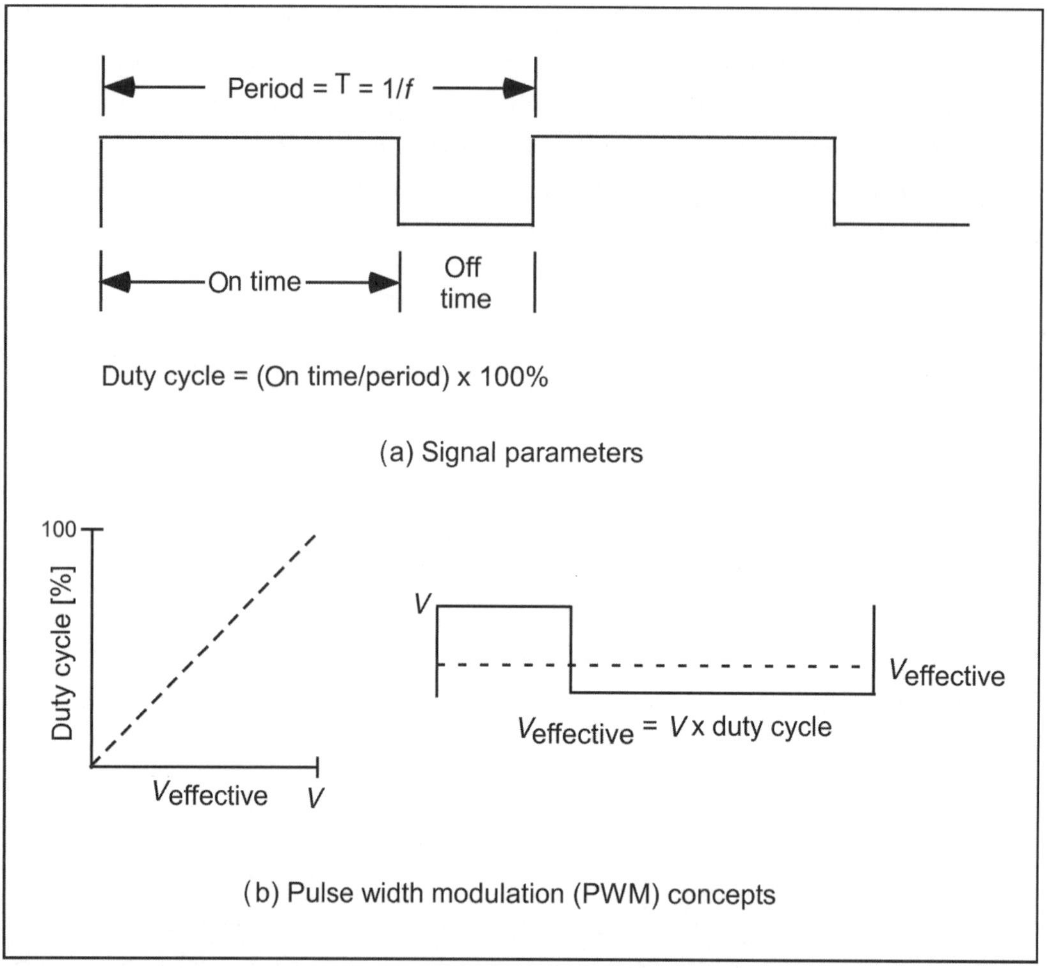

FIGURE 3.5: Timing concepts

Applications

Most microcontrollers are equipped with a multichannel timing system. The channels within the timing system may be configured to

- Measure parameters of input signals such as period and duty cycle.
- Generate precision output pulses or repetitive signals.
- Count incoming pulses present in an input signal.
- Generate PWM signals.

3.6 PORT SYSTEMS

Microcontrollers are equipped with a series of ports to provide access to the world beyond the microcontroller. Frequently these ports are organized as eight-bit input/output ports as shown in Figure 3.1. Usually a port register is equipped with an accompanying data direction register. This is used to set the direction (input or output) for a given port pin.

Although the ports are used by the input and output of digital signals, many have alternate functions such as analog-to-digital conversion, serial communication, and network interfacing. These features are briefly described in the next several sections and in additional detail later in the textbook.

Even when equipped with alternate functions, often there are not enough external pins to provide access to all microcontroller features. To alleviate this challenge, some microcontroller ports are equipped with time multiplexed capability. This simply means that the function of a given port alternates at a prescribed interval. For example, expansion ports provided to route the data and address ports outside the microcontroller may employ time multiplexed features. To deinterleave the time multiplexed port data, external data latches are required. A common latch used for this application is the 74HC573 (octal, three-state, D-type transparent latch). Expansion ports are also provided to route control signals from inside the microcontroller to external components.

3.7 ANALOG-TO-DIGITAL CONVERTERS

Many microcontrollers are equipped with analog-to-digital conversion (ADC) subsystems. This subsystem converts continuously varying analog signals from the outside world into a binary representation suitable for use by the microcontroller. These converters commonly have 8–10-bit resolution. Therefore, a continuous signal is converted to a series of digital snapshots of the analog signal. You as the system designer must determine how often to initiate an ADC conversion for a given application. Later in the text we provide detailed information to aid in the design of a system using an ADC.

3.8 COMMUNICATION SYSTEMS

When we consider communication between two systems, the taxonomy consists of parallel and serial communications. Simply put, a parallel communication method utilizes multiple channels, bus wires, to send and receive multiple streams of data simultaneously, compared to a serial communication method where only a single stream of data is sent and received at a time.

The most obvious advantage of a parallel communication method over a serial communication method, provided that the communication rate is equal, is the speed of data transfer. Using the multiple number of connections, the same amount of data can be sent and received quicker, proportional to the number of connections, than a single connection used in a serial communication method. The disadvantage of a parallel communication method is the hardware and software cost to enable the fast data transfer. Typically, parallel communication techniques are used for short distance communication within and outside of a microcontroller. For a long distance communication, a serial communication technique is used to send and receive data.

3.8.1 Serial Communications

There are two different types of serial communications available: synchronous communication and asynchronous communication. The key challenge in serial communications is to maintain synchronization between the transmitter and the receiver. The asynchronous communication method uses a start and stop bit protocol to synchronize a transmitter and receiver. The start and stop synchronization bits are embedded in each transmitted signal byte. The advantage of the asynchronous communication is that it is inexpensive, but the disadvantage is that the data transmission rates are typically slower than a synchronous serial communication system due to its overhead (start and stop bits). Asynchronous serial communication subsystems are referred to as the UART (universal asynchronous receiver transmitter) or the SCI (serial communications interface).

The synchronous serial communication uses a synchronized clock to send and receive each bit. The synchronous serial communication receiver is much more simple compared to an asynchronous serial communication receiver module. It may be viewed as a synchronous 16-bit shift register with an eight-bit half residing in the transmitter and the other eight-bit half residing in the receiver. The disadvantage of the synchronous serial communication compared to the asynchronous communication method is the task of synchronizing the transmitter and receiver clocks. This is usually accomplished with an additional clock line linking the transmitter and receiver. For a long distance communication, the synchronous serial communication technique is not recommended. For a short distance communication, the synchronous communication system can considerably improve the bit throughput over the asynchronous communication system. A synchronous serial communication system is referred to as the SPI (serial peripheral interface).

3.8.2 Terminology

To better understand serial communication concepts, some terminology must be first introduced. Space constraints do not allow us to include an exhaustive list of terms associated with serial communication techniques. We only present those that one would readily encounter in the technical documentation and literature.

- *Simplex mode:* In this mode, the serial communication is accomplished by transmitting data in one direction at a time.
- *Duplex mode:* In this serial communication mode, data can be transmitted and received from both ends of the communication link at the same time.
- *BAUD rate:* The rate of bits sent or received. It describes the number of bits communicated per second.

- *ASCII code:* The American Standard Code for Information Interchange code is used in communication to encode alphabets, numbers, punctuation, and control characters using a seven-bit representation. ASCII is a subset of the international Unicode standard.
- *Bit time:* The time required to transmit or receive a single bit.
- *Serial line code:* A specific encoding mechanism used to transmit and receive information.

3.8.3 Serial Communication Signals

In this section, we consider five different serial communication line codes. It is important that both the transmitter and receiver use a common line code, BAUD rate, and parity setting for proper communications. The different line codes are illustrated in Figure 3.6.

- *Non-return-zero (NRZ) line code:* Bit one is represented as voltage high and bit zero is represented as zero voltage.
- *Non-return-zero-inverted (NRZI) line code:* Bit change is represented with voltage high and no change is represented as zero voltage.
- *Return-zero (RZ) line code:* Bit zero is represented with voltage zero and bit one is represented with voltage high for the first half bit time and the rest of the bit time is represented with voltage zero.

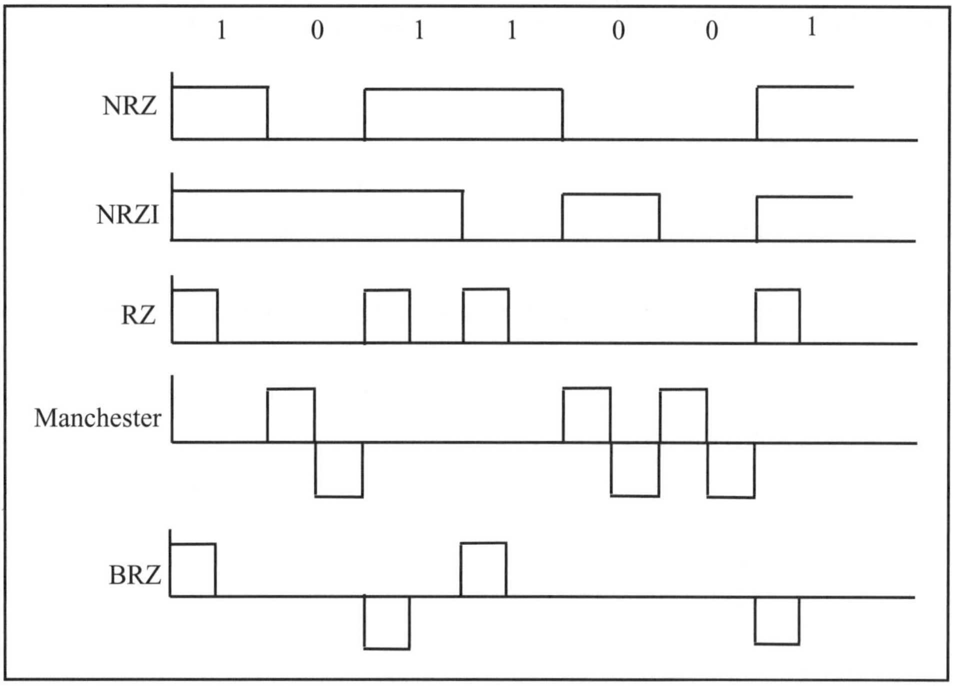

FIGURE 3.6: Serial line codes

- *Manchester line code:* Bit one is represented with voltage zero and bit zero is represented with voltage one for the first half of a bit time and the rest of the bit time is the opposite voltage level of the first half of the bit time.
- *Bipolar-return-zero (BPRZ) line code:* Bit zero is represented by voltage zero and bit one is represented by either voltage high and voltage low for the first half of a bit time and the rest of the bit time is represented by voltage zero. The representation for bit one alternates between voltage high and voltage low.

3.8.4 Handshake Mechanisms

To ensure the robust serial communication with minimal errors, several handshake mechanisms have been developed. For asynchronous serial communications, each data frame is constructed with start bits, data bits, a parity bit, and stop bits. The start bits inform a receiver that a data frame has arrived. Typically, one or two bits are used as the start bits. When the transmitter is idle it typically provides a logic high idle signal on the serial communication output pin. A start bit is typically a logic low. The transition from idle (logic high) to a start bit (logic low) allows the receiver to detect the start of a new incoming data frame.

Data bits can be either eight or nine bits. The ASCII format uses seven bits to represent a character. The eighth bit is used for parity. A nine-bit format is used when the data to be transmitted is eight bits in width (e.g. the output from an eight-bit analog-to-digital converter). In this case, the ninth bit in the data portion of the frame is the parity bit.

The parity bit improves the accuracy of serial communication by providing the receiver the ability to detect the presence of a single bit error in transmission. There are two different types of parity bits: an even parity bit and an odd parity bit. When an even parity bit mode is used, the parity bit is used to make the number of logic high data bits and the parity bit to be an even number. When an odd parity bit mode is used, the total number of logic high bits in the data bits and the parity bit must be an odd number. If a receiver detects a parity error, it may notify the transmitter to resend the data. If more robust error correction capability is required, additional parity bits may be used.

Similar to the start bits, the stop bits inform a receiver of the end of a data frame. Start, data, parity, and stop bits are used by a receiver to ensure correct reception of a frame and synchronization between transmitter and receiver.

3.8.5 RS-232 Protocol

Many peripheral devices are equipped to communicate with an RS-232 compatible interface. The "RS-232" is an Electronic Industry Association (EIA) standard formally designated EIA-232-D. This standard specifies the different aspects of a serial communication interface including

- electrical specifications,
- functional signal specification,
- mechanical specification, and
- procedural specifications.

A variety of chips are available to translate microcontroller compatible signals to RS-232 compatible signals. The RS-232 standard represents a logic high with a −10 VDC level and a logic low with a +10 VDC level. These chips are equipped to provide interfacing for a two-way (transmit and receive) communication system. That is, the output serial bit stream from a microcontroller to a RS-232 compatible peripheral device must be converted to a RS-232 signal. The serial bit stream returning from the peripheral device must be converted from the RS-232 format back to microcontroller compatible signal levels.

3.9 INTERRUPT SYSTEM

A program consists of a number of program steps that are executed in sequence. The normal execution of a program step follows the fetch, decode, execute sequence as shown in Figure 3.7. Sometimes this normal sequence of events must be interrupted to respond to high priority faults and status both inside and outside the microcontroller. When these higher priority events occur,

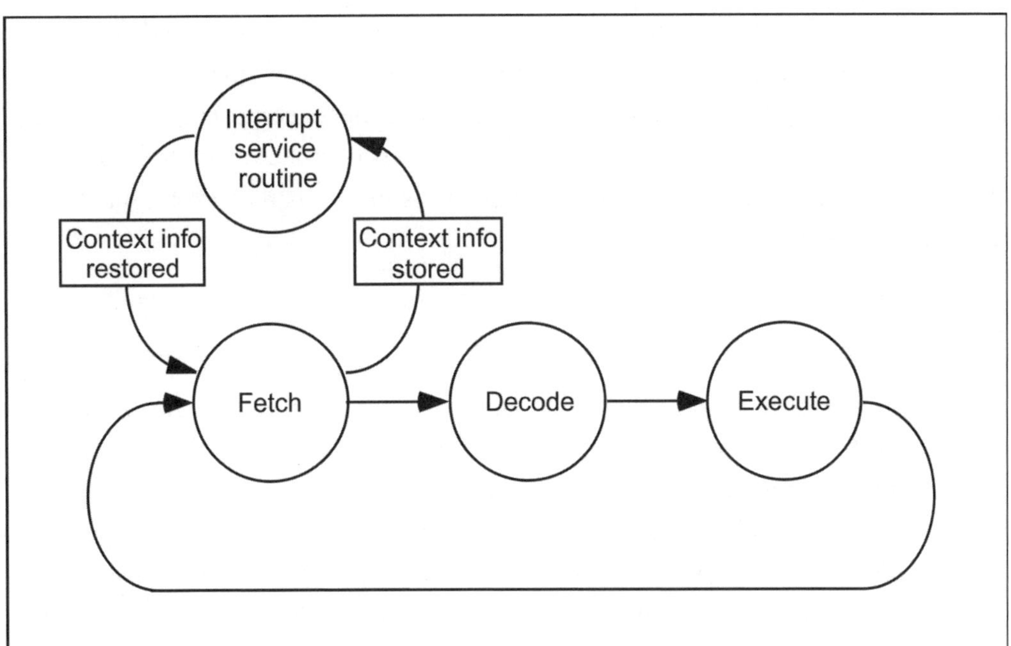

FIGURE 3.7: Interrupt concepts

the microcontroller must temporarily suspend normal operation and execute event specific actions. These actions are commonly called an interrupt service routine. Once the higher priority event has been serviced, the microcontroller returns and continues processing the normal program. It must be emphasized that although the interrupt temporarily suspends the operation of the normal sequence of events, an orderly transition to and from the interrupt must be provided. This is accomplished by the use of a stack. A stack is a temporary storage location set aside as a portion of RAM memory.

When an interrupt occurs the microcontroller will finish executing the current instruction. It will then place context information, key register values and the return address to the normal program, on the stack. The interrupt service routine (ISR) will then be executed. The ISR consists of the microcontroller activity required by the interrupt. When the ISR is finished, control will return to the program that was executing when the interrupt occurred. The context information previously stored on the stack will be used to restore the controller to its preinterrupt configuration.

There is a wide variety of interrupts available with most microcontrollers. Some are generated by hardware or software malfunctions. Others are available to the system designer. Two very useful interrupts are the real time interrupt (RTI) and the external interrupt request (IRQ). The RTI interrupt provides a regular, periodic interruption in the main program flow. This is useful for updating a real time clock or checking critical system status such as the system battery level. The IRQ is an external controller hardware pin. When this pin is asserted an interrupt is generated and the associated interrupt service routine is executed. This is a useful method of allowing an external system hardware component to alert the microcontroller that a malfunction has occurred. For example, if the microcontroller is controlling an access gate and something is preventing the gate from opening, the microcontroller needs to be alerted of this condition. If not, the microcontroller will continue to issue an open control signal to the gate's motor and cause the motor to be damaged.

3.10 SPEED

A novice system designer might believe that the best microcontroller is the one that has the fastest operating speed and equipped with the most number of features. As mentioned earlier, it is the system designer's responsibility to find the most economical microcontroller with the proper features for a given application. Microcontrollers are available in a wide range of operating speeds. In general, you should use the lowest acceptable speed for a given application. This is because that the power consumption of a microcontroller is directly proportional to its operational speed. Since many microcontroller applications are battery powered, conserving power and hence extending battery life is essential.

3.11 CHOOSING A MICROCONTROLLER FOR A SPECIFIC DESIGN

In this section, we introduce a systematic approach to choosing a specific microcontroller for a given design. Rather than presenting a dry list of procedures, we use a case study of developing a microcontroller-based pulse oximetry system to present this material.

Pulse oximetry: Pulse oximetry is a method of transcutaneously measuring the saturated oxygen ($\%S_aO_2$) content of arterial blood using an ear or finger clip as illustrated in Figure 3.8(a). This technique of measuring saturated oxygen content of blood is based on the concept of measuring the relative amount of oxygen carried by the hemoglobin in the red blood cells within the arteries. As can be seen in Figure 3.8(b) the light absorption properties of oxygenated hemoglobin (HbO_2) is different from hemoglobin (Hb). When probed at two different wavelengths

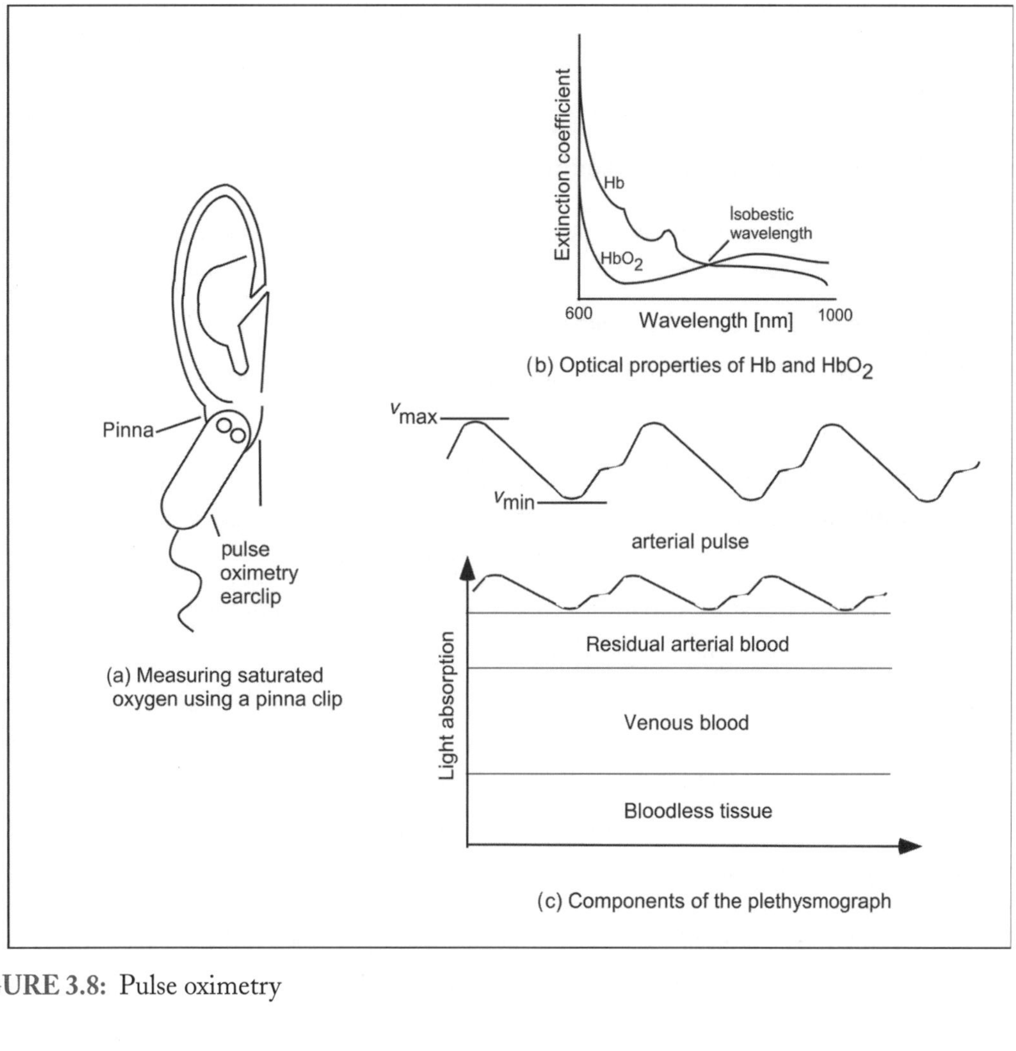

FIGURE 3.8: Pulse oximetry

the light transmittance may be measured and used to determine $\%S_aO_2$ (percent saturated oxygen).

The pulse oximetry clip contains two light-emitting diode (LED) sources and a light detector. One LED emits light at a red wavelength (approximately 660 nm) while the other emits at an infrared wavelength (approximately 940 nm). The transmission of light through the pinna (ear lobe) or finger tip is sequentially measured at the two wavelengths resulting in a waveform as shown in Figure 3.8(c). The maximum and minimum voltage of the resulting waveform at each wavelength is determined and used to determine the saturated oxygen measurement using an expression similar to

$$\%S_aO_2 = -(100R/3) + (340/3)$$

where

$$R = \ln[V_{max}R(n)/V_{min}R(n)]/\ln[V_{max}IR(n)/V_{min}IR(n)]$$

3.11.1 System Requirements

To develop a microcontroller-based system to host a pulse oximetry system, the following microcontroller subsystems would be required:

- Two output timing channels to issue the control signals to assert the red and infrared LEDs.
- A single analog-to-digital conversion channel to collect data from the detector.
- An eight-bit output port and two additional output control signals for a liquid crystal display.
- An eight-input port to interface a keypad (optional).
- A memory system large enough to host the algorithm used to control the data collection sequence, analyze the data, and display the information.
- A serial communication capability to interface the pulse oximeter to a central computer at an intensive care unit control desk.

Figure 3.9 summarizes these requirements. With this list of requirements, a suitable microcontroller may be chosen. As you might suspect, there will be many, good candidate controllers to satisfy these requirements.

To summarize, these are some of the considerations for choosing an appropriate microcontroller for a specific application:

- Number of ports: Enough ports and external pins to support application?
- Required subsystems: Equipped with the proper subsystems for a specific application?

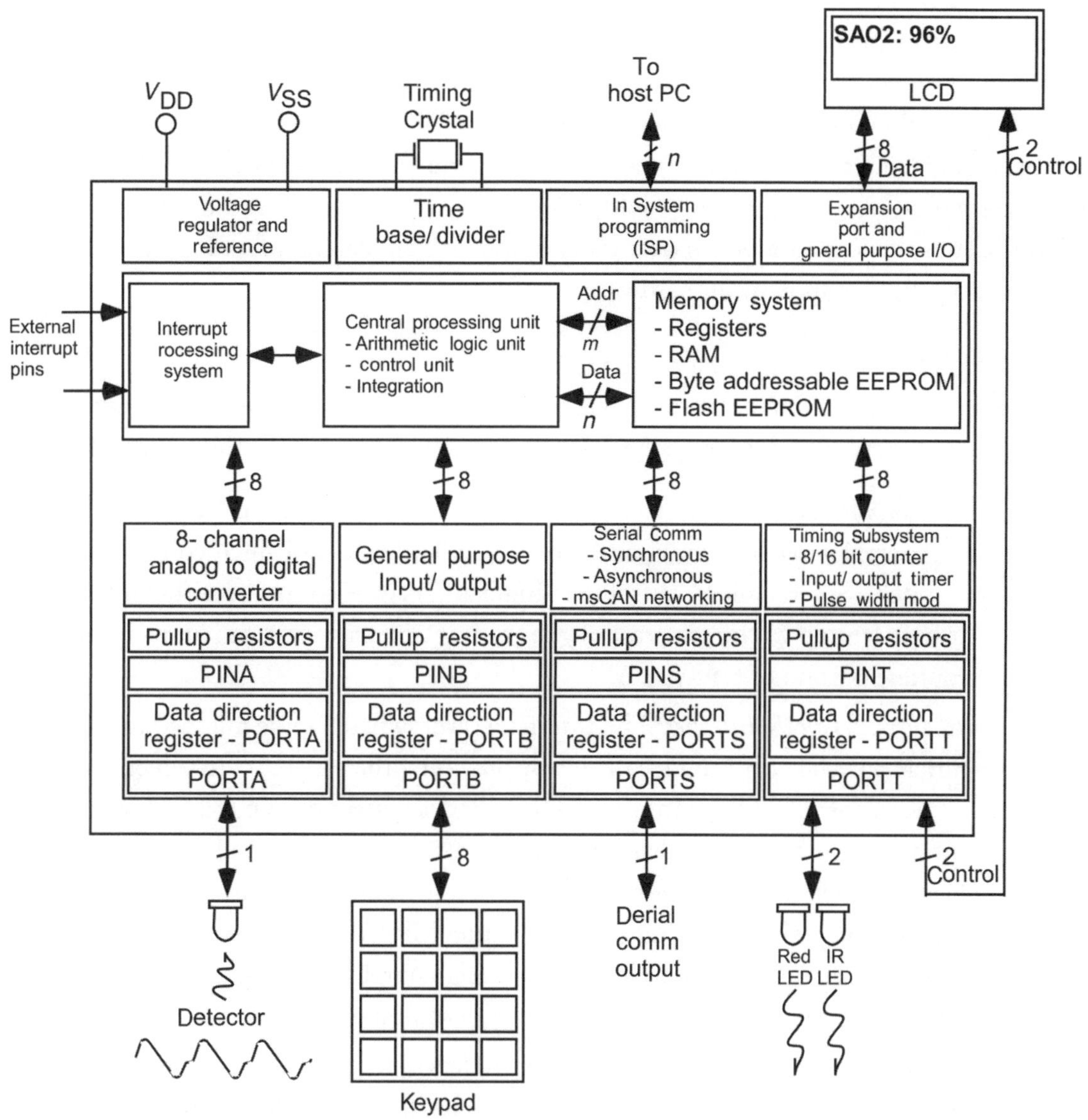

FIGURE 3.9: Pulse oximetry block diagram

- Memory size and type: Memory large enough to support application algorithm?
- Clock speed: Does the processor operate at a sufficient speed to complete the algorithm in the required time?
- Power dissipation: Is a low power dissipation processor required for a specific application?
- Special feature set: Fuzzy logic? Signal processing features? etc.

3.12 MICROCONTROLLER VENDORS

We will not attempt to list all of the microcontroller manufacturers here or their products. As you might imagine the product lines are constantly evolving. Rather we refer you to the Internet. A simple search on the keyword microcontroller will provide a multipage listing of available manufacturers. There are also a number of good websites dedicated to providing the latest information on new technology and products. We also recommend using the Internet to find the latest information on support tools: compilers, emulators, programming tools, and related microcontroller application hardware.

3.13 CUTTING EDGE TECHNOLOGY

With the brief review of microcontroller features and subsystems complete, let us take a brief look at some of the cutting edge subsystems and trends occurring in microcontroller technology.

- *FPGA and microcontroller combined processors:* It is becoming common place to develop a design employing both programmable logic hardware and microcontrollers in the same system. Several microcontroller manufacturers have come out with joint development kits. It is up to the system designer to implement system functions in the most appropriate portion of the system.
- *Microcontrollers on the Internet:* The Internet is everywhere! Some microcontroller manufacturers have taken advantage of this by providing the capability to directly interface their microcontroller to the Internet. This allows a system design to employ the Internet infrastructure to connect distributed microcontrollers. This would be well suited for applications such as providing security and safety monitoring within a large building or complex with existing Internet infrastructure. Microcontroller-based data collection nodes can be distributed throughout a building complex and then relay the collected information via the Internet infrastructure to a central processor.
- *Data links:* As mentioned earlier, microcontrollers may be networked together in a simple but powerful configuration. These microcontroller networks have become common place in a wide variety of commercial and military products.
- *USB controllers:* The universal serial bus (USB) has become a common method of connecting peripheral devices to a PC. Several manufacturers have provided USB support for their microcontrollers. This allows for the easy interface of a microcontroller to a PC for data exchange and peripheral development.

3.14 SUMMARY

In this chapter we have provided an overview of the systems commonly available in microcontrollers. We have investigated memory technology, ADC converters, timing systems, serial communication, and cutting edge features. This chapter was meant only as an overview. Additional information on several of these subsystems is provided later in the text.

BIBLIOGRAPHY

D Pack and S Barrett, 68HC12 Microcontroller: Theory and Applications, Prentice-Hall, Upper Saddle River, NJ, 2002.

S Barrett and D Pack, Embedded Systems Design with the 68HC12 and HCS12, Prentice-Hall, Upper Saddle River, NJ, 2005.

H Huang, The HCS12/9S12: An Introduction to Software and Hardware Interfacing, Thomson, Clifton Park, NY, 2006.

M Predko, Programming and Customizing PICmicro Microcontrollers, McGraw-Hill/TAB Electronics, 2000.

D Clark, Programming and Customizing the OOPic Microcontroller: The Official OOPic Handbook, McGraw-Hill, 2003.

M Predko, Programming and Customizing the 8051 Microcontroller, McGraw-Hill/TAB Electronics, 1999.

I Mackenzie, The 8051 Microcontroller, Prentice-Hall, Upper Saddle River, NJ, 1998.

D Gadre, Programming and Customizing the AVR Microcontroller, McGraw-Hill/TAB Electronics, 2000.

A Williams, Microcontroller Projects Using the Basic Stamp, CMP Books, 2002.

T Van Sickle, Programming Microcontrollers in C, CMP Books, 2002.

J Skroder, Using the M68HC11 Microcontroller: A Guide to Interfacing and Programming. Prentice-Hall Incorporated, Upper Saddle River, NJ, 1996.

G Lipovski, Embedded Microcontroller Interfacing for M-CORE Systems, Academic Press, 2000.

J Hennessy and D Patterson, Computer Architecture A Quantitative Approach, Morgan Kaufman, San Francisco CA, 2003.

D Patterson and J Hennessy, Computer Orgination and Design The Hardware Software Interface, Morgan Kaufman, San Francisco CA, 1994.

CHAPTER PROBLEMS

- **Fundamental**
  1. *Question*: What is the relationship between the width of a microcontroller's data and address bus and its ability to interface with memory?
  2. *Question*: What is the difference between linear and paged memory?
  3. *Question*: Describe the different types of memory commonly available on a microcontroller. What is each type of memory commonly used for?

- **Advanced**
  1. *Question*: Briefly describe the three basic common hardware CPU architectures found within a microcontroller.
  2. *Question*: Describe how to select a specific microcontroller for a given application.
  3. *Question*: Describe electrical parameters associated with a microcontroller and their importance in interfacing external devices.
- **Challenging**
  1. *Question*: From memory, sketch a block diagram of a microcontroller and its subsystems. Provide a brief explanation of the function of each subsystem and a possible application for each.
  2. *Question*: Research three cutting edge features of microcontroller technology. Write a one page point paper on each feature.

CHAPTER 4

Timing Subsystem

Objectives: After reading this chapter, the reader should be able to

- Compute the frequency and the period of a periodic signal using a microcontroller.
- Explain functional components of a microcontroller timer system.
- Describe the procedure to capture incoming signal events.
- Describe the procedure to generate time critical output signals.
- Design an embedded system to interface a microcontroller with external devices requiring time critical signals.

The focus of this chapter is to introduce timing related topics of microcontrollers. In the next section, we first present some fundamental concepts required to discuss more complex topics that follow in Section 5.2. We complete the chapter with a set of timer related applications in Section 5.3.

4.1 BACKGROUND THEORY

One of the most important reasons for using microcontrollers in embedded systems is the capabilities of microcontrollers to perform time related tasks. In a simple application, one can program a microcontroller system to turn on or turn off an external device at a programmed time. In a more involved application, we can use a microcontroller to generate complex digital waveforms with varying pulse widths to control the speed of a DC motor. Before we delve into the microcontroller technologies used to perform time related tasks, we first need to define some terms.

4.1.1 Frequency

Consider a signal $x(t)$ that repeats itself. We call this signal periodic with period T, if it satisfies

$$x(t) = x(t + T).$$

We find periodic signals all around us. For example, periodic signals are used to transfer power to your homes and generate telephone ring tones. To measure the frequency of a periodic signal, we need to agree on the time interval within which we count the number of times a particular event repeats. The time period commonly used is the second, and the number of times an event occurs is defined using the Hertz (Hz) unit. For example, a sinusoidal signal with a 60 Hz frequency means that a full cycle of a sinusoid signal repeats itself 60 times each second. In fact such periodic signals are used to turn on light bulbs in your homes; light bulbs turn on and off 60 times per second. Why do not we see lights flickering? Humans can only detect periodic signals up to approximately 28 Hz. Another example you are already familiar with is your television. A television in the United States displays 30 image frames per second. The event in this application is an image and we say the frame rate is 30 Hz. You are also familiar with the clock speed on your computer. It seems only a couple of years ago when we thought 133 MHz (133 times 1 million cycles per second) clock was fairly fast. The normal clock speed of a computer today is 2.5 GHz (2.5 times 1 billion cycles per second).

4.1.2 Period

The flip side of a frequency is a period. If an event occurs with a rate of 1 Hz, the period of that event is 1 s. To find a period, given a frequency, or vice versa, we simply need to remember their inverse relationship $f = \frac{1}{T}$ where f and T represent a frequency and the corresponding period, respectively. Both periods and frequencies of signals are often used to specify timing constraints of embedded systems. For example, when your car is on a wintery road and slipping, the engineers who designed your car configured the antislippage unit to react within some millisecond period, say 20 ms. The constraint then forces the design team that monitors the slippage to program their monitoring system to check a slippage at a rate of 50 Hz.

4.1.3 Duty Cycle

In many applications, periodic pulses are used as control signals. A good example is the use of a periodic pulse to control a servo motor. To control the direction and sometimes the speed of a motor, a periodic pulse signal with a changing duty cycle over time is used. The periodic pulse signal shown in Figure 4.1 frame (a) is on for 50% of the signal period and off for the rest of the period. The pulse shown in frame (b) is on for only 25% of the same period as the signal in frame (a) and off for 75% of the period. The duty cycle is defined as the percentage of one period for which a signal is on. Therefore, we call the signal shown in frame (a) in Figure 4.1 as a periodic pulse signal with a 50% duty cycle and the corresponding signal in frame (b), a periodic pulse signal with a 25% duty cycle. In Section 5.3, we discuss how the variation of the duty cycle may be used to directly control the speed of a DC motor.

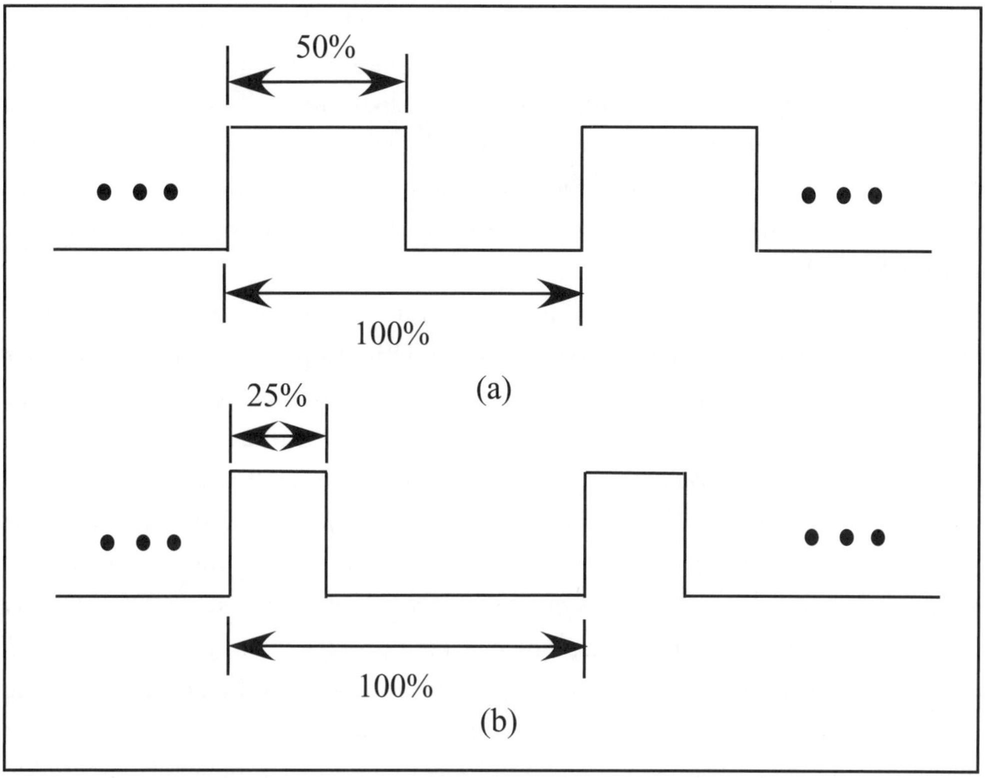

FIGURE 4.1: Two signals with the same period but different duty cycles. Frame (a) shows a periodic signal with a 50% duty cycle and frame (b) displays a periodic signal with a 25% duty cycle

4.2 TIMER SYSTEM

Although the sophistication of the timer unit varies from microcontroller to microcontroller, most microcontrollers have a built-in timer unit. In this section, we consider timer functions that one can find in most microcontrollers. The overall goal of a timer system in a microcontroller is twofold (1) detect and capture external timing events and (2) generate timing events to control and access external systems.

4.2.1 Hardware

Over the years, the timing hardware on microcontrollers has evolved to provide sophisticated input and output capabilities. The hardware starts with a crystal time base. The crystal's frequency of a microcontroller is used to generate a baseline clock signal. Each operation of a microcontroller is then governed by the clock signal; an instruction is fetched, decoded, and executed in synchronization with the clock.

For a timer system, the system clock is used to update the contents of a special register called a free-running counter. The job of a free-running counter is to count up (increment) each time it sees a rising edge (or a falling edge) of a clock signal. Thus, if a clock is running at the rate of 2 MHz, the free-running counter will count up each 0.5 *usecs*. All other timer related units reference the contents of the free-running counter to perform input and output time related activities: measurement of time periods, capture of timing events, and generation of time related signals.

For input time related activities, all microcontrollers typically have timer hardware components that detect signal logic changes on one or more input ports. Such components rely on a free-running counter to capture external event times. We can use such ability to measure the period of an incoming signal, the width of a pulse, and the time of a signal logic change.

For output timer functions, a microcontroller uses a comparator, a free-running counter, logic switches, and special purpose registers to generate time related signals on one or more output ports. A comparator checks the value of the free-running counter for a match with the contents of another special purpose register where a programmer stores a specified time value in terms of the free-running counter value. The checking process is executed at each clock cycle and when a match occurs, the corresponding hardware system induces a programmed logic change on a programmed output port pin. Using such capability, one can generate a simple logic change at a designated time incident, a pulse with a desired time width, or a pulse width modulated signal to control servo or Direct Current (DC) motors.

For both input and output time related activities, a typical microcontroller has associated interrupt capabilities. That is, when a desired input signal appears, an interrupt system can be triggered to capture the current free-running counter value and use the value for a desired purpose. Similarly, output timer related interrupts are used to service tasks when an output event occurs.

Usually, the timer hardware of a microcontroller demands a setup of the most number of registers. A programmer must initialize the speed of the free-running counter, program the desired logic signal to be detected for an input timer related activity, configure the desired logic signal for an output timer related activity, specify a physical input or output port, set up time related registers, set up flags associated with a timer port, and configure all registers associated with interrupts, if needed. A novice student usually takes some additional time to master the workings of a micrcontroller timer system, but the mastery is critical for any embedded system applications.

4.2.2 Operation

Figure 4.2 shows a generic architecture of a microcontroller timer system. A crystal is used to generate a periodic clock signal which in turn controls all timing activities of the microcontroller.

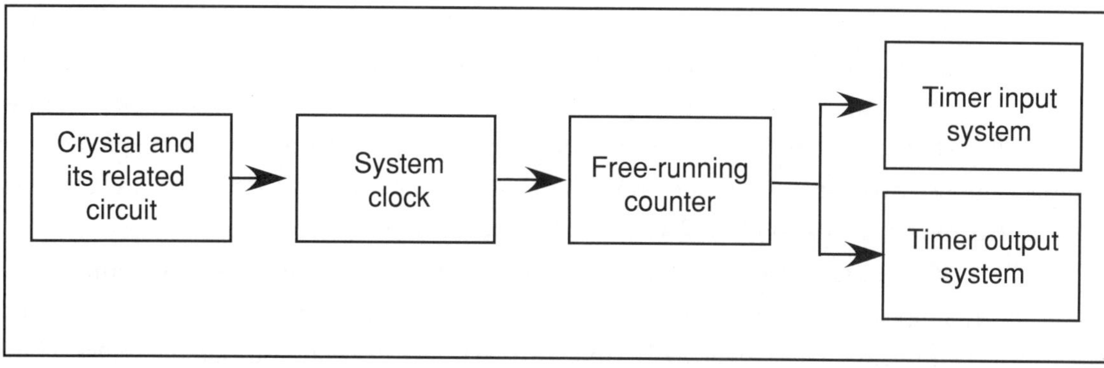

FIGURE 4.2: A generic diagram of a timer system for a microcontroller

Note that in the figure we call a unit that captures external events as the timer input system while the system that generates events for external devices as the timer output system. Before we continue, we need to define what is a timing event. For the input timer system, a timing event is a change of logic state on one of the microcontroller input pins. Figure 4.3 shows how the timer input system uses a free-running counter to capture the time when the signal state changes on the external pin. As shown in Figure 4.3, the contents of a free-running counter are used to capture the relative time of the event. In addition to the capability of simply capturing

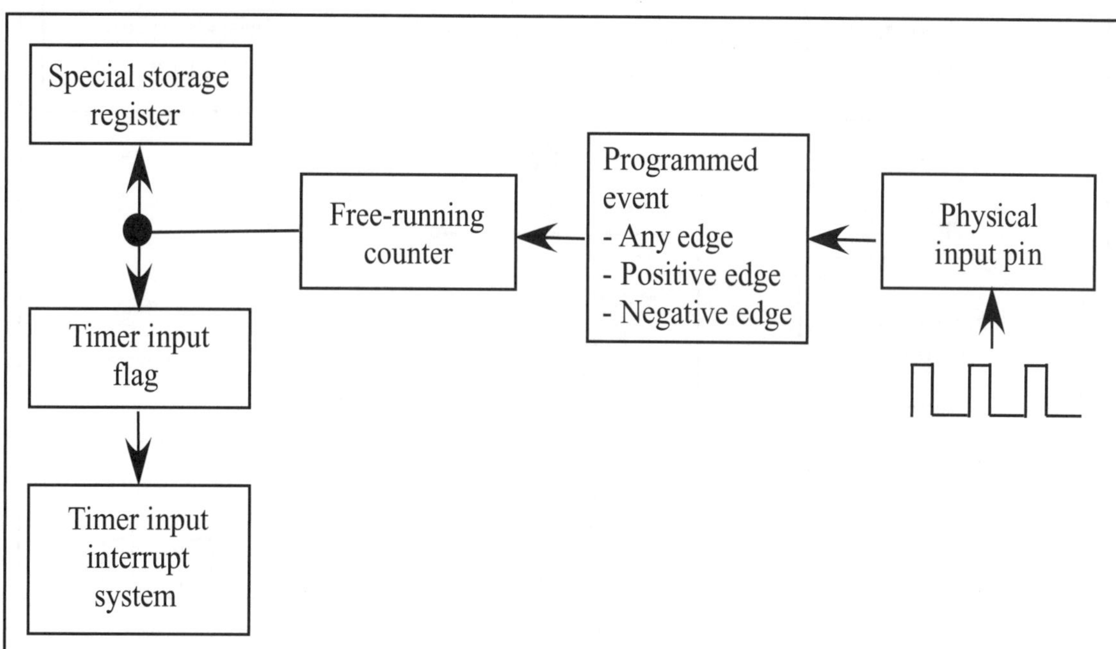

FIGURE 4.3: A diagram of a timer input system

the time of an incoming event, a microcontroller can capture the times of appearance for rising and falling edges to compute the period and the frequency of a periodic signal. To find the frequency of a periodic pulse, we can capture two consecutive rising edges and use the time between the events to compute the corresponding period and subsequently the frequency. For example, suppose two rising edges shown in Figure 4.3 occur at times 1.0 s and 1.2 s. One can then easily compute the period of 0.2 s and the frequency of 5 Hz.

You can also use the timer input system to measure the pulse width of an aperiodic signal. For example, suppose that the times for the rising edge and the falling edge of an incoming signal are 1.5 s and 1.6 s, respectively. We can use these values to easily compute the pulse width of 0.1 s.

The second overall goal of the timer system is to generate signals to control external devices. An event again simply means a change of logic states on an output pin of a microcontroller at a specified time. Now consider Figure 4.4. Suppose an external device connected to the microcontroller requires a pulse signal to turn itself on. Again suppose the particular pulse the external device needs is 2 ms wide. In such situations, we can use the free-running counter value to synchronize the time of desired logic state changes. Naturally, extending the same capability, we can also generate a periodic pulse with a fixed duty cycle or a varying duty cycle.

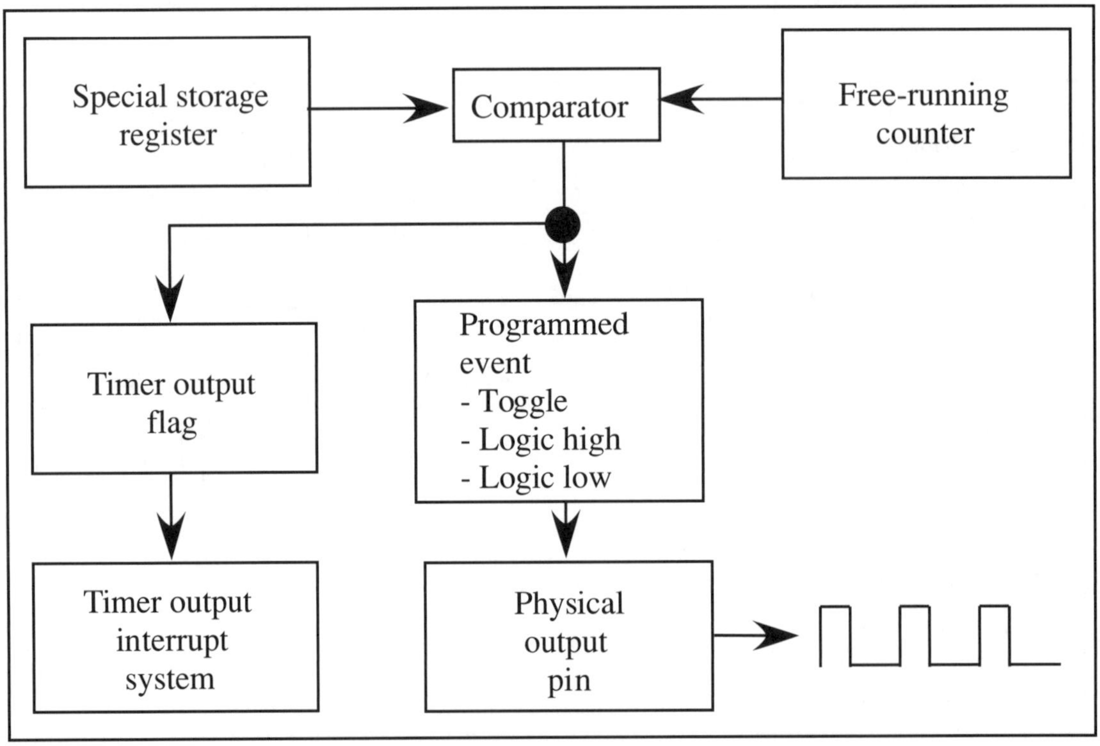

FIGURE 4.4: A diagram of a timer output system

For the timer input and output systems to function properly, a typical microcontroller requires you to set up a number of special registers. For example, for the timer input system, one must program an event (logic state high or logic state low), configure a physical input pin, and set up parameters for the related interrupt, if used. When an event occurs, another special register is used to capture the free-running counter value. Similarly, for the timer output system, one must configure a physical output pin, program an event, and set the timing value to special registers to dictate when the programmed event should appear on the output pin. Typically, an output timer system will also have an associated interrupt that system one can utilize.

From the examples we discussed above, you may have wondered how a microcontroller can be used to compute absolute times from the relative free-running counter values, say 1.5 s and 1.6 s. The simple answer is that we cannot do so directly. A programmer must use the relative system clock values and derive the absolute time values. Suppose your microcontroller is clocked by a 2-MHz signal and the system clock uses a 16-bit free-running counter. For such a system, each clock period represents 0.5 ms and it takes approximately 32.78 ms to count from 0 to 64k (65 536). The timer input system then uses the clock values to compute frequencies, periods, and pulse widths. For example, suppose again you want to measure a pulse width of an aperiodic signal. If the rising edge and the falling edge occurred at count values $0010 and $0114, can you find the pulse width when the free-running counter is counting at 2 MHz? Recall that the $ symbol represents that the following value is in a hexadecimal form. Let us first convert the two values into their corresponding decimal values, 276 and 16. The pulse width of the signal in the number of counter value is 260. Since we already know how long it takes for the system to count one, we can readily compute the pulse width as $260 \times 0.5\ usec = 130\ usec$.

Our calculations do not take into account time increments lasting longer than the rollover time of the counter. When a counter rolls over from its maximum value back to zero, a flag is set to notify the processor of this event. The rollover events may be counted to correctly determine the overall elapsed time of an event.

4.3 APPLICATIONS

In this section, we consider some important uses of the timer system of a microcontroller such as to measure an input signal timing event, to count the number of external signal occurrences, to generate timed signals, and finally, to generate pulse width modulated signals. We first start with a case of measuring the time duration of an incoming signal.

4.3.1 Measuring External Timing Event

In many applications, we are interested in measuring the elapsed time of an external event using a microcontroller. For example, an application may call for measuring the frequency of an incoming signal. Using the hardware and functional units discussed in the previous sections, we

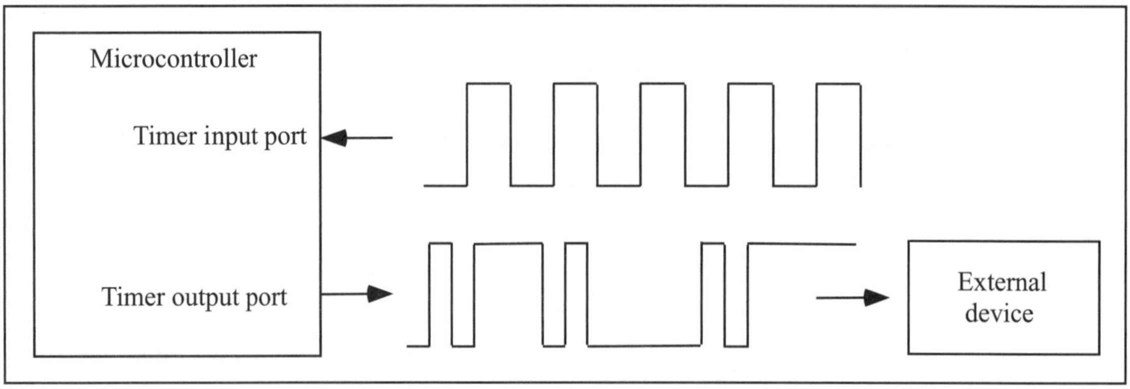

FIGURE 4.5: Use of the timer input and output systems of a microcontroller. The signal on top is fed into a timer input port. The captured signal is subsequently used to compute the input signal frequency. The signal at the bottom is generated using the timer output system. The signal is used to control an external device.

now present a procedure to accomplish the task. Figure 4.5 shows an incoming periodic signal to our microcontroller. We assume that the microcontroller has the basic input signal capturing capabilities.

The first necessary step for the current task is to turn on the timer system. A microcontroller usually does not turn on all of its functional systems including a timer system to reduce power consumption. Therefore, a programmer must provide the controller with an instruction to do so. In addition to a separate timer module, many microcontroller manufacturers allow a programmer to choose the rate of the clock that governs the overall functions of a timer module.

Once the timer is turned on and the clock rate is selected, a programmer must configure the physical port to which the incoming signal arrives. This step is done using a special input timer port configuration register. The next step is to program the event to capture. In our current example, we should capture two consecutive rising edges or falling edges. Again, the programming portion is done by storing an appropriate set up value to a special register.

Now that the input timer system is configured appropriately, you have two options to accomplish the task. The first one is the use of a polling technique; the microcontroller continuously polls a flag, which holds a logic high signal when a programmed event occurs on the physical pin. Once the microcontroller detects the flag, it needs to clear the flag and record the time when the flag was set using another special register that captures the time of the associated free-running counter value. The program needs to continue to wait for the next flag which indicates the end of one period of the incoming signal. A programmer then needs to record the newly acquired captured time represented in the form of a free-running counter value again. The period of the signal can now be computed by computing the time difference between the two captured event times, and, based on the clock speed of the microcontroller, the programmer can compute the actual time changes and consequently the frequency of the signal. In many cases, a microcontroller cannot afford the time to poll for one event. Such situation introduces the second method: interrupt systems. Most microcontroller manufacturers have developed built-in interrupt systems with their timer input modules. Instead of continuously polling for a flag, a microcontroller performs other tasks and relies on its interrupt system to detect the programmed event. The task of computing the period and the frequency is the same as the first method, except that the microcontroller will not be tied down to constantly checking the flag, increasing the efficient use of the microcontroller resources. To use interrupt systems, of course, we must pay the price by appropriately configuring the interrupt systems to be triggered when a desired event is detected. Typically, additional registers must be configured, and a special program called an interrupt service routine must be written.

Suppose that for an input capture scenario the two captured times for the two rising edges are \$1000 and \$5000, respectively. Note that these values are not absolute times but the representations of times reflected as the values of the free-running counter. The period of the signal is \$4000 or 16384 in a decimal form. If we assume that the timer clock runs at 10 MHz, the period of the signal is 1.6384 ms, and the corresponding frequency of the signal is approximately 610.35 Hz.

4.3.2 Counting Events

The same capability of measuring the period of a signal can also be used to simply count external events. Suppose we want to count the number of logic changes of an incoming signal for a given period of time. Again, we can use the polling technique or the interrupt technique to accomplish the task. For both techniques, the initial steps of turning on a timer and configuring a physical input port pin are the same. In this application, the programmed event should be any logic state changes instead of looking for a rising or a falling edge as we have done in the previous section. If the polling technique is used, at each event detection, the corresponding flag must be cleared and a counter must be updated as events occur within the specified time window. If the interrupt technique is used, one must write an interrupt service routine within which the flag is cleared and a counter is updated.

4.3.3 Generating Timing Signals to Interface External Devices

In the previous two sections, we considered two applications of capturing external incoming signals. In this subsection and the next one, we consider how a microcontroller can generate time critical signals for external devices.

Suppose in this application, we want to send a signal shown in Figure 4.5 to turn on an external device. The timing signal is arbitrary but the application will show that one can generate any desired signal using the timer output system within the speed capability of the microcontroller.

Similar to the use of the timer input system, one must first turn on the timer system and configure a physical pin as a timer output pin using special registers. In addition, one also needs to program the desired external event using another special register associated with the timer output system. To generate the signal shown in Figure 4.5, one must compute the time required between the rising and the falling edges. Suppose that the external device looks for a pulse which is 2 ms wide. To generate the desired pulse, one must first program the logic state for the particular pin to be low and set the time value using a special register with respect to the contents of the free-running counter. As was mentioned in Section 5.2, at each clock cycle, the special register contents are compared with the contents of the free-running counter and when a match occurs, the programmed logic state appears on the designated hardware pin. Once the rising edge is generated, the program then must reconfigure the event to be a falling edge (logic state low) and change the contents of the special register to be compared with the free running counter. For the particular example in Figure 4.5, let us assume that the main clock runs at 2 MHz, the free-running counter is a 16-bit counter, and the name of the special register (16-bit register) where we can put appropriate values is output timer register. To generate the desired pulse, we can put $0000 first, and after the rising edge has been generated, we need to change the program event to a falling edge and put $0FA0 or 4000 in decimal to the output timer register. As was the case with the input timer system module, we can use output timer system interrupts to generate the desired signals as well.

4.3.4 Industrial Implementation Case Study (PWM)

In this section, we discuss a well-known method to control the speed of a DC motor using a pulse width modulated (PWM) signal. The underlying concept is as follows. If we turn on a DC motor and provide the required voltage, the motor will run at its maximum speed. Suppose we turn the motor on and off rapidly, by applying a periodic signal. The motor at some point cannot react fast enough to the changes of the voltage values and will run at the speed proportional to the average time the motor was turned on. By changing the duty cycle, we can control the speed of a DC motor as we desire. Suppose again we want to generate a speed profile shown in Figure 4.6. As shown in the figure, we want to accelerate the speed, maintain the speed, and decelerate the speed for a fixed amount of time.

The first task necessary is again to turn on the timer system, configure a physical port, and program the event to be a rising edge. As a part of the initialization process, we need to put $0000 to the output timer register we discussed in the previous subsection. Once the rising edge is generated, the program then needs to modify the event to a falling edge and change

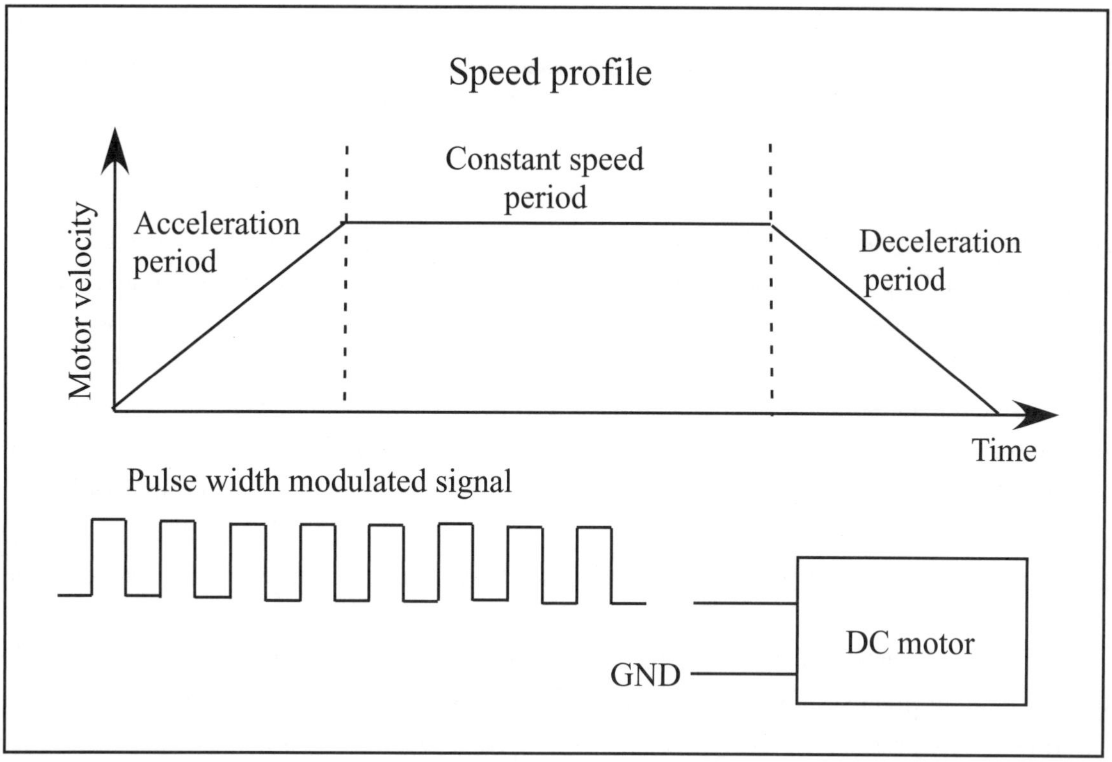

FIGURE 4.6: The figure shows the speed profile of a DC motor over time when a pulse width modulated signal is applied to the motor

the contents of the output timer register to a value proportional to a desired duty cycle. For example, if we want to start off with 25% duty cycle, we need to input \$4000 to the register, provided that we are using a 16-bit free-running counter. Once the falling edge is generated, we now need to go back and change the event to be a rising edge and the contents of the output timer counter value back to \$0000. If we want to continue to generate a 25% duty cycle signal, then we must repeat the process indefinitely. Note that we are using the time for a free-running counter to count from \$0000 to \$FFFF as one period.

Now suppose we want to increase the duty cycle to 50% over 1 s and that the clock is running at 2 MHz. This means that the free-running counter counts from \$0000 to \$FFFF every 32.768 ms, and the free-running counter will count from \$0000 to \$FFFF approximately 30.51 times over the period of 1 s. That is, we need to increase the pulse width from \$4000 to \$8000 in approximately 30 turns, or approximately 546 clock counts every 32.768 ms.

For both examples we illustrated in this and the previous sections, many microcontrollers have a built-in timer output system that can control logic states for more than one physical pin.

For such microcontrollers, the implementation of the applications we showed becomes easier by dedicating two output timer modules to generate desired signals: one is assigned to turn the logic state to high while the other one is responsible to turn the logic state to low.

4.4 SUMMARY

In this chapter, we considered a microcontroller timer system. We introduced some terminology for timer related topics, discussed typical functions of a timer subsystem, studied timer hardware operations, and considered some applications where the timer subsystem of a microcontroller can be used. A typical microcontroller has a timer system that can capture timings for incoming signals and generate outgoing signals at desired times. By taking advantage of the capabilities, we can measure the period of an incoming periodic signal, the width of a pulse generated by an external device, and the time when a rising or a falling edge appears on an input pin. In addition, we can generate a periodic signal with varying duty cycles, any aperiodic signal, and force logic state changes on output pins. All these capabilities allow us to easily interface a microcontroller with external devices.

One of the important reasons to use a microcontroller in an application is the readily useable built-in timer system. Such system makes time critical interface of a microcontroller to external devices possible. To fully apply the timer system capabilities, one must become familiar with the particular assembly language for a microcontroller. The primary reason for the use of assembly language in the current context is that a programmer has a full control of each clock cycle operation; the programmer determines precisely when each instruction is executed.

BIBLIOGRAPHY

K Short, Embedded Microprocessor Systems Design: An Introduction Using the INTEL 80C188EB, Prentice Hall, Upper Saddle River, NJ, 1998.

F Driscoll, R Coughlin, and R Villanucci, Data Acquisition and Process Control with the M68HC11 Microcontroller, 2nd ed., Prentice Hall, Upper Saddle River, NJ, 2000.

T Morton, Embedded Microcontrollers, Prentice Hall, Upper Saddle River, NJ, 2001.

CHAPTER PROBLEMS

- **Fundamental**

  1. *Question*: Given an eight-bit free-running counter and the system clock rate of 24 MHz, find the time required for the counter to count from zero to its maximum value.

  2. *Question*: If we desire to generate periodic signals with periods ranging from 125 ns to 500 ms, what is the minimum frequency of the system clock?

- **Advanced**
  1. *Question*: Describe how you can compute the period of an incoming signal with varying duty cycles.
  2. *Question*: Describe how one can generate an aperiodic pulse with a pulse width of 2 min?
- **Challenging**
  1. *Question*: Design a microcontroller system to control a sprinkler controller that performs the following tasks. We assume that your microcontroller runs with 10 MHz clock and it has a 16-bit free-running counter. The sprinkler controller system controls two different zones by turning sprinklers within each zone on and off. To turn on the sprinklers of a zone, the controller needs to receive a 152.589 Hz PWM signal from your microcontroller. To turn off the sprinklers of the same zone, the controller needs to receive the PWM signal with a different duty cycle.
     - (a) Your microcontroller needs to provide the PWM signal with 10% duty cycle for 10 ms to turn on the sprinklers in zone one.
     - (b) After 15 min, your microcontroller must send the PWM signal with 15% duty cycle for 10 ms to turn off the sprinklers in zone one.
     - (c) After 15 min, your microcontroller must send the PWM signal with 20% duty cycle for 10 ms to turn on the sprinklers in zone two.
     - (d) After 15 min, your microcontroller must send the PWM signal with 25% duty cycle for 10 ms to turn off the sprinklers in zone two.

CHAPTER 5

Analog-to-Digital Conversion

Objectives: After reading this chapter, the reader should be able to

- Explain the difference between analog and digital signals.
- Illustrate the analog-to-digital conversion process.
- Assess the quality of analog-to-digital conversion using the metrics of sampling rate, quantization levels, number of bits used for encoding, and dynamic range.
- Design signal conditioning circuits to interface sensors with analog-to-digital converters.

In this chapter we present the process of analog-to-digital conversion, a process extensively used by microcontrollers to bring in signals from their environments for analysis, testing, and manipulation in digital forms. Since we cannot decouple the process completely from the hardware, we therefore include discussions on various analog-to-digital converters (ADCs) when it is appropriate. The conversion process is so essential for a large number of microcontroller applications that most of all commercial microcontrollers come with one or more built-in analog-to-digital converters (ADCs). For this reason we believe it is important for practicing engineers to have a firm grasp of the current subject.

We present the fundamental concepts associated with the conversion process in the first section. In the following section, we discuss the conversion process itself, followed by the presentation of a number of hardware implementations of the process in Section 5.4. We show readers some applications of the ADC in Section 5.5 and conclude the chapter in Section 5.6. For those who want to delve further into the project, we provide a list of reading references in Section 5.7.

5.1 BACKGROUND THEORY

Before we show the conversion process, we need to make you familiar with underlying theories that support the process. We assume the reader has some background knowledge in algebra and basic engineering. We start with some definitions on analog and digital signals.

5.1.1 Analog Signals versus Digital Signals

Signals contain information. For example, when you hear a siren from a fire truck, you are hearing a set of sinusoidal signals with different frequencies repeating over time. The signals convey that there is an emergency nearby and the truck needs to get there quickly. Consider another example. Suppose you are measuring and recording the temperature in your backyard over 24 h. If you plot the recorded values over the time axis, you have a signal that contains the temperature information of your backyard on the recorded day. Now consider yet another example. Suppose we put light emitting diodes all over your body. We then ask you to run on a treadmill while filming your movements using a video camera. If we plot the movements of the diodes over time, they will show how your elbows, shoulders, knees, hips, and feet, for example, move over time when you run. We consider the plots as signals containing information.

From our discussion, we see that a signal is a collection of values representing the state of a physical variable. The collection can be as small as only one value or can have as many values as you wish. In engineering, we usually arrange the values in order, for example over time or over a spatial axis, to display the information. The time and the spatial variables are called independent variables, since they are not affected by the physical variables of interests. For example, we measure the temperature change over time. The temperature measured is dependent on the time, not the other way around. Figure 5.1 shows an altitude trajectory of a bird flying over a short period of time. The signal shows how the altitude of the bird changes continuously.

Figure 5.2 shows a grayscale image of a six-legged robot. The image captured the light intensities of the scene using a CCD camera. If we move from left to right of the image and observe the intensity changes, we can find vertical edges, by detecting signal intensity changes in the spatial axis. The same analysis can be performed as we move from top to bottom of the image.

Analog signals are those whose physical variable values change continuously over their independent variables. Most physical variables, your speech, movement of stars, and the music you hear at a concert are analog signals, signals that we find all around us. The example of observing the movement of your body parts using the light emitting diodes is an analog signal. Digital signals, however, have their physical variables defined only for discrete instances over their independent variables. Although it may look continuous to human eyes, the photo example shown in Figure 5.2 is a discrete signal since pixels that make up a camera image cannot capture all space within the camera frame. The image is only a finite composition of intensity values seen by a discrete number of pixels.

Digital signals are important since all signals represented in digital systems, computers, PDAs, and microcontrollers, are in digital forms. The important task is how to faithfully

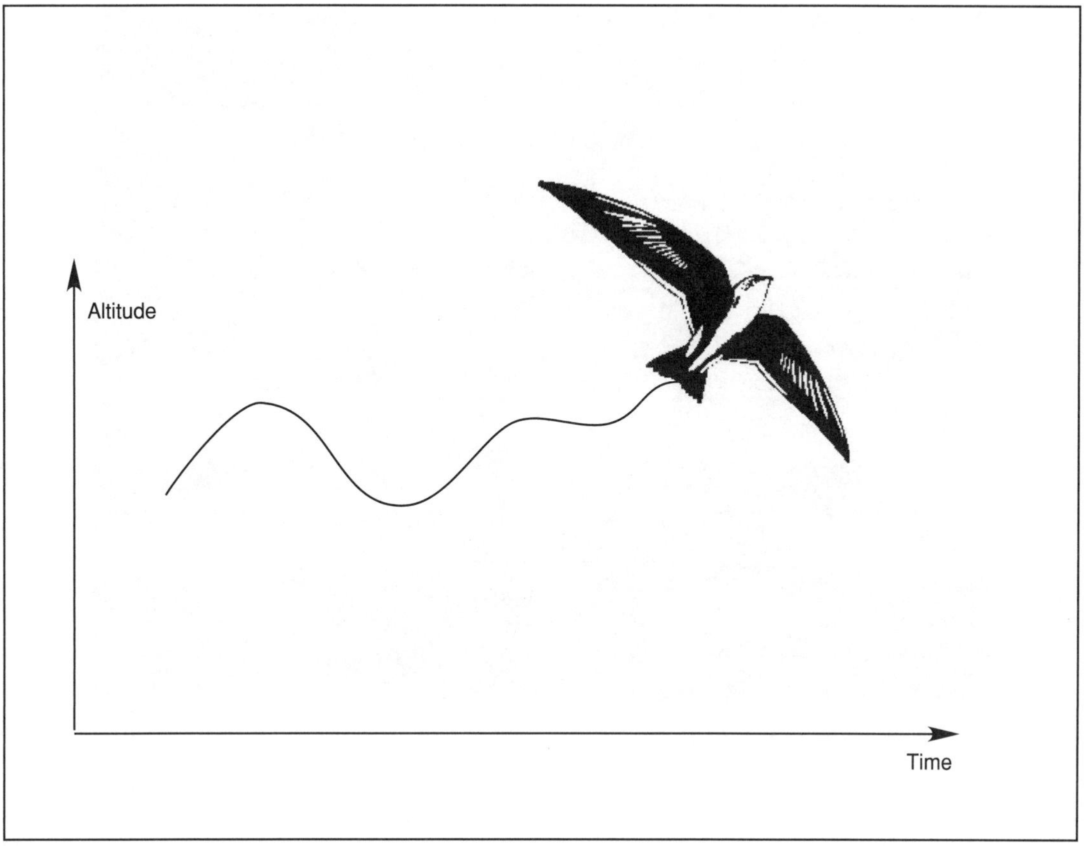

FIGURE 5.1: Altitude trajectory generated by a flying bird

represent analog signals using digital signals. For example, human voices must be converted to a digital form before they can be routed by digital switching circuits in telephone communication systems. Similarly, commands to robots must be converted to a digital form before robots can understand the command.

As shown in the examples above, we live in an analog world; that is, physical variables are analog signals. It is precisely this reason why the analog-to-digital conversion is so very important in any digital system that interacts with an analog environment.

5.1.2 Sampling, Quantization, and Encoding

In this subsection, we present three important processes associated with the ADC. We first start with the subject of sampling. Imagine yourself as a photographer in an Olympic diving stadium. Your job is taking a sequence of pictures of divers jumping off from a diving board 10 m above the surface of the diving pool. Your goal is to put the sequence of pictures together

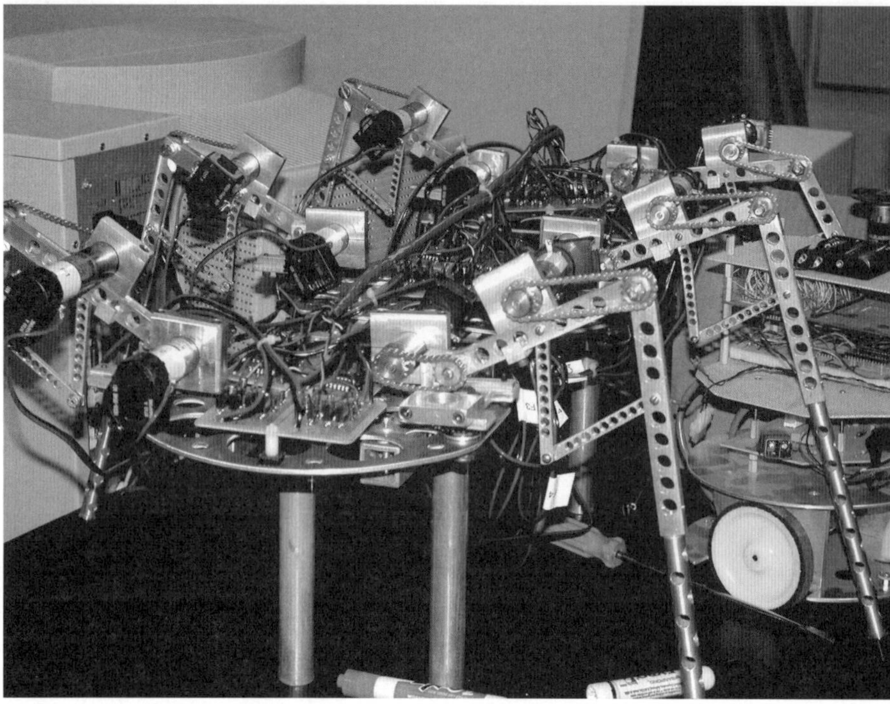

FIGURE 5.2: A photo of a walking robot

to reconstruct the motion of each diver. The sequence of pictures makes up samples of divers' motions. If a diver tries a complex motion and you want to faithfully reconstruct his motion, you must take enough pictures from the start to the end of the dive. If a diver makes a simple routine dive, you only need to take few pictures over the period of the dive. Two very different cases of motions generated by a diver are shown in Figure 5.3. The same time sequence is used to capture samples for both motions. As can be seen from the figure, frame (a) motion cannot be regenerated from the samples while the motion shown in frame (b) can clearly be reconstructed from the same number of samples used to capture both motions.

Sampling is the process of taking 'snap shots' of a signal over time. Naturally, when we sample a signal, we want to sample it in an optimal fashion such that we can capture the essence of the signal while minimizing the use of resources. In essence, we want to minimize the number of samples while faithfully reconstructing the original signal from the samples. As can be deduced from our discussion above, the rate of change in a signal determines the number of samples required to faithfully reconstruct the signal, provided that all adjacent samples are captured with the same sample timing intervals.

Harry Nyquist from Bell Laboratory studied the sampling process and derived a criterion that determines the minimum sampling rate for any continuous analog signals. His, now famous,

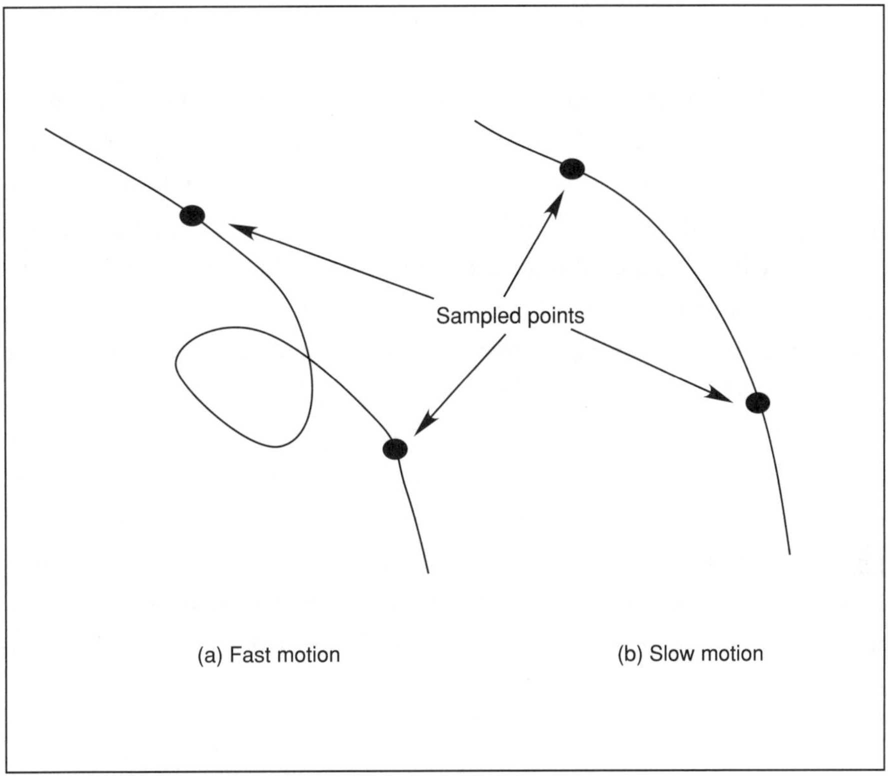

FIGURE 5.3: Two divers jumping off the platforms

minimum sampling rate is known as the Nyquist sampling rate, which states that one must sample a signal at least twice as fast as the highest frequency content of the signal of interest. For example, if we are dealing with the human voice signal that spans from about 20 Hz to 4 kHz, the Nyquist sample theorem tells us that we must sample the signal at least at 8 kHz, 8000 snap shots every second. Engineers who work for telephone companies must deal with such issues. For further study on the Nyquist sampling rate, refer to Pack and Barrett (2002) listed in the References section.

Sampling is important since when we want to represent an analog signal in a digital system, such as a computer, we must use the appropriate sampling rate to capture the analog signal for a faithful representation in digital systems.

Now that we understand the sampling process, let us move on to the second process of the analog-to-digital conversion, quantization. Each digital system has a set number of bits it uses as the basic units to represent data. A bit is the most basic unit where single binary information, one or zero, is represented. A nibble is made up of four bits put together and a byte of eight bits.

In the previous section, we tacitly avoided the discussion of the form of captured signal samples. When a signal is sampled, digital systems need some means to represent the captured samples. The quantization of a sampled signal is how the signal is represented as one of the quantization levels.

Suppose you have a single bit to represent an incoming signal. You only have two numbers, 0 and 1. You may say that you can distinguish only low from high. Suppose you have two bits. You can represent four different levels, 00, 01, 10, and 11. What if you have three bits? Yes, you now can represent eight different levels: 000, 001, 010, 011, 100, 101, 110, and 111. Think of it as follows. When you had two bits, you were able to represent four different levels. If we add one more bit, that bit can be one or zero, making the total possibilities eight. Similar discussion can lead us to conclude that given n bits, we have 2^n different numbers or levels one can represent.

Figure 5.4 shows how n bits are used to quantize a range of values. In many digital systems, the incoming signals are voltage signals. The voltage signals are first obtained from physical signals with the help of transducers, such as microphones, angle sensors, and infrared sensors. The voltage signals are then conditioned to map their range with the input range of a digital system, typically 0–5 V. In Figure 5.4, n bits allow you to divide the range of a digital system into 2^n different quantization levels. As can be seen from the figure, the more quantization levels means the better mapping of an incoming signal to its true value. If we only had a single

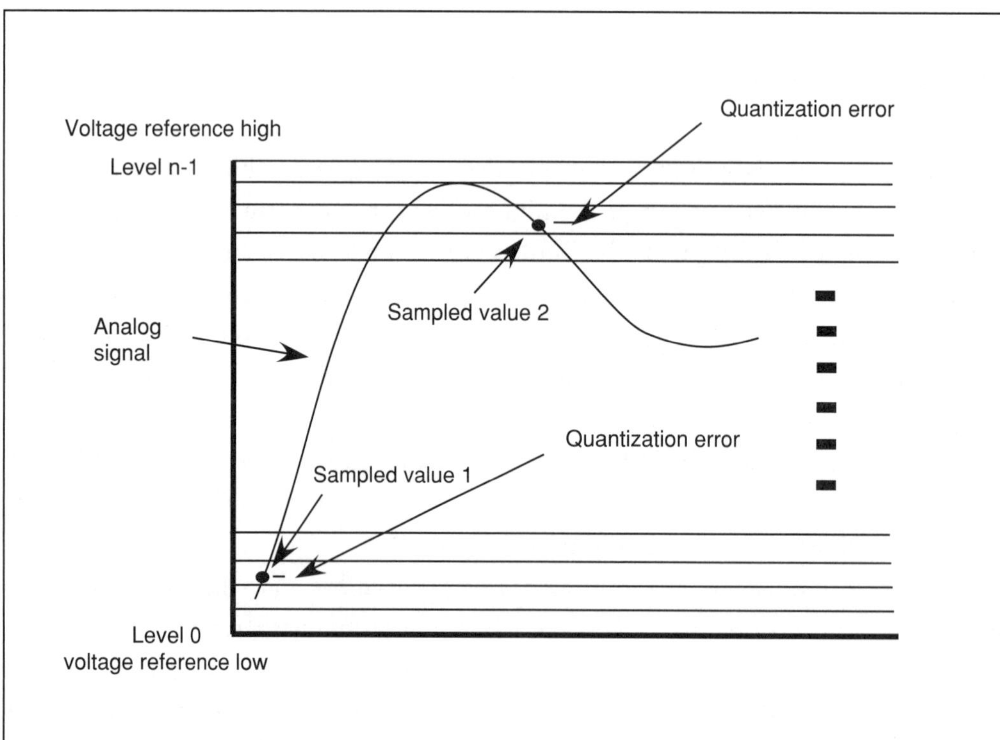

FIGURE 5.4: Quantization

bit, we can only represent level 0 and level 1. Any analog signal value in between the range had to be mapped either as level 0 or level 1, not many choices. Now imagine what happens as we increase the number of bits available for the quantization levels.

What happens when the available number of bits is eight? How many different quantization levels are available now? Yes, 256. How about 10, 11, 12? Notice also that as the number of bits used for the quantization levels increases the 'distance' between two adjacent levels decreases with a factor of a polynomial. Now resolution.

The encoding process involves converting a quantized signal into a digital binary number. Suppose again we are using eight bits to quantize a sampled analog signal. The quantization levels are determined by the eight bits and each sampled signal is quantized as one of the 256 quantization levels. Consider the two sampled signals shown in Figure 5.4.

The first sample is mapped to quantization level 1 and the second one is mapped to quantization level 198. Note the amount of quantization error introduced for both samples. Now consider Figure 5.5. The same signal is sampled at the same time but quantized using a

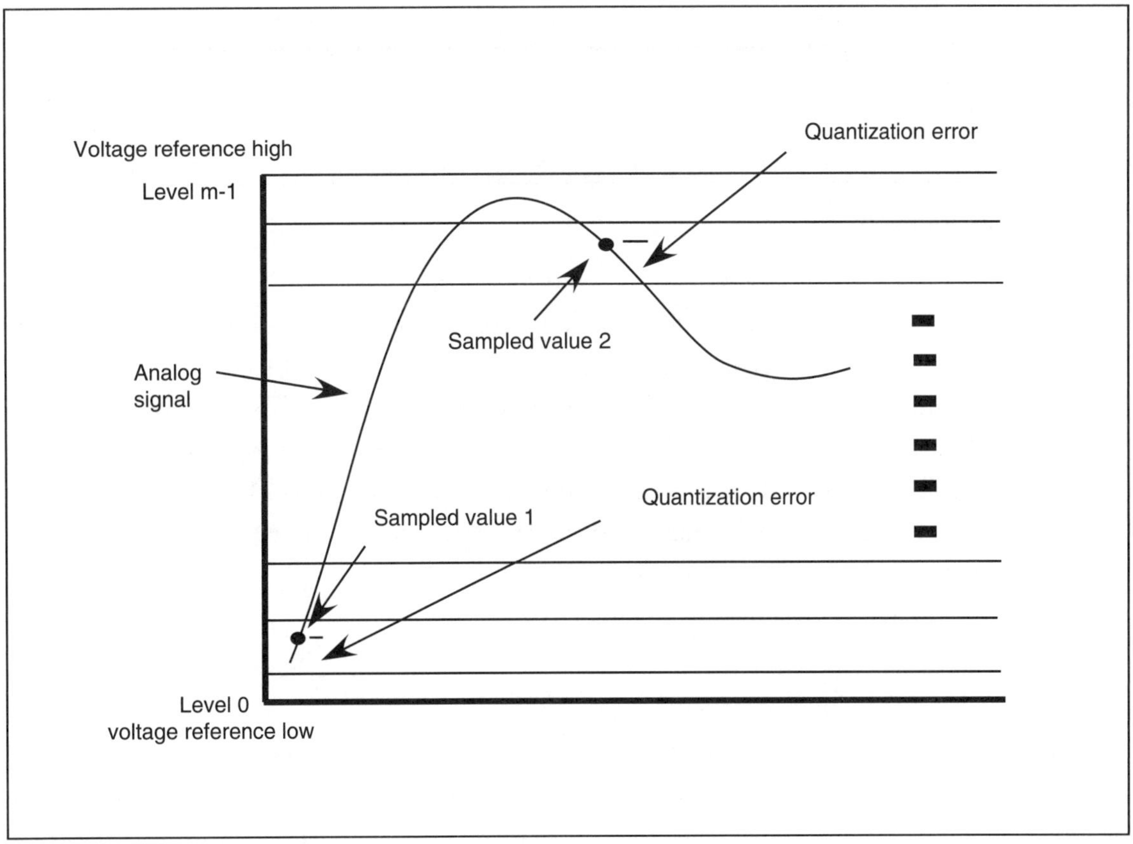

FIGURE 5.5: Quantization with fewer bits

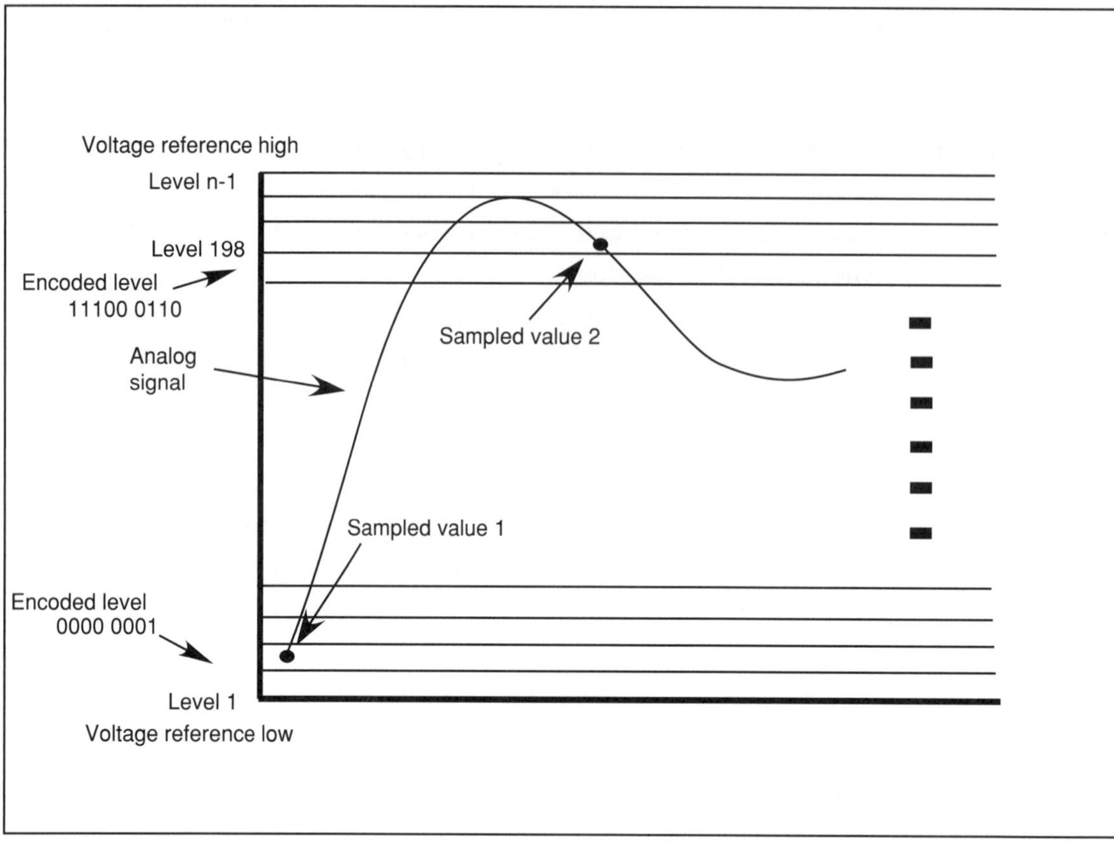

FIGURE 5.6: Encoding

lesser number of bits. Note that the quantization error is inversely proportional to the number of bits used to quantize the signal.

Once a sampled signal is quantized, the encoding process involves representing the quantization level with the available bits. Thus, for the first sample, the encoded sampled value is 0000 0001, while the encoded sampled value for the second sample is 1100 0110. As a result of the encoding process, sampled analog signals are now represented as a set of binary numbers. Thus, the encoding is the last necessary step to represent a sampled analog signal into its corresponding digital form, shown in Figure 5.6.

5.1.3 Resolution and Data Rate

You hear the term 'resolution' often when people discuss computer screens, TV screens, and digital audio signals. Resolution is a measure used to quantize an analog signal. In fact, resolution is nothing but the 'distance' between two adjacent quantization levels we discussed earlier.

Suppose again we have a range of 5 V and one bit to represent an analog signal. The resolution in this case is 2.5 V, a very poor resolution. You can imagine how your TV screen will look if you had only two levels to represent each pixel, black and white. The maximum error, called the resolution error, is 2.5 V for the current case, 50% of the total range of the input signal. Suppose you now have four bits to represent quantization levels. The resolution now becomes 1.25 V or 25% of the input range. Suppose you have 20 bits for quantization levels. The resolution now becomes 4.77×10^{-6}, 9.54×10^{-5}% of the total range. The discussion we presented simply illustrates that as we increase the available number of quantization levels within a range, the distance between adjacent levels decreases, reducing the quantization error of a sampled signal. As the number grows, the error decreases, making the representation of a sampled analog signal the more accurate in the corresponding digital form. The number of bits used for the quantization is directly proportional to the resolution of a system. You now should understand the technical background when you watch high definition television broadcasting.

Now let us move onto the discussion of the data rate. The definition of the data rate is the amount of data generated by a system per some time unit. Typically, the number of bits or the number of bytes per second is used as the data rate of a system. We just saw that the more bits we use for the quantization levels, the more accurate we can represent a sampled analog signal. Why not use the maximum number of bits current technologies can offer for all digital systems, when we convert analog signals to digital counter parts? It has to do with the cost involved. In particular, suppose you are working for a telephone company and your switching system must accommodate 100 000 customers. For each individual phone conversation, suppose the company use 8 kHz sampling rate and you are using 10 bits for the quantization levels for each sampled signal.[1] If all customers are making out of town calls, what is the number of bits your switching system must process to accommodate all calls? The answer will be $100\,000 \times 8000 \times 10$ or eight billion bits per every second! You will need some major computing power to meet the requirement. For such reasons, when designers make decisions on the number of bits used for the quantization levels, they must consider the computational burden the selection will produce on the computational capabilities of a digital system.

You will also encounter the term dynamic range when you consider finding appropriate analog-to-digital converters. The dynamic range is a measure used to describe the signal-to-noise ratio. The unit used for the measurement is Decibel (dB), which is the strength of a signal with respect to a reference signal. The greater the dB number the stronger is the signal compared to a noise signal. The definition of the dynamic range is $20 log\, 2^b$ where b is the number

[1]For the sake of our discussion, we ignore other overheads involved in processing a phone call such as multiplexing, demultiplexing, and serial-to-parallel conversion.

of bits used to convert analog signals to digital signals. Typically, you will find 8–12 bits used in commercial analog-to-digital converters, translating the dynamic range from $20 log 2^8$ dB to $20 log 2^{12}$ dB.

5.2 ANALOG-TO-DIGITAL CONVERSION PROCESS

The goal of the ADC process is to accurately represent analog signals as digital signals. Toward this end, three signal processing procedures, sampling, quantization, and encoding, described in the previous sections must be combined together. Before the ADC process takes place, we first need to convert a physical signal into an electrical signal with the help of a transducer. A transducer is an electrical and mechanical system that converts physical signals into electrical signals or electrical signals to physical signals. Depending on the purpose, we categorize a transducer as an input transducer or an output transducer. If the conversion is from physical to electrical, we call it an input transducer. The mouse, the keyboard, and the microphone for your personal computer all fall under this category. A camera, an infrared sensor, and a temperature sensor are also input transducers. The output transducer converts electrical signals to physical signals. The computer screen and the printer for your computer are output transducers. Speakers and electrical motors are also output transducers. Therefore, transducers play the central part for digital systems to operate in our physical world by transforming physical signals to and from electrical signals.

In addition to transducers, we also need a signal conditioning circuitry before we apply the ADC process or its opposite process of digital-to-analog conversion. The signal conditioning circuitry is called the transducer interface. The objective of the transducer interface circuit is to scale and shift the electrical signal range to map the output of the input transducer to the input of the ADC. Figure 5.7 shows the transducer interface circuit using an input transducer.

The output of the input transducer is first scaled by constant K. In the figure, we use a microphone as the input transducer whose output ranges from -5 VDC to $+5$ VDC. The input to the analog-to-digital converter ranges from 0 VDC to 5 VDC. The box with constant K maps the output range of the input transducer to the input range of the converter. Naturally, we need to multiply all input signals by $1/2$ to accommodate the mapping. Once the range has been mapped, the signal now needs to be shifted. Note that the scale factor maps the output range of the input transducer as -2.5 VDC to $+2.5$ VDC instead of 0 VDC to 5 VDC. The second portion of the circuit shifts the range by 2.5 VDC, thereby completing the correct mapping. Actual implementation of the circuit components is accomplished using amplifiers with some feedback loops. We refer interested readers to Thomas and Rosa (2003) listed in the References section. We can go through the same exercise for the circuit that connects a digital-to-analog converter and an output transducer, which we leave as an exercise for the readers.

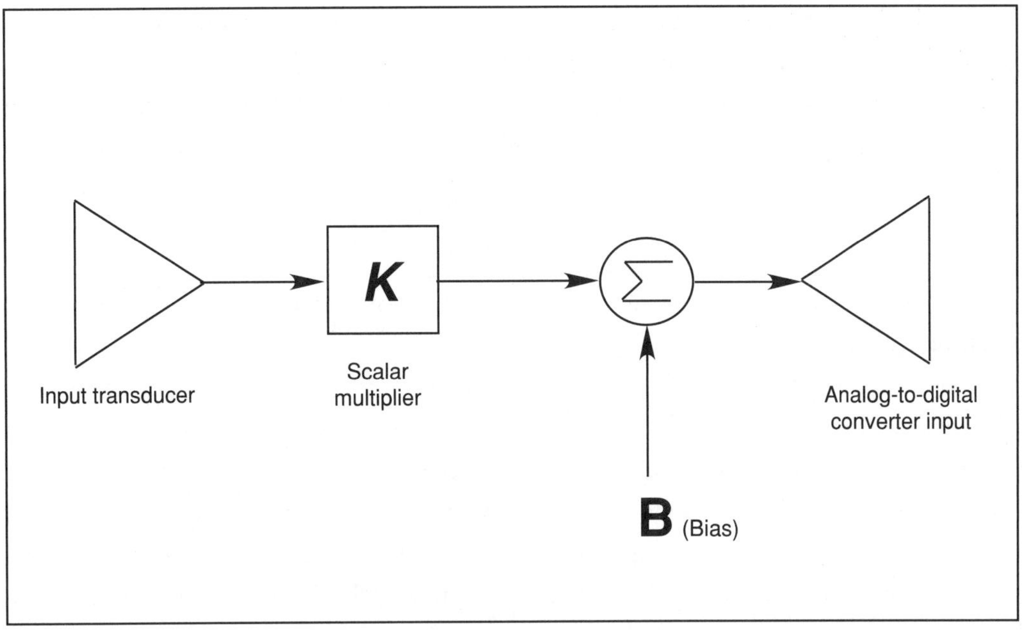

FIGURE 5.7: A block diagram of the signal conditioning for an analog-to-digital converter. The range of the sensor voltage output is mapped to the analog-to-digital converter input voltage range. The scalar multipler maps the magnitudes of the two ranges and the bias voltage is used to align two limits.

Once a physical signal has been converted to its corresponding electrical signal with the help of an input transducer and the output of the transducer mapped correctly to the input of the analog-to-digital converter, the ADC process can start. The first step of the ADC process is the sampling of the analog signal. When selecting a converter, one must consider the type of physical signal that is being converted to properly ensure the sampling rate. As discussed in the previous section, using the proper sampling rate is the first step that determines whether an analog signal will be represented correctly in digital systems. What this means for the reader is to select an analog-to-digital converter that can handle a required conversion rate. Since most microcontrollers now come with a built-in analog-to-digital converter, one must study the user manual portion discussing the conversion rate and make sure that the required sampling rate for the application falls under the advertised conversion rate. For example, if you need to convert the signal representing a person's blood pressure, the sampling rate with 100 Hz (100 samples per second) will suffice. However, if you are dealing with human voice, you need at least an 8-kHz sampling rate capacity [Enderle *et al.* (2000)].

Once the analog signal has been sampled, the quantization process takes place. For this process, again one must decide how much quantization error can be allowed. At one extreme where you are only concerned with finding out only two states, say on and off, quantization error of 1 V is not important. We can operate safely with two bits with a maximum quantization error of 1.25 V. However, if we can only operate with a maximum quantization error of 0.01 V, we need to choose a converter with at minimum 10 bits (about 9 mV). To determine the number of bits and its corresponding maximum quantization error, we use the following equation.

$$Resolution = \frac{range}{2^b}$$

Thus, we can determine the number of bits that will meet the error requirement using the equation above.

Once the quantization level has been determined, we can now encode it using the available bits. As seen in the previous section, the process is a simple conversion of a decimal number (quantization level) to a binary number. Note that the binary number should use all available bits. For example, quantization level four using eight bits is converted as 0000 0100, not 100.

In summary, the analog-to-digital conversion process has three steps for completion: sampling, quantization, and encoding. In the next section, we delve into four different technologies used to implement the analog-to-digital conversion process.

5.3 ADC CONVERSION TECHNOLOGIES

In this section, we briefly discuss four different types of technologies used in analog-to-digital converters: the successive-approximation conversion, the integration conversion, the counter-based conversion, and the parallel conversion. For a detailed discussion, we refer the interested readers to Pack and Barrett (2002). For the following discussion of different analog-to-digital conversion technologies, see Figure 5.8. One can group all existing technologies into two different categories: direct conversion versus indirect conversion. The successive-approximation conversion, counter-based conversion, and parallel conversion use techniques to directly convert analog input signals into digital counter parts, belonging to the direct conversion category, while the integration method uses conversion time as the means to indirectly measure the magnitude of the incoming analog signals, an indirect approach.

5.3.1 Successive-Approximation

The successive-approximation technique uses a digital-to-analog converter, a controller, and a comparator to perform the ADC process. Starting from the most significant bit down to the least significant bit, the controller turns on each bit at a time and generates an analog signal, with the help of the digital-to-analog converter, to be compared with the original input analog signal. Based on the result of the comparison, the controller changes or leaves the current bit and turns on the next most significant bit. The process continues until decisions are made for

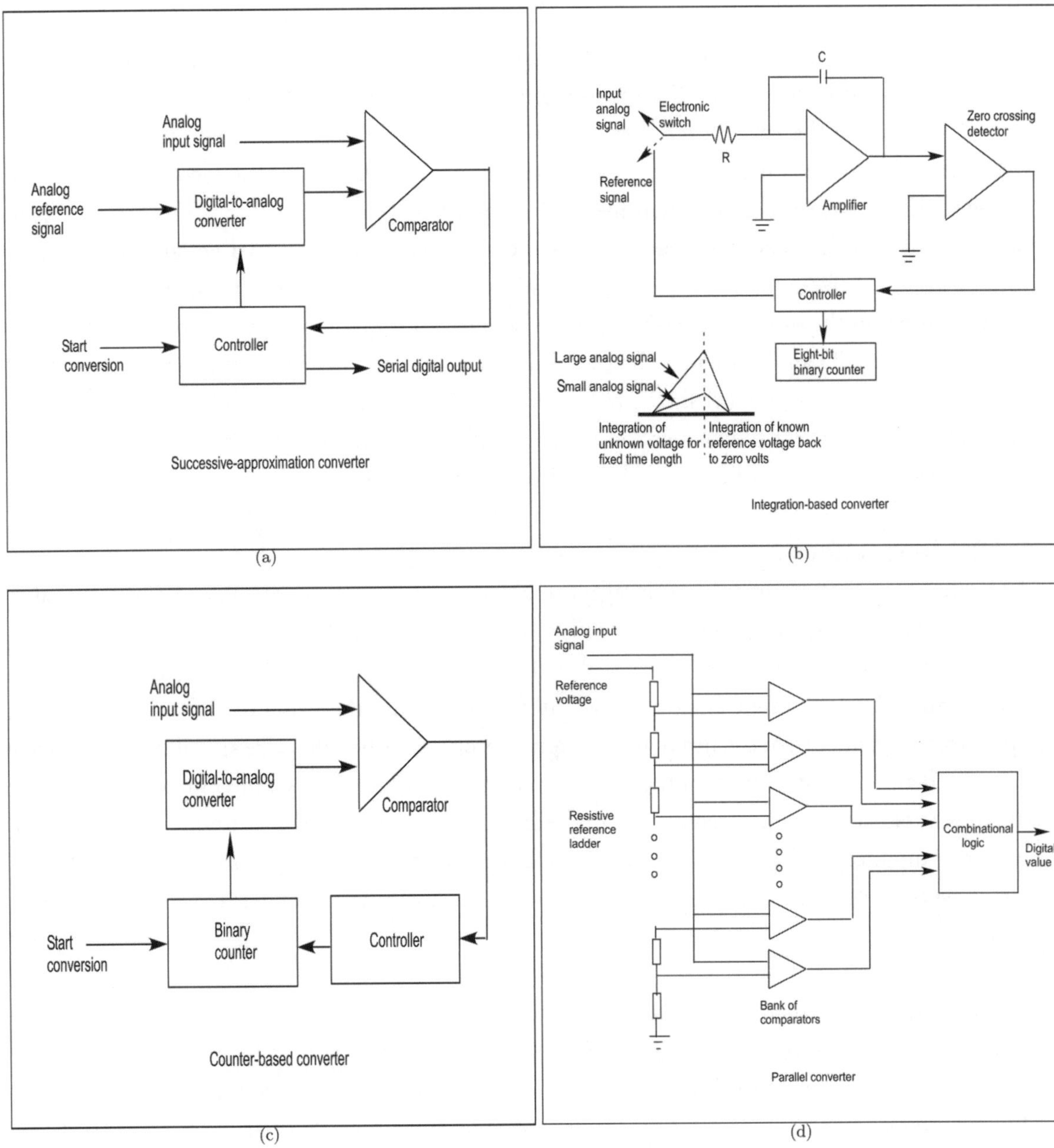

FIGURE 5.8: Four different technologies used to convert analog signals to digital signals. Frame (a) shows the successive-approximation converter, frame (b) illustrates an integration-based converter, frame (c) depicts the counter-based converter, and frame (d) shows a parallel analog-to-digital converter

all available bits. One can consider the process similar to a game children play often. One child thinks of a number in the range of zero to ten and asks another child to guess the number within *n* turns. The first child will tell the second child whether a guessed number is higher or lower than the answer at the end of each turn. The optimal strategy in such a situation is to guess the middle number in the range, say five. If the answer is higher than five, the second guess should be eight. If the answer is lower than five, the second guess should be three. The strategy is to narrow down to the answer by partitioning the available range into two equal parts at every turn. The successive-approximation method works similarly in that the most significant bit is used to partition the original input range of an analog-to-digital converter into two halves, the second most significant bit divides the remaining half into two quarters of the input range, and so forth. Figure 5.8, frame (a) shows the architecture of this type of converter. The advantage of this technique is that the conversion time is uniform for any input, but the disadvantage of the technology is the use of complex hardware for implementation.

5.3.2 Integration

The integration technique uses an integrator, a comparator, and a controller to convert analog signals to digital signals. A sampled analog signal is integrated over a fixed period of time, say *n* clock cycles of the digital system. Another fixed reference signal is integrated over time and compared to the input analog signal integrated. While the value of the reference signal integrated is smaller than the input analog signal integrated, the reference signal is continuously integrated and the time for the integration is measured. When the two integrated values equal, the measured time is converted to a digital encoded value. Frame (b) of Figure 5.8 shows the system components of the converter. One disadvantage of this technique is the varying time for the conversion process. A small analog value will take less time to convert compared to a large value.

5.3.3 Counter-Based Conversion

The third technique used in analog-to-digital converter is the counter-based conversion. This conversion is performed by using a counter, and a digital-to-analog converter, and a comparator. The counter starts at zero and counts up. As the counter counts up, the corresponding value is converted to an analog value and compared to an input analog signal. As long as the input analog signal is greater than the signal generated by the digital-to-analog converter, the counter counts up and the process continues. When the comparator detects that the signal from the digital-to-analog converter is greater than the input analog signal, the counter value is converted to a digital value representing the sampled analog signal. Frame (c) shows the overall architecture of this converter. Similar to the converter based on the integration principle, the disadvantage of this type of converter is the varying conversion time.

5.3.4 Parallel Conversion

The last technique allows the quickest conversion time among the techniques we discussed. The parallel conversion uses a large number of comparators and circuitry to simultaneously measure the input signal and convert it to a digital value. The obvious disadvantage of this technique is the cost involved in building the circuitry. Frame (d) shows the architecture of the converter.

5.4 APPLICATIONS

The ADC process is ubiquitous where digital systems work within our analog world. In this section, we present some application examples of analog-to-digital converters.

5.4.1 Signal Processing

More and more sensors now generate digital outputs using analog-to-digital converters. A digital compass outputs digital values corresponding to directions of the compass. Infrared sensors convert the intensity of received infrared light to digital values. Often these sensors are used in mobile robots to avoid obstacles, detect heat sources, and detect ground targets. Voice recognition systems use built-in analog-to-digital converters to sample, quantize, and encode human speech signals.

5.4.2 Signal Conditioning for ATD Converters

In addition to the transducers and transducer interface circuits described in Section 5.2, typical analog-to-digital converters also contain two more signal conditioning circuits. The first one is a low pass filter to filter out high frequency signals after the sampling is completed and before the quantization is done. All signals can be represented in both the time and frequency domains. We can draw a sinusoidal signal with respect to the time axis or with respect to the frequency axis. The convenience provided with the frequency domain representation of a signal is that we now can see the frequency components that contribute to make up the overall signal. The low pass filter can be used to remove any undesired high frequency noise that can cause aliasing effects. When an aliasing effect is not removed during the analog-to-digital conversion process, the reconstructed analog signal from the digital representations can be significantly different from the original analog signal. Due to space constraints, we do not examine frequency representation of signals and the aliasing effects to frequency components of signals in this chapter. Interested readers are referred to Discrete-Time Signal Processing by Oppenheim and Schafer (1999) listed in the Bibliography section.

The second circuit is not necessary if the input analog signal is slow moving such as the change of ambient temperature of a room, but required if the input signal changes quickly such as the arm motion of a baseball pitcher. The circuit is called a track-hold or sample and hold circuit which consists of latches to hold the captured signal magnitude. The primary purpose of the circuit is to maintain the sampled signal magnitude while the rest of the analog-to-digital process is being processed. We do not want the sample to vary during the conversion process.

5.4.3 Digital-to-Analog Conversion

Once a signal is acquired to a digital system with the help of the analog-to digital conversion process and has been processed, frequently the processed signal is converted back to another analog signal. A simple example of such a conversion occurs in digital audio processing. Human voice is converted to a digital signal, modified, processed, and converted back to an analog signal for people to hear. The process to convert digital signals to analog signals is completed by a digital-to-analog converter. The most used technique to convert digital signals to analog signals is the summation method shown in Figure 5.9.

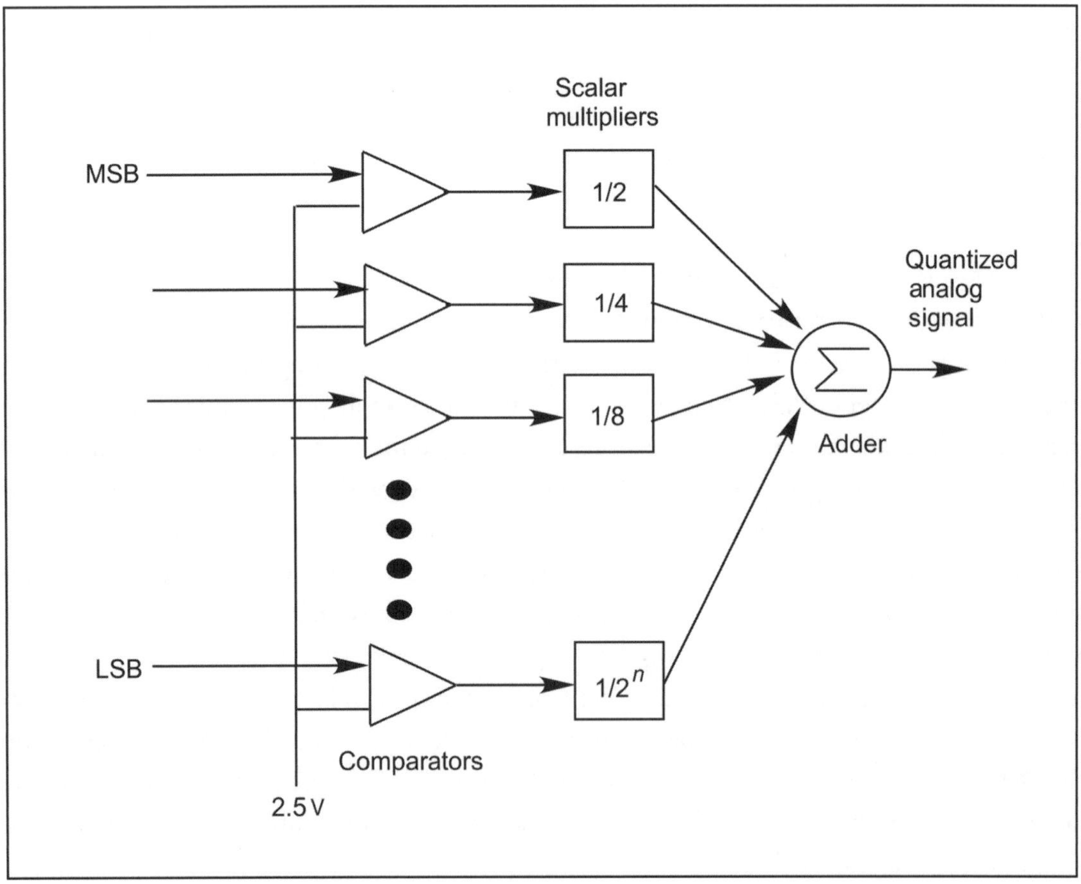

FIGURE 5.9: A summation method to convert a digital signal into a quantized analog signal. Comparators are used to clean up incoming signals and the resulting values are multiplied by a scalar multiplier and the results are added to generate the output signal. For the final analog signal, the quantized analog signal should be connected to a low pass filter followed by a transducer interface circuit

With the summation method of digital-to-analog conversion, a digital signal, represented by a set of ones and zeros, enters the digital-to-analog converter from the most significant bit to the least significant bit. For each bit, a comparator checks its logic state, high or low, to produce a clean digital bit, represented by a voltage level. Typically, in a microcontroller context, the voltage level is +5 or 0 V to represent logic one or logic zero, respectively. The voltage is then multiplied by a scalar value based on its significant position of the digital signal as shown in Figure 5.9. Once all bits for the signal have been processed, the resulting voltage levels are summed together to produce the final analog voltage value. Notice that the production of a desired analog signal may involve further signal conditioning such as a low pass filter to 'smooth' the quantized analog signal and a transducer interface circuit to match the output of the digital-to-analog converter to the input of an output transducer.

5.4.4 Industrial Implementation: Digital Cameras

The way we take pictures have changed significantly due to the explosion of digital cameras and their abilities to capture scenes as well as the analog cameras due in part by the advancement of data acquisition technologies. One of these technologies is the analog-to-digital converter. The goal of the photography is to capture the 3D scene onto a 2D space as accurately as possible. In the past, the surface of a 2D film was exposed to light reflected from a variety of objects in a scene. By controlling the exposure time to the film and later by processing it in a dark room, we acquired pictures.

A digital camera uses the same process but a set of two-dimensional photo receptors replace the film. Each receptor stores the amount of light it receives in a capacitor. In some digital cameras, three photo receptors are used to capture the light intensities in three capacitors to store red, green, and blue intensities received by the receptors. Other digital cameras use a monochrome receptor with rotating color filters. Once the amount of light intensity is captured in a capacitor, the voltage associated with the charge is converted to a digital value using an analog-to-digital converter.

The quality of a digital image is often evaluated using the resolution and the dynamic range. The resolution for a camera is the number of pixels used to capture an image. For example, a camera with 1000 by 2000 pixels has 2 mega pixels of resolution. Once an image is captured, a set of low pass antialiasing filters is used to smooth out the picture. If filters are not used color aliasing effects may result. The dynamic range of a digital camera describes the ability of the analog-to-digital converter to represent light intensity captured in a capacitor into a digital number in the presence of system noise. The system noise comes from lens, electronic noise, and receptors. A bigger number for a camera means it is capable of showing a small change of light intensity.

5.5 SUMMARY

In this chapter we presented the differences between analog and digital signals and used this knowledge to discuss three subprocessing steps involved in analog to digital converters: sampling, quantization, and encoding. We also presented the quantization errors and the data rate associated with the ADC process. The dynamic range of an analog-to-digital converter, one of the measures to describe a conversion process, was also presented. Two different categories exist to represent technologies used to convert analog signals to their corresponding digital signals: direct approach and indirect approaches. For the direct approach, we presented the successive-approximation, counter-based, and parallel conversion techniques. For the indirect approach, the integration-based conversion technique was discussed. The application section briefly touched upon the use of the analog-to-digital conversion technologies and gave an industry example of the analog-to-digital conversion process in digital cameras.

BIBLIOGRAPHY

R Thomas and A Rosa, The Analysis and Design of Linear Circuits, 4th ed., Wiley & Sons, New York, 2003.

D Pack and S Barrett, 68HC12 Microcontroller: Theory and Applications, Prentice Hall, Upper Saddle River, NJ, 07458, 2002.

A Oppenheim and R Schafer, Discrete-time Signal Processing, 2nd ed., Prentice Hall, Upper Saddle River, NJ, 1999.

J Enderle, S Blanchard, and J Bronzino, Introduction to Biomedical Engineering, Academic Press, 2000.

CHAPTER PROBLEMS

- **Fundamental**
  1. *Question*: Given a sinusoid with 500-Hz frequency, what should be the minimum sampling frequency for an analog-to-digital converter, if we want to faithfully reconstruct the analog signal after the conversion?
  2. *Question*: If 12 bits are used to quantize a sampled signal, what is the number of available quantized levels?
- **Advanced**
  1. *Question*: Given the 12-V input range of an analog-to-digital converter and the resolution of 0.125 V, what should be the minimum number of bits used for the conversion?

2. *Question*: Perform a trade-off study on the four technologies used for the analog-to-digital conversion. Use cost, conversion time, and accuracy as the list of criteria.

- **Challenging**
  1. *Question*: Investigate the analog-to-digital converters in your audio system. Find the sampling rate, the quantization bits, and the technique used for the conversion.

CHAPTER 6

Networked Microcontrollers

Objectives: After reading this chapter, the reader should be able to

- Describe the purpose and objectives of computer networks.
- Explain microcontroller networks.
- Describe Controller Area Networks (CANs) and Byte Data Link Controller (BDLC) networks.

In this chapter, we introduce an important subfield of microcontroller technologies, networked microcontrollers. Over the past few years, two different emerging technologies became the enablers of creating powerful network systems which harbor resources of all network members. The first technology comes from the computer industry where the cost for computation continues to decline. Today, we can easily find an inexpensive microcontroller or microprocessor with amazing computational capacity for which we had to pay a high price to acquire only a few years ago. The second enabling technology arose from the sensor industry. The costs for sensors, as for the microcontroller, came down so much that extracting information from multiple locations simultaneously became a common application of these sensors. Recently, the two technology advancements have come together to create a network of microcontrollers to provide efficiency, effectiveness, redundancy, and communication. Although connecting a number of computers together to create a computer networked system is not a new idea (computers in the 1970s were connected together to share printers), it is a fairly recent event to put together a set of microcontrollers together to create a microcontroller network. The pioneer of such networks is the automobile industry where networked microcontrollers in cars are used to enhance the experience of riders. In this chapter we present the technologies behind creating a networked computers, focusing mainly on the microcontroller network. The first section provides fundamental knowledge required to create a networked computer system, the second section presents specific microcontroller network technologies, and the final section discusses an automobile controller network application familiar to most of the readers.

6.1 BACKGROUND THEORY

In this section we briefly discuss important network concepts necessary for the rest of the chapter.

6.1.1 Designing Computer Networks

The main motivation of a networked computer system is to share the resources of the network members when connected together. Hence, the design of a network is dictated by the system requirements. For example, one may want to create a small network of computers that share computational burden with minimal communication. The primary purpose of another computer network may be the communication among the network members. Yet for another network of computers, the main purpose of the network can be the sharing of hardware resources such as printers, scanners, and fax machines. Designers of different computer networks, which we discuss in the next subsection, must consider the use and purpose of networks as the networks are created.

6.1.2 Types of Networks and Protocols

There are several types of computer networks. The most well-known computer network is the world wide web, which is often called the Internet. This giant network connects computers all around the globe and continues to play the main role in the vast amount of information exchange everyday. The world wide web is considered a Wide Area Network (WAN) since it connects computers that are located at a great distance from each other. The smaller version of the WAN is the Local Area Network (LAN). This network connects computers in a relatively small area such as the one within a company or an organization. People also refer to a LAN as an intranet to distinguish it from the Internet. The computers associated with a LAN are usually located inside of fire walls and are accessed by a computer outside of the network through servers. A computer network that is even smaller than a LAN is called a Small Area Network (SAN). With the emerging wireless technologies, one can easily find such a network within a home, an office, and a commercial shop.

In addition to the aforementioned three computer networks, the microcontroller industry developed Controller Area Networks (CANs) and Byte Data Link Controller (BDLC) networks, which we discuss in the next section.

All computer networks must have protocols for communicating data and commands among the nodes. A protocol enforces a set of rules that must be followed by the network computers to send and receive data and commands. A protocol specifies communication requirements that include the timing of bits transmitted and received, the data length, and the error checking mechanism. We do not have the space to discuss in-depth the protocols used for the microcontroller networks in this chapter. We refer interested readers to the Bibliography section at the end of this chapter.

6.2 MICROCONTROLLER NETWORKS

The automobile industry is the main player which has advanced microcontroller networks. Over the years, the industry has pushed microcontroller technologies to control fuel injection systems, antilock brakes, and four-wheel drive systems. Today, a typical automobile contains more than ten microcontrollers connected together to coordinate the engine performance, the brake system, the navigational system, the communication system, the display system, and the status monitoring system. The monitoring system checks a variety of automobile states including fluid levels and tire pressures. The success in the automobile industry has helped the microcontroller controller networks to be adopted in other industries since the early 1990s. Today, we can find microcontroller networks in trains, medical equipment, home audio systems, home theater systems, communication systems, military systems, and security systems. In this section, we present two primary microcontroller networks: the Controller Area Networks and the Byte Data Link Controller network.

6.2.1 Controller Area Network

Among the microcontroller networks, the Controller Area Network (CAN) is the most commonly used microcontroller network in the industry. The network was initially developed for automobile applications, but has been adopted to a variety of systems used in military and robotics applications. The CAN network protocol uses three layers of the International Standards Organization/Open Systems Interconnect (ISO/OSI) model: the object layer, the transfer layer, and the physical layer. The three layers manage handling of messages, interface between layers, and application programs on microcontrollers. The attractive feature of the CAN network protocol is that messages do not contain originating and destination addresses. Instead, an identifier is embedded in each message, which means that a node can be connected to the network without altering any existing software or hardware on the network. It also means that multiple nodes can act upon the same message, allowing multicasting capabilities.

The protocol uses a simple arbitration technique for communication conflicts and a sophisticated error detection mechanism to inform the sender of the message that generated errors. The actual message consists of a message identifier, a message, and a length specifier. Any member of a CAN network can send and request messages over the network. Four types of messages exist: a data message, a data request message, an error message, and an overload message.

The data message, called a data frame, starts with a single bit that is used to synchronize the rest of the message for the receiving microcontoller. The second part of the frame holds the identifier number of a message. Receiving microcontrollers use the identifier number to either accept or disregard the rest of a message. The third part of the message is allocated as the arbitration field which is used to distinguish data from a request. The next two parts of a message specifies the data length in bytes and the actual data. Following the data section, a message contains the Cyclical Redundancy Check (CRC) field used to test the validity of the data received. An end of message field concludes the message frame. Reference Figure 6.1.

A data request message is generated by a receiving microcontroller to send a request to retransmit a data frame. The overall structure of the data request message is the same as the one for the data message except that the data field is empty and the arbitration field indicates a data request. An error frame, made of a number of error flags, indicates that either an active or passive error occurred on the network. An active error frame is generated by the transmitting microcontroller and a passive error frame is generated by the receiving microcontroller. Five different errors can trigger the generation of an error frame: a bit error, a stuff error, a CRC error, a form error and an acknowledgement error. A bit error is detected when a transmitting node detects inconsistency between the bit transmitted and the one it monitors on the network. The stuff error occurs when one or more of the designated bits with the same logic states are not detected. The CRC error is issued when the computed and received CRC values do not agree, the form error occurs when invalid bits form a frame subfield, and the acknowledgement error occurs when an acknowledgement from a receiving node is not received. Finally, an overload frame is generated when too many valid frames exist on the network for a receiving node to process, resulting a time delay.

Bit Timing

For proper transmission and reception of messages, timing requirements must be consistently met. For the CAN bus, the bit time is governed by dividing the duration period for each bit into four segments: the synchronization segment, the propagation time segment, phase buffer segment 1, and phase buffer segment 2. Figure 6.1 shows the four segments that are used for each bit duration.

The synchronization segment looks for an edge to synchronize nodes on the bus. The propagation time segment is used to accept transmitter/receiver delay and signal propagation time within the bus. The phase segment 1 and the phase segment 2 are lengthened and shortened to resynchronize bits on the bus. The actual logic state of a bit is measured after the phase buffer segment 1 (PHASE_SEG1).

We briefly presented the CAN protocol in this section. For further details on the protocol, we refer the reader to Barrett and Pack (2005) and the CAN Version 2.0 protocol standard listed in the References section.

6.2.2 BDLC Networks

The Byte Data Link Controller (BDLC) network uses the Society of Automotive Engineers (SAE) J1850 protocol. The protocol is equivalent to serial data communications at a low speed, typically less than 125 kb/s. The BDLC network adopted a variable pulse width bit format, noise

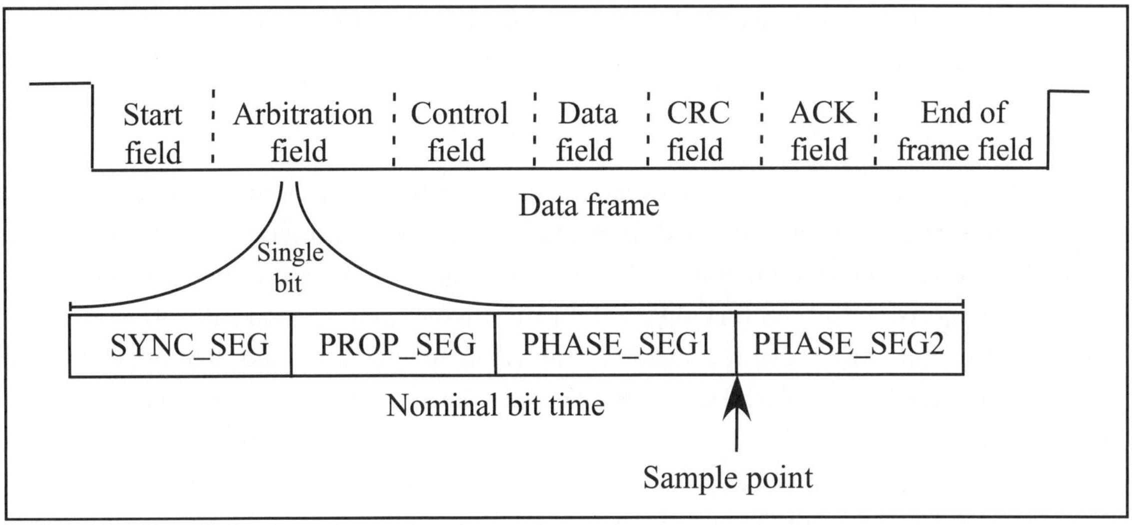

FIGURE 6.1: Nominal bit time segments

filters, a collision detection mechanism, and a cyclical redundancy check to accurately transfer and receive messages within the network. Each message contains a start of frame symbol, actual data (message priority, message ID, followed by actual data), cyclical redundancy check byte, and an end-of-data symbol. Over the past decade, the SAE J1850 protocol and the CAN protocol have competed to dominate the controller area networks. Today, the CAN protocol has gained significantly larger microcontroller network applications among industry users. Typically, one will find a CAN network in an automobile manufactured in Europe and a BDLC network in an automobile manufactured by General Motors, Ford, and Chrysler.

6.2.3 Customized Networks

One can certainly create his or her own microcontroller network with a custom-made protocol. A typical microcontroller contains multiple input/output ports, one of which can be easily used as a network communication port. If the number of microcontrollers involved in a network is small, a customized protocol can easily be adopted for such a network. For example, a three nodes network with a customized protocol is often assigned as a senior design project at our institutions. Typically, a token-passing technique among the members is used to regulate the transmission of messages which contain a transmitting node identification number, a receiving identification number, and a message.

6.3 APPLICATIONS

In this section, we present two different applications of microcontroller networks. A form of the first application is found in most of the automobiles manufactured after the year 2000. The second application demonstrates the effective use of a microcontroller network for a nonautomobile application.

6.3.1 Automobiles

Ever since the benefits of using an inexpensive microcontroller to monitor and enhance automobile performance were discovered, more and more microcontrollers are incorporated in automobile designs and production. Today, a typical car contains more than ten microcontrollers working to control the fuel injection for the engine, monitor and control the antilock brake system, check tire pressures, maintain the cabin temperature, and communicate with Global Positioning System (GPS) satellites, to name a few. Figure 6.2 shows a typical set up for an automobile where multiple microcontrollers are embedded to control and monitor different automobile subsystems individually.

Increasingly, the automobile designers found it necessary to send information from one microcontroller to another, especially the data gathered by microcontrollers that are in charge of sensors to microcontrollers that control subsystems and the display system, discovering the need for a network. The need evolved and developed into microcontroller networks, an example shown in Figure 6.3. Robert Bosch from Germany developed the Controller Area Network (CAN) for the German automobile industry in late 1980s.

Using the protocol, the microcontrollers responsible for automobile subsystems communicate among themselves to enhance the automobile performance and improve the comfort

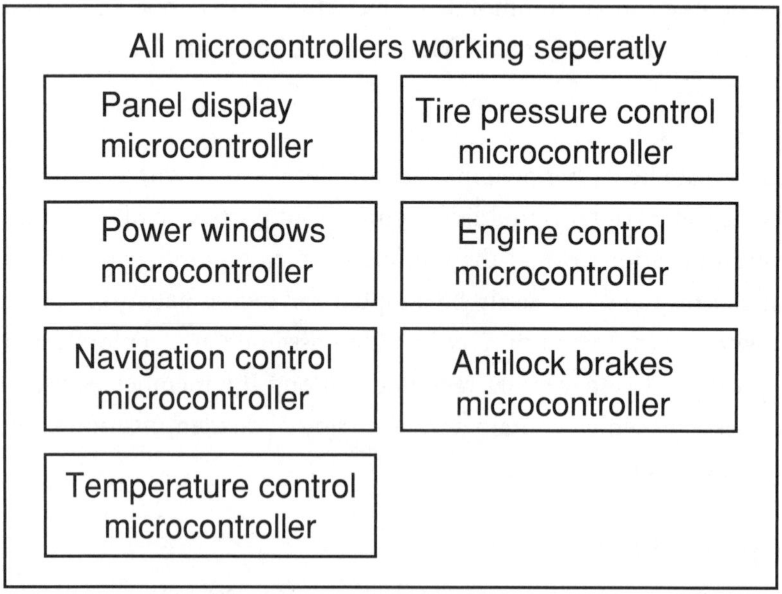

FIGURE 6.2: Multiple microcontrollers working separately

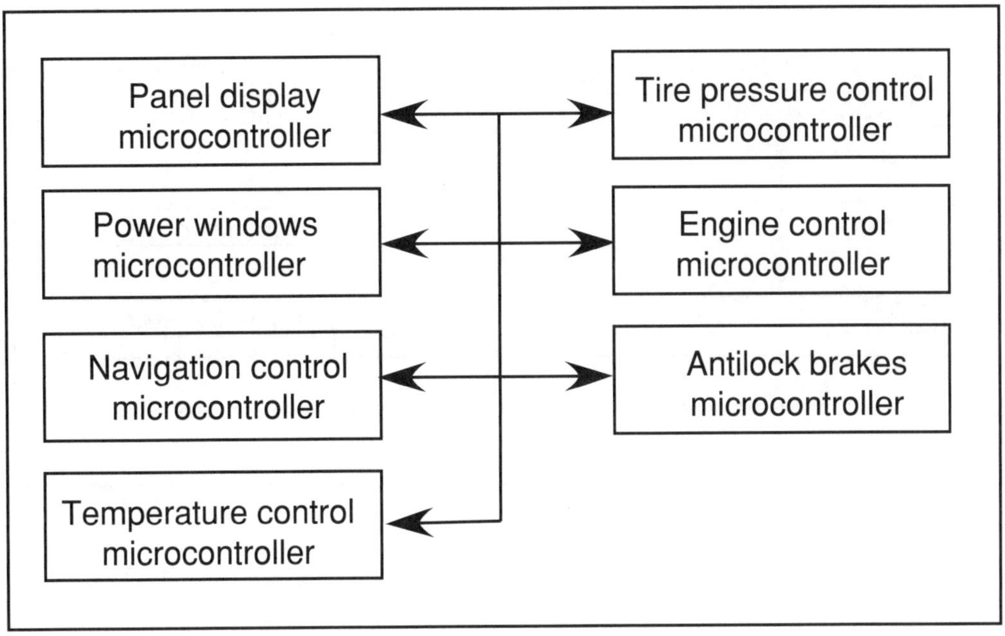

FIGURE 6.3: Multiple microcontrollers working separately

of the riders. For example, the front panel display microcontroller continuously polls microcontrollers to update the cabin temperature, outside temperature, engine performance, and tire slippage. If a tire slippage is detected, due to snow on the road, for example, the information is not only relayed to the driver but also used in the antilock brake system, if the driver chose to brake. Such communication among the microcontrollers makes an automobile safer for people and improves the automobile performance.

6.3.2 Mobile Robots

In this section we describe a CAN application used in a mobile robot application. The CAN network connects multiple microcontrollers that collect sensor data, manage CAN bus activities, and control servo motors. Sensor data are collected and processed by a microcontroller. Based on the sensor data reflecting robot motion, proper control commands are sent to another microcontroller that generates pulse width modulation signals to control servo motors of the mobile robot. The bus activities are supervised by yet another microcontroller that monitors the 'health' of each microcontroller on the controller area network. Figure 6.4 shows the CAN used for the application. The CAN system is used by a mobile robot to track and follow a moving target. For a complete description of the system see Fierro *et al.* (2005) cited in the Bibliography section.

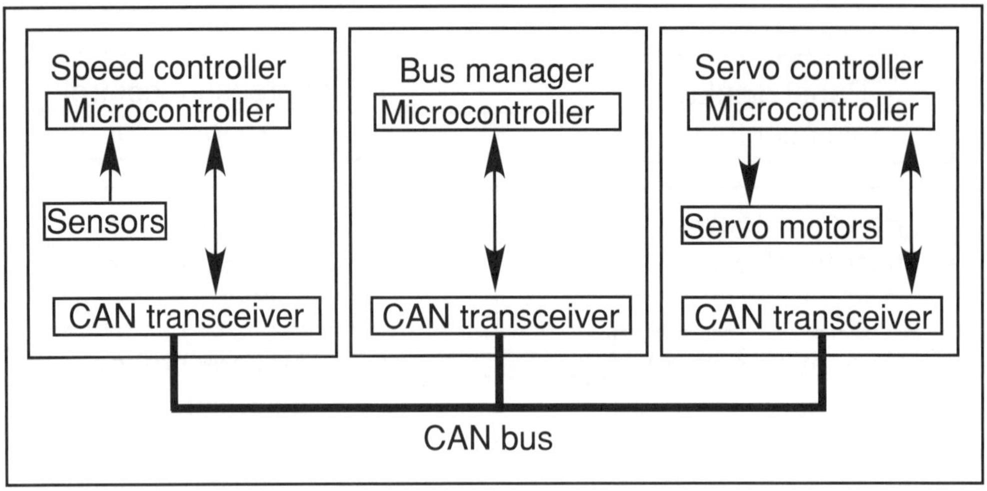

FIGURE 6.4: A mobile robot controller area network

6.4 SUMMARY

In this chapter, we provided a brief description of different computer networks including WAN, LAN, and SAN. In addition, two microcontroller networks, CAN and BDLC are presented. The evolution of these microcontroller networks has heavily been influenced by the automobile industries. The BDLC networks are widely accepted in the United States automobile industries while the CAN networks are adopted in most of the European automobile industries. The two applications show the power of the microcontroller networks. We can safely expect that the number of microcontroller networks will continually grow as the price of microcontrollers drops and the networking technologies continue to improve.

BIBLIOGRAPHY

R Fierro, J Clark, D Hougen, and S Commuri, A multi-robot testbed for biologically-inspired cooperative control, in multi-robot systems. From Swarms to Intelligent Automata, Volume III, L. E. Parker, F. E. Schneider, and A. C. Schultz (eds.), Springer, pp. 171–182, 2005.

CAN-Bosch Controller Area Network (CAN) Version 2.0, Protocol Standard, BCANPSV2.0/D, Rev. 3, Motorola.

The msCAN on the MC9S12DP256 Compared with the msCAN on the HC12 Family, AN2011/D, Rev. 1, 01/2002, Motorola.

Motorola Scalable Controller Area Network (msCAN) Interrupts, AN2283/D, Rev. 0, 08/2002, Motorola.

BDLC Reference Manual-HC08 and HC12 MCUs, BDLCRM/D, Motorola.

S Barrett and D Pack, Embedded Systems Design with the 68HC12 and HCS12, Prentice-Hall, Upper Saddle River, NJ, 2005.

CHAPTER PROBLEMS

- **Fundamental**
  1. *Question*: Discuss the advantages and disadvantages of a networked computer compared to an isolated computer.
  2. *Question*: Give a short definition of a network communication protocol.
- **Advanced**
  1. *Question*: Describe the advantage of not specifying a destination node address and an origination node address of a CAN data frame.
  2. *Question*: Describe the mechanism used in the CAN protocol to synchronize an incoming bit from a CAN bus.
- **Challenging**
  1. *Question*: What are the main differences between the BDLC and the CAN microcontroller networks?
  2. *Question*: Describe the technique used to send and receive messages in a controller area network without specifying starting and destination addresses in messages.

CHAPTER 7

Operating Parameters and Interfacing

Objectives: After reading this chapter, the reader should be able to

- Describe the voltage and current parameters for an HC CMOS type microcontroller.
- Apply the voltage and current parameters toward properly interfacing input and output devices to a microcontroller.
- Describe the special concerns that must be followed when a microcontroller is used to interface to a high power DC or AC device.
- Discuss the requirement for an optical-based interface.
- Describe how to control the speed and direction of a DC motor.

In this chapter, we introduce you to the extremely important concepts of the operating envelope for a microcontroller. We begin by reviewing the voltage and current electrical parameters for an HC CMOS based microcontroller. We then show how to apply this information in properly interfacing input and output devices to a microcontroller. We then discuss the special considerations for controlling a high power DC or AC load such as a motor and introduce the concept of an optical interface. We close the chapter by providing a detailed example of controlling the speed and direction of a high power DC motor.

7.1 OPERATING PARAMETERS

Microcontrollers are electronic devices which have precisely defined operating parameters. As long as the microcontroller is used within its defined parameters it should continue to operate correctly. However, if the microcontroller is used outside of its specified electrical parameters, spurious results may result. Any time a device is connected to a microcontroller, careful interface analysis must be performed. Most microcontrollers are members of the "HC," or high-speed CMOS, family of chips. As long as all components in a system are also of the "HC" family, electrical interface issues are minimal. If the microcontroller is connected to some component not in the "HC" family, electrical interface analysis must be completed. Manufacturers readily provide the electrical characteristic data necessary to complete this analysis in the data support documentation.

There are eight different electrical specifications required for electrical interface analysis. The electrical parameters are

- V_{OH}: the lowest guaranteed output voltage for a logic high,
- V_{OL}: the highest guaranteed output voltage for a logic low,
- I_{OH}: the output current for a V_{OH} logic high,
- I_{OL}: the output current for a V_{OL} logic low,
- V_{IH}: the lowest input voltage guaranteed to be recognized as a logic high,
- V_{IL}: the highest input voltage guaranteed to be recognized as a logic low,
- I_{IH}: the input current for a V_{IH} logic high, and
- I_{IL}: the input current for a V_{IL} logic low.

These electrical characteristics are required for both the microcontroller and the external components. Typical values for a microcontroller in the HC CMOS family assuming V_{DD} = 5.0 V and V_{SS} = 0 V are provided below. The minus sign on several of the currents indicates a current flow out of the device. A positive current indicates current flow into the device.

- $V_{OH} = 4.2$ V,
- $V_{OL} = 0.4$ V,
- $I_{OH} = -0.8$ mA,
- $I_{OL} = 1.6$ mA,
- $V_{IH} = 3.5$ V,
- $V_{IL} = 1.0$ V,
- $I_{IH} = 10$ μA, and
- $I_{IL} = -10$ μA.

If external circuitry is connected such that the microcontroller acts as a current source (current leaving microcontroller) or current sink (current entering microcontroller), the voltage parameters listed above will also be affected.

In the current source case, an output voltage V_{OH} is provided at the output pin of the microcontroller when the load connected to this pin draws a current of I_{OH}. If a load draws more current from the output pin than the I_{OH} specification, the value of V_{OH} is reduced. If the load current becomes too high, the value of V_{OH} falls below the value of V_{IH} for the subsequent logic circuit stage and is not recognized as an acceptable logic high signal. When this occurs erratic and unpredictable circuit behavior results.

In the sink case, an output voltage V_{OL} is provided at the output pin of the microcontroller when the load connected to this pin delivers a current of I_{OL} to this logic pin. If a load delivers more current to the output pin of the microcontroller than the I_{OL} specification, the value of V_{OL} increases. If the load current becomes too high, the value of V_{OL} rises above the value of V_{IL} for the subsequent logic circuit stage and is not recognized as an acceptable logic low signal. As before, when this occurs, erratic and unpredictable circuit behavior results.

For convenience this information is illustrated in Figure 7.1. In (a) we have provided an illustration of the direction of current flow from the HC device and also a comparison of voltage levels. As a reminder current flowing out of a device is considered a negative current (source case) while current flowing into the device is considered positive current (sink case). The magnitude of the voltage and current for HC CMOS devices is provided in (b). As more current is sinked or sourced from a microcontroller pin the voltage will be pulled up or pulled down, respectively (c). If input and output devices are improperly interfaced to the microcontroller these loading conditions may become excessive and voltages will not be properly interpreted as the correct logic levels. In this chapter, we will discuss how to properly apply this information to correctly interface input and output devices to a microcontroller.

7.2 INPUT DEVICES

In this section we will discuss how to properly interface input devices to a microcontroller. We will start with the most basic input component, a simple on/off switch.

7.2.1 Switches

Switches come in a variety of types. As a system designer it is up to you to choose the appropriate switch for a specific application. Switch varieties commonly used in microcontroller applications are illustrated in Figure 7.2(a). Here is a brief summary of the different types:

- *Slide switch:* A slide switch has two different positions: on and off. The switch is manually moved to one position or the other. For microcontroller applications slide switches are available that fit in the profile of a common integrated circuit size dual inline package (DIP). A bank of four or eight DIP switches in a single package is commonly available.
- *Momentary contact pushbutton switch:* A momentary contact pushbutton switch comes in two varieties normally closed (NC) and normally open (NO). A normally open switch as its name implies does not normally provide an electrical connection between its contacts. When the pushbutton portion of the switch is depressed the connection between the two switch contacts is made. The connection is held as long as the switch is depressed. When the switch is released the connection is opened. The converse is true

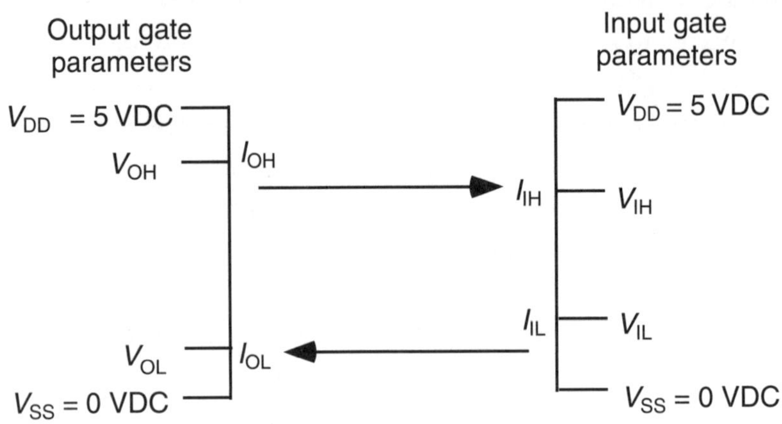

(a) Voltage and current electrical parameters

Output parameters	Input parameters
$V_{OH} = 4.2$ V	$V_{IH} = 3.5$ V
$V_{OL} = 0.4$ V	$V_{IL} = 1.0$ V
$I_{OH} = -0.8$ mA	$I_{IH} = 10\ \mu$A
$I_{OL} = 1.6$ mA	$I_{IL} = -10\ \mu$A

(b) HC CMOS voltage and current parameters

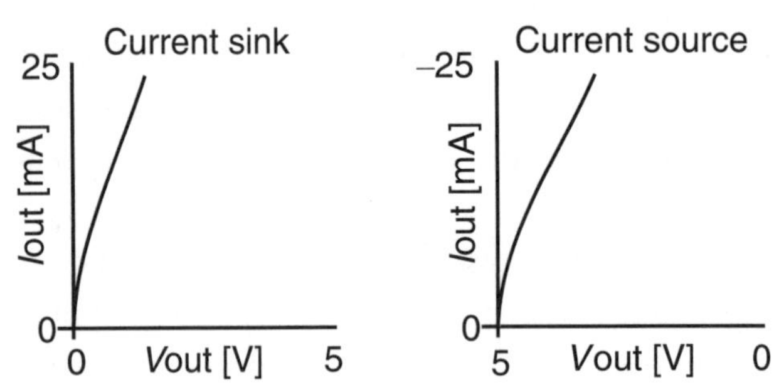

(c) CMOS loading curves

FIGURE 7.1: Electrical voltage and current parameters

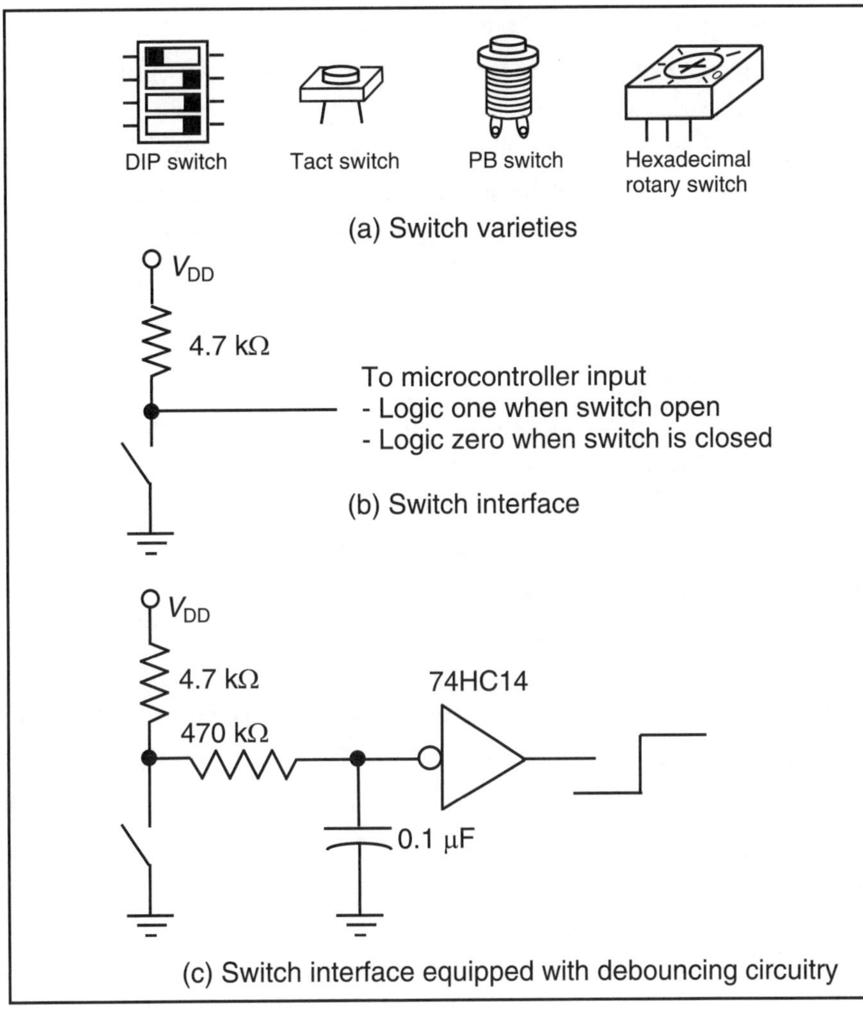

FIGURE 7.2: Switch interface

for a normally closed switch. For microcontroller applications pushbutton switches are available in a small tact type switch configuration.

- *Push on/push off switches:* These type of switches are also available in normally open and normally closed configurations. For the normally open configuration, the switch is depressed to make connection between the two switch contacts. The pushbutton must be depressed again to release the connection.
- *Hexadecimal rotary switches:* Small profile rotary switches are available for microcontroller applications. These switches commonly have 16 rotary switch positions. As the switch is rotated to each position a unique four-bit binary code is provided at the switch contacts.

A common switch interface is provided in Figure 7.2(b). This interface allows a logic one or zero to be properly introduced to a microcontroller input port pin. The basic interface consists of the switch in series with a current limiting resistor. The node between the switch and the resistor is provided to the microcontroller input pin. In the configuration shown, the resistor pulls the microcontroller input up to the supply voltage V_{DD}. When the switch is closed, the node is grounded and a logic zero is provided to the microcontroller input pin. To reverse the logic of the switch configuration the position of the resistor and the switch is simply reversed.

7.2.2 Switch Debouncing

Mechanical switches do not make a clean transition from one position (on) to another (off). When a switch is moved from one position to another it makes and breaks contact multiple times. This activity may go on for tens of milliseconds. A microcontroller is relatively fast as compared to the action of the switch. Therefore, the microcontroller is able to recognize each switch bounce as a separate and erroneous transition.

To correct the switch bounce phenomena additional external hardware components may be used or software techniques may be employed. A hardware debounce circuit is illustrated in Figure 7.2(c). The node between the switch and the limiting resistor of the basic switch circuit is fed to a low pass filter (LPF) formed by the 470 kΩ resistor and the capacitor. The LPF prevents abrupt changes (bounces) in the input signal from the microcontroller. The LPF is followed by a 74HC14 Schmitt Trigger which is simply an inverter equipped with hysteresis. This further limits the affects of switch bouncing.

Switches may also be debounced using software techniques. This is accomplished by inserting a 30–50 ms lockout delay in the function responding to port pin changes. The delay prevents the microcontroller from responding to the multiple switch transitions related to bouncing.

7.2.3 Keypads

A keypad is simply an extension of the simple switch configuration. A typical keypad configuration and interface is provided in Figure 7.3. As we can see the keypad is simply multiple switches in the same package. A hexadecimal keypad is provided in the figure. A single row of keypad switches are asserted by the microcontroller and then the host keypad port is immediately read. If a switch has been depressed, the keypad pin corresponding to the column the switch is in will also be asserted. The combination of row and column assertion can be decoded to determine which key has been pressed as illustrated in the table. Keypad rows are continually asserted one after the other in sequence.

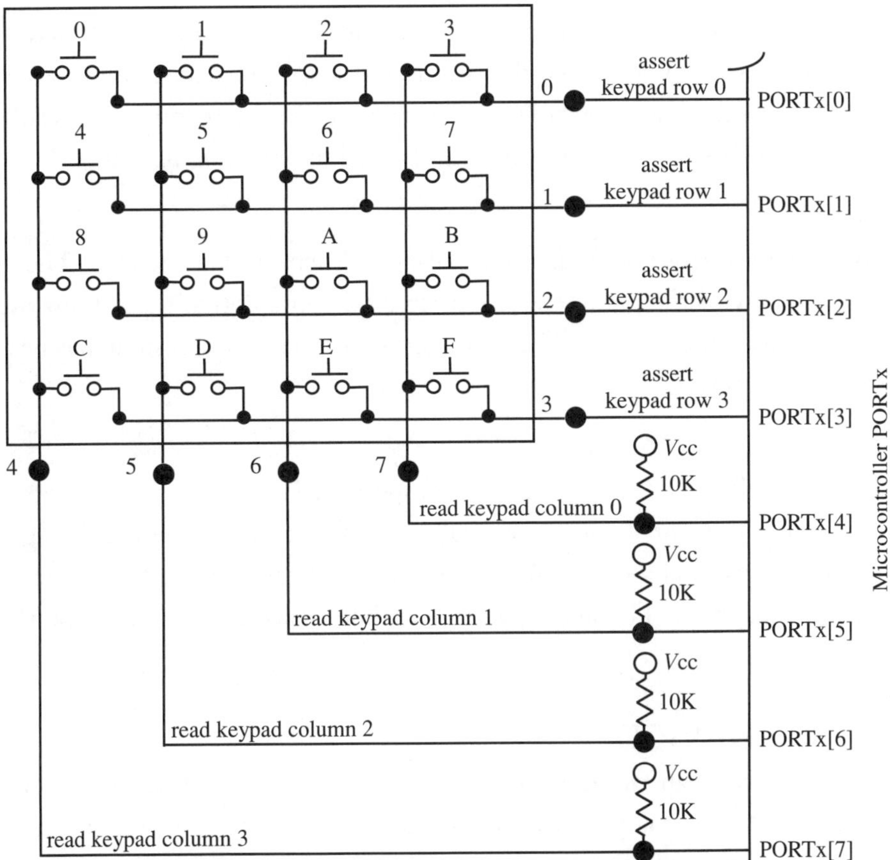

Key pressed	Row asserted 3 2 1 0	Column response 7 6 5 4	Row/column combination (PORTx[7:0])
0	1 1 1 0	1 1 1 0	0xEE
1	1 1 1 0	1 1 0 1	0xDE
2	1 1 1 0	1 0 1 1	0xBE
3	1 1 1 0	0 1 1 1	0x7E
4	1 1 0 1	1 1 1 0	0xED
5	1 1 0 1	1 1 0 1	0xDD
6	1 1 0 1	1 0 1 1	0xBD
7	1 1 0 1	0 1 1 1	0x7D
8	1 0 1 1	1 1 1 0	0xEB
9	1 0 1 1	1 1 0 1	0xDB
A	1 0 1 1	1 0 1 1	0xBB
B	1 0 1 1	0 1 1 1	0x7B
C	0 1 1 1	1 1 1 0	0xE7
D	0 1 1 1	1 1 0 1	0xD7
E	0 1 1 1	1 0 1 1	0xB7
F	0 1 1 1	0 1 1 1	0x77
None	X X X X	1 1 1 1	0xXF

FIGURE 7.3: Keypad interface

7.2.4 Sensors

A microcontroller is typically used in control applications where data is collected, the data is assimilated and processed by the host algorithm, and a control decision and accompanying signals are provided by the microcontroller. Input data for the microcontroller is collected by a complement of input sensors. These sensors may be digital or analog in nature.

Digital Sensors

Digital sensors provide a series of digital logic pulses with sensor data encoded. The sensor data may be encoded in any of the parameters associated with the digital pulse train such as duty cycle, frequency, period, or pulse rate. The input portion of the timing system may be configured to measure these parameters.

Analog Sensors

Analog sensors provide a DC voltage that is proportional to the physical parameter being measured. As discussed in the analog to digital conversion chapter, the analog signal may be first preprocessed by external analog hardware such that it falls within the voltage references of the conversion subsystem. The analog voltage is then converted to a corresponding binary representation.

7.3 OUTPUT DEVICES

As previously mentioned, an external device should not be connected to a microcontroller without first performing careful interface analysis to insure that the voltage, current, and timing requirements of the microcontroller and the external device are met. In this section we describe interface considerations for a wide variety of external devices as shown in Figure 7.4.

7.3.1 Light-Emitting Diodes (LEDs)

An LED is typically used as a logic indicator to indicate the presence of a logic one or a logic zero at a specific pin of a microcontroller. An LED has two leads: the anode or positive lead and the cathode or negative lead. To properly bias an LED, the anode lead must be biased at a level approximately 1.7–2.2 V higher than the cathode lead. This specification is known as the forward voltage (V_f) of the LED. The LED current must also be limited to a safe level known as the forward current (I_f). The diode voltage and current specifications are usually provided by the manufacturer. An example of an LED biasing circuit is provided in Figure 7.4(a). A logic one is provided by the microcontroller to the input of the inverter. The inverter provides a logic zero at its output which provides a virtual ground at the cathode of the LED. Therefore, the proper voltage biasing for the LED is provided. The resistor (R) provides the current limiting for the LED. Its value may be calculated using $R = (V_{DD} - V_{DIODE})/I_{DIODE}$. It is important to note that a 7404 inverter must be used due to its capability to safely sink 16 mA of current.

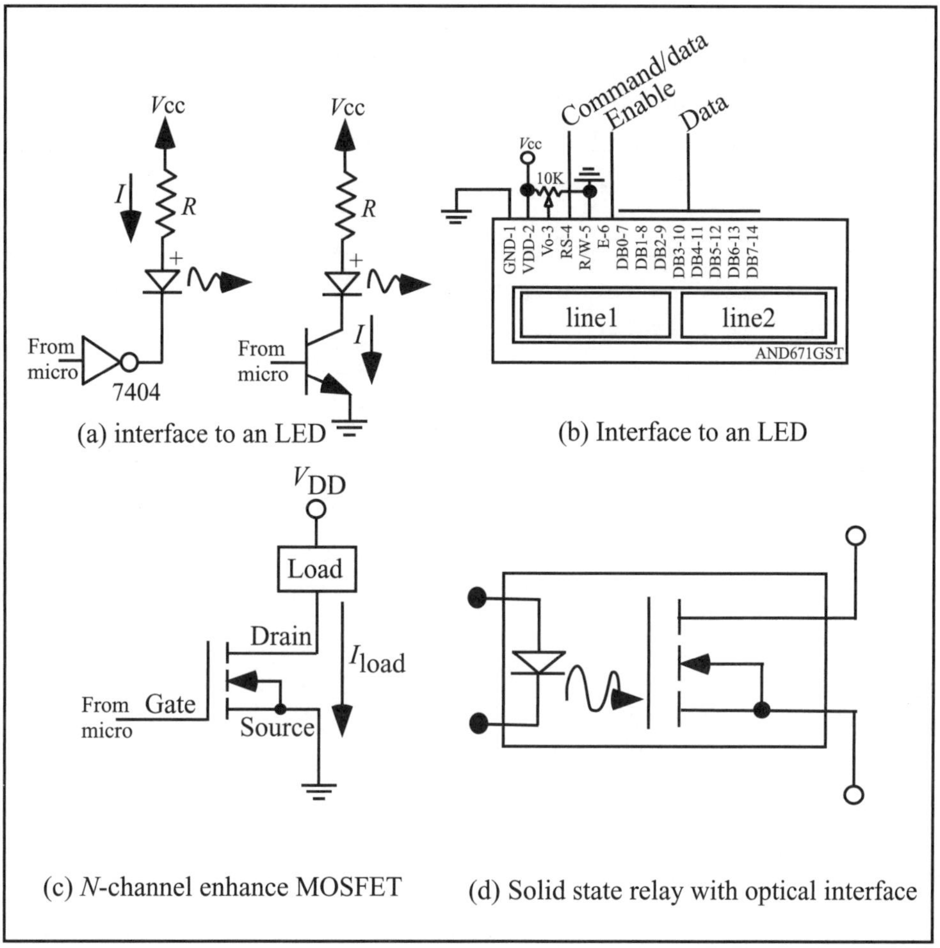

FIGURE 7.4: Output devices

Alternately, an NPN transistor such as a 2N2222 may be used in place of the inverter as shown in the figure.

7.3.2 Liquid Crystal Display (LCD)

A LCD is an output device to display text information as shown in Figure 7.4(b). LCDs come in a wide variety of configurations including multicharacter, multiline format. A 16×2 LCD format is common. That is, it has the capability of displaying two lines of 16 characters each. The characters are sent to the LCD via ASCII format a single character at a time. For a parallel configured LCD an eight-bit data path and two lines are required between the microcontroller and the LCD. A small microcontroller mounted to the back panel of the LCD translates the ASCII data characters and control signals to properly display the characters. LCDs are configured for either parallel or serial data transmission format.

Some sample C code is provided below to send data and control signals to an LCD. In this specific example an AND671GST 1 × 16 character LCD was connected to an Atmel ATmega8535 eight-bit microcontroller. One eight-bit port and two extra control lines are required to connect the microcontroller to the LCD. Note: The initialization sequence for the LCD is specified within the manufacturer's technical data.

```
//***********************************************************************
//LCD_Init: initialization for an LCD connected in the following manner:
//LCD: AND671GST 1x16 character display
//LCD configured as two 8 character lines in a 1x16 array
//LCD data bus (pin 14-pin7)   ATMEL ATmega8535: PORTC
//LCD RS (pin 4)  ATMEL ATmega8535: PORTA[7]
//LCD E  (pin 6)  ATMEL ATmega8535: PORTA[6]
//***********************************************************************

void LCD_Init(void)
{
delay_5ms();
delay_5ms();
delay_5ms();
                              // output command string to initialize LCD
putcommand(0x38);             //function set 8-bit
delay_5ms();
putcommand(0x38);             //function set 8-bit
putcommand(0x38);             //function set 8-bit
putcommand(0x38);             //one line, 5x7 char
putcommand(0x0C);             //display on
putcommand(0x01);             //display clear-1.64 ms
putcommand(0x06);             //entry mode set
putcommand(0x00);             //clear display, cursor at home
putcommand(0x00);             //clear display, cursor at home
}

//***********************************************************************
//putchar:prints specified ASCII character to LCD
//***********************************************************************

void putchar(unsigned char c)
{
DDRC=0xff;                    //set PORTC as output
DDRA=DDRA|0xC0;               //make PORTA[7:6] output
PORTC=c;
```

```
PORTA= PORTA|0x80;                  //RS=1
PORTA= PORTA|0x40;                  //E=1
PORTA= PORTA&0xbf;                  //E=0
delay_5ms();
}

//*********************************************************************
//performs specified LCD related command
//*********************************************************************

void putcommand(unsigned char d)
{
DDRC=0xff;                          //set PORTC as output
DDRA=DDRA|0xC0;                     //make PORTA[7:6] output
PORTA= PORTA&0x7f;                  //RS=0
PORTC=d;
PORTA= PORTA|0x40;                  //E=1
PORTA= PORTA&0xbf;                  //E=0
delay_5ms();
}

//*********************************************************************
```

7.3.3 DC Devices

A number of direct current devices may be controlled with an electronic switching device such as a MOSFET. Specifically, an *N*-channel enhancement MOSFET (metal oxide semiconductor field effect transistor) may be used to switch a high current load on and off (such as a motor) using a low current control signal from a microcontroller as shown in Figure 7.4(c). The low current control signal from the microcontroller is connected to the gate of the MOSFET. The MOSFET switches the high current load on and off consistent with the control signal. The high current load is connected between the load supply and the MOSFET drain. It is important to note that the load supply voltage and the microcontroller supply voltage do not have to be at the same value. When the control signal on the MOSFET gate is logic high, the load current flows from drain to source. When the control signal applied to the gate is logic low, no load current flows. Thus the high power load is turned on and off by the low power control signal from the microcontroller.

Often the MOSFET is used to control a high power motor load. A motor is a notorious source of noise. To isolate the microcontroller from the motor noise an optical isolator may be used as an interface as shown in Figure 7.4(d). The link between the control signal from the microcontroller to the high power load is via an optical link contained within a Solid State Relay (SSR). The SSR is properly biased using techniques discussed previously.

7.3.4 AC Devices

In a similar manner, a high power alternating current (AC) load may be switched on and off using a low power control signal from the microcontroller. In this case a SSR is used as the switching device. SSRs are available to switch a high power DC or AC load [Crydom].

7.4 APPLICATION: DC MOTOR SPEED AND DIRECTION CONTROL

Often a microcontroller is used to control a high power motor load. To properly interface the motor to the microcontroller we must be familiar with the different types of motor technologies. Motor types are illustrated in Figure 7.5.

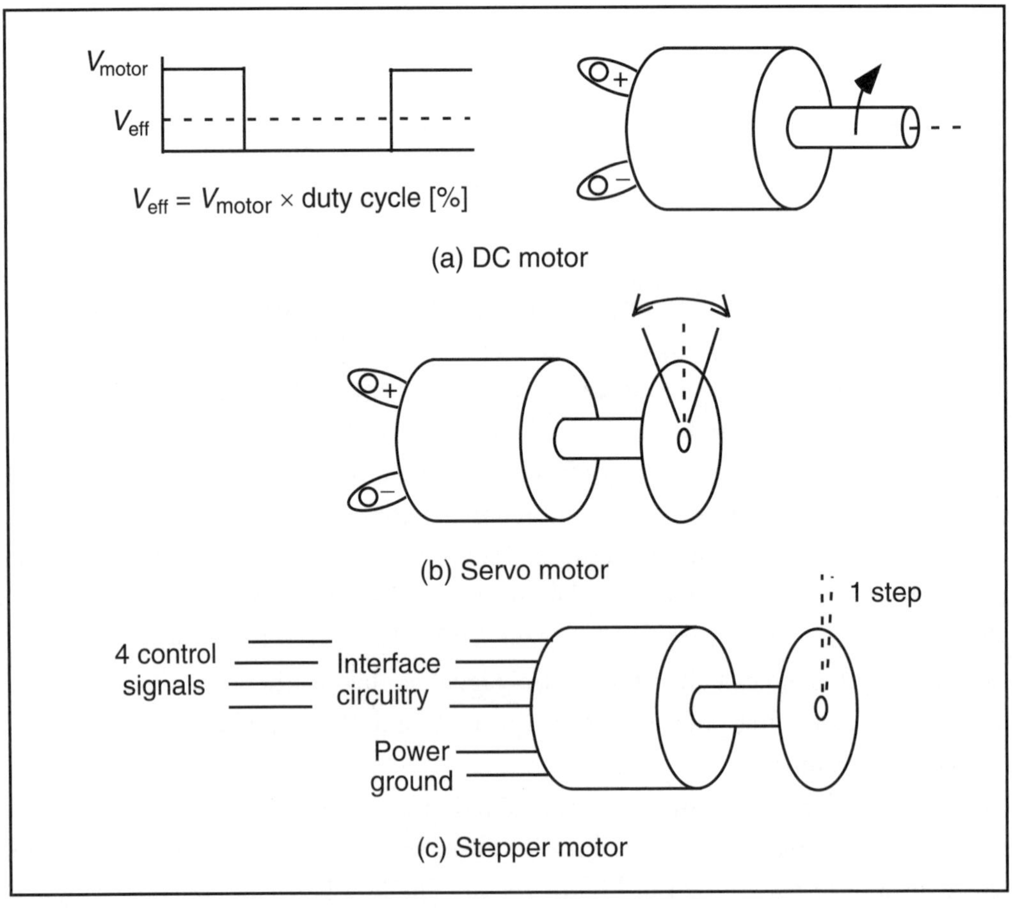

FIGURE 7.5: Motor types

- *DC motor:* A DC motor has a positive and negative terminal. When a DC power supply of suitable current rating is applied to the motor it will rotate. If the polarity of the supply is switched with reference to the motor terminals, the motor will rotate in the opposite direction. The speed of the motor is roughly proportional to the applied voltage up to the rated voltage of the motor.
- *Servo motor:* A servo motor provides a precision angular rotation for an applied pulse width modulation duty cycle. As the duty cycle of the applied signal is varied, the angular displacement of the motor also varies. This type of motor is used to change mechanical positions such as the steering angle of a wheel.
- *Stepper motor:* A stepper motor as its name implies provides an incremental step change in rotation (typically 2.5 degree per step) for a step change in control signal sequence. The motor is typically controlled by a two or four wire interface. For the four wire stepper motor, the microcontroller provides a four-bit control sequence to rotate the motor clockwise. To turn the motor counterclockwise, the control sequence is reversed. The low power control signals are interfaced to the motor via MOSFETs or power transistors to provide for the proper voltage and current requirements of the pulse sequence.

7.4.1 Motor Operating Parameters

Space does not allow a full discussion of all motor types. We will concentrate on the DC motor. As previously mentioned the motor speed may be varied by changing the applied voltage. This is difficult to do with a digital control signal. However, PWM control signal techniques discussed earlier in the chapter may be combined with a MOSFET interface to precisely control the motor speed. The duty cycle of the PWM signal will also be the percentage of the motor supply voltage applied to the motor and hence the percentage of rated full speed at which the motor will rotate. The interface circuit to accomplish this type of control is shown in Figure 7.6. Various portions of this interface circuit have been previously discussed. The resistor R_G, typically 10 kΩ, is provided to discharge the MOSFET gate when no voltage is applied to the gate. For an inductive load a reversed biased protection diode must be provided across the load. The interface circuit shown allows the motor to rotate in a given direction. As previously mentioned, to rotate the motor in the opposite direction the motor polarity must be reversed. This may be accomplished with a high power switching network called an H-bridge specifically designed for this purpose. See Pack and Barrett (2002) for more information on this topic.

7.5 SUMMARY

In this chapter we have discussed the voltage and current operating parameters for an HC CMOS type microcontroller. We discussed how this information may be applied to properly

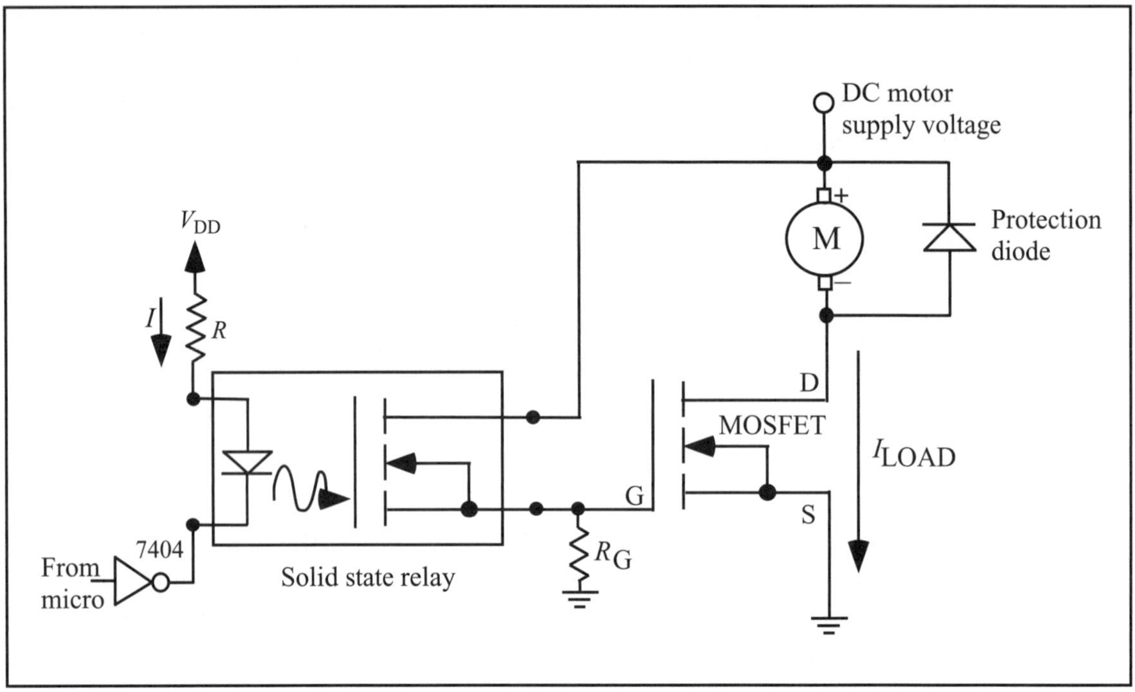

FIGURE 7.6: DC motor interface

design an interface for common input and output circuits. It must be emphasized that a properly designed interface allows the microcontroller to operate properly within its parameter envelope. If due to a poor interface design, a microcontroller is used outside its prescribed operating parameters, spurious and incorrect logic values will result. We also discussed the concept of interfacing a motor to a microcontroller using PWM techniques coupled with high power MOSFET or SSR switching devices.

BIBLIOGRAPHY

D Pack and S Barrett, 68HC12 Microcontroller: Theory and Applications. Prentice-Hall, Upper Saddle River, NJ, 2002.

S Barrett and D Pack, Embedded Systems Design with the 68HC12 and HCS12. Prentice-Hall, Upper Saddle River, NJ, 2004.

Crydom Corporation, 2320 Paseo de las Americas, Suite 201, San Diego, CA (www.crydom .com).

CHAPTER PROBLEMS

- **Fundamental**
  1. *Question*: What will happen if a microcontroller is used outside of its prescribed operating envelope?

2. *Question*: Discuss the difference between the terms "sink" and "source" as related to current loading of a microcontroller.

- **Advanced**
  1. *Question*: In your own words provide a brief description of each of the microcontroller electrical parameters provided in Figure 9.1(b).
  2. *Question*: What is switch bounce? Describe two techniques to minimize switch bounce.
  3. *Question*: What are the two purposes of the resistor in Figure 9.2(a)?
- **Challenging**
  1. *Question*: How would the table provided in Figure 9.3 change if rows were individually asserted with a logic one instead of a logic zero?
  2. *Question*: What is the purpose of the 7404 inverter in Figure 9.4(a) in terms of current sinking?

Index

Author Biographies

Steven F. Barrett received the BS Electronic Engineering Technology from the University of Nebraska at Omaha in 1979, the M.E.E.E. from the University of Idaho at Moscow in 1986, and the Ph.D. from The University of Texas at Austin in 1993. He was formally an active duty faculty member with the United States Air Force Academy, Colorado and is now an Associate Professor of Electrical and Computer Engineering, University of Wyoming. He is a member of IEEE (senior) and Tau Beta Pi (chief faculty advisor). His research interests include digital and analog image processing, computer-assisted laser surgery, and embedded controller systems. He is a registered Professional Engineer in Wyoming and Colorado. He co-wrote with Dr. Daniel Pack "68HC12 Microprocessor: Theory and Application," Prentice-Hall, 2002; "Embedded Systems Design and Applications with the 68HC12 and HS12," Prentice-Hall, 2005; and "Microcontroller Fundamentals for Engineers and Scientists," Morgan-Claypool Publishers, 2006. In 2004, Barrett was named "Wyoming Professor of the Year" by the Carnegie Foundation for the Advancement of Teaching. Email: steveb@uwyo.edu

Daniel J. Pack is a Professor in the Department of Electrical Engineering at the United States Air Force Academy, CO. He received the Bachelor of Science degree in Electrical Engineering in 1988, the Master of Science degree in Engineering Sciences in 1990, and the Ph.D. degree in Electrical Engineering in 1995 from Arizona State University, Harvard University, and Purdue University, respectively. He was a visiting scholar at Massachusetts Institute of Technology-Lincoln Laboratory. He co-authored two textbooks on microcontrollers and embedded systems and authored over 70 journal and conference papers. He is a member of Eta Kappa Nu, Tau Beta Pi (faculty advisor), IEEE (senior), and ASEE. He is a registered Professional Engineer in Colorado. In 2005, Pack was named "Colorado Professor of the Year" by the Carnegie Foundation for the Advancement of Teaching. His research interests include cooperative UAVs, intelligent control, automatic target recognition, and robotics. Email: daniel.pack@usafa.edu

High-Speed Digital System Design

Justin Davis
Mississippi State University

MORGAN & CLAYPOOL PUBLISHERS

I would like to dedicate this book:

To my parents for your lifelong dedication to me.

To my friends for supporting my morale.

To Georgia Tech for training me to be a helluva engineer.

To my academic colleagues for accepting me into your world and opening doors to amazing possibilities for me.

ABSTRACT

High-Speed Digital System Design bridges the gap from theory to implementation in the real world. Systems with clock speeds in low megahertz range qualify for high-speed. Proper design results in quality digital transmissions and lowers the chance for errors. This book is for computer and electrical engineers who may or may not have learned electromagnetic theory. The presentation style allows readers to quickly begin designing their own high-speed systems and diagnosing existing designs for errors. After studying this book, readers will be able to:

- Design the power distribution system for a printed circuit board to minimize noise
- Plan the layers of a PCB for signals, power, and ground to maximize signal quality and minimize noise
- Include test structures in the printed circuit board to easily diagnose manufacturing mistakes
- Choose the best PCB design parameters such a trace width, height, and routed path to ensure the most stable characteristic impedance
- Determine the correct termination to minimize reflections
- Predict the delay caused by a given PCB trace
- Minimize driver power consumption using AC terminations
- Compensate for discontinuities along a PCB trace
- Use pre-emphasis and equalization techniques to counteract lossy transmission lines
- Determine the amount of crosstalk between two traces
- Diagnose existing PCBs to determine the sources of errors

KEYWORDS

Digital design, Computer engineering, Circuits, Printed circuit board, High-speed

Contents

CHAPTER 1

PCB Planning for High-speed Systems

This chapter assumes that the reader is familiar with analog components to analyze simple circuits, basic printed circuit board (PCB) design, and digital circuits. The purpose of this chapter is to set up a printed circuit board environment which will enable the best signal quality when routing traces.

1.1 LEARNING OBJECTIVES

After reading this chapter, you will be able to perform the following tasks:

- Design the power distribution system for a printed circuit board (PCB) to minimize noise.
- Plan the layers of a PCB for signals, power, and ground to maximize signal quality and minimize noise.
- Include test structures in the PCB to easily diagnose manufacturing mistakes.
- Determine the ideal size for vias to minimize impact on signal quality.

1.2 MULTILAYERED POWER DISTRIBUTION SYSTEM

The power system in a digital system should provide stable voltage references to all logic devices. Digital devices are typically very noisy and inject that noise into the power system. The power supply can filter some of this noise at low frequency, but higher frequency noise must be filtered using on-board, on-package, and on-die passive components.

The most important concept of this section is Ohm's law as it applies to inductors. Inductors are seen as short circuits as long as a steady current is flowing through them. As the current changes, the inductors act to resist that change. The result is a voltage difference across the inductor. The main power problem is that all real wires have a finite, but small inductance. This inductance does not matter in a circuit unless a large amount of changing current is flowing through them. In an ac signal, current flows in one direction, then turns around and flows in the other direction. The faster this occurs, the more of a voltage change will be present across the inductor. Therefore, as the signal frequency increases, the more normal wires act like inductors.

A power supply is designed to produce a specific voltage regardless if it provides a small amount of current or a large amount of current. As the current demand increases, the power supply must provide more current to maintain a steady voltage. In a typical digital circuit, the current demand changes proportionally to how fast the gates are switching. Therefore, at higher frequency of operation the current demand changes faster. This creates a large change in current through the wires from the power supply to the logic devices. The inductance of those wires will then become a problem in maintaining a steady voltage across the entire digital circuit. Those wires include all the metal that the current flows through such as the leads within a chip, the pins of a package, the traces/planes on the circuit board, and the cables leading to the power supply. This parasitic inductance is the bane of all power distribution systems.

In an ideal circuit, a uniform voltage is supplied to every logic device. This implies zero impedance through the wiring supplying those devices. In a realistic circuit, this wiring will have finite impedance which can cause differences in the voltage seen at each device. Therefore, the goal of designing a good power system is to minimize the impedance in the path from power on each gate on the die of a circuit to the power supply, and then back to ground on each gate. This implies a low-impedance path from the power rail to the ground rail as well.

Power supplies have very low impedance; however, the wires, cables, and circuit board traces which connect to logic devices do not. This cabling is called the *power distribution wiring*. The resistive element of the relatively large impedance can be compensated by adding sense wires to the end of the distribution wiring for feedback to the power supply. Alternatively, increasing the size of those wires or traces will decrease the resistance.

The inductance in power distribution wiring cannot be compensated by increasing the size of the wires or implementing sense wires. The inductance in the wiring from the power supply to the circuit board will slow the response of the power supply to changes in power demand. If the power demand increases without supplying more current to the circuit board, the voltage seen at the logic devices will decrease. The opposite happens when the power demand decreases: the voltage seen at the logic devices will increase. Typical logic devices are only rated to accept a difference in voltage by $\pm 5\%$, so the power supply still needs to provide stable power/ground voltage levels.

The first possible solution is to reduce the rate of change of the current demand. This can be accomplished by slowing the clock rate or the slew rate of the logic devices; however, by definition high-speed circuits will operate above a few kilohertz making this option not possible. The alternative solution is to use board-level bypass capacitors to provide/store extra current.

Bypass Capacitors

Bypass capacitors are capacitors which connect to both the power rail and the ground rail. Installing bypass capacitors on the circuit board can provide extra power when the power supply cannot react fast enough; however, this is not limitless energy. The capacitors store power based on their capacitance (higher capacitance means more stored power). With a good design, the capacitors will be able to supply this extra power until the power supply can compensate for the change. The capacitors can operate in the opposite way as well: storing power when the power demand decreases.

The bypass capacitors have a significant limitation. The capacitors have a parasitic inductance and parasitic resistance mainly resulting from the wire leads of the package. As a rule of thumb, larger capacitors have a larger parasitic inductance. This can be modeled as a small inductor on either side of the capacitor, but typically these inductors are combined into one. This will limit the rate at which current can be provided by the capacitor. Therefore, large capacitors provide more power, but at a slower rate. Small capacitors provide only a little power, but do so very quickly. As a result of this, bypass capacitors are usually tiered on a circuit board from very large capacitors to very small capacitors.

The inductance in the wire leads of the capacitors is based on the package. A short wire has less inductance, so short wire leads will have less inductance. Therefore, the smallest package should be chosen for a given value of capacitance. Surface-mount chip capacitors are the smallest PCB capacitors available. For capacitance values of 2.2 μF to 0.001 μF, X7R, X5R, or NP0 type capacitors are usually used for their small inductance (typically less than 2 nH). For larger capacitance values, low-inductance electrolytic capacitors are used.

The leads of a capacitor also have a slight resistance, but it is very small. This is called the effective series resistance (ESR) and can be modeled as one resistor in series with the capacitor. This resistance is only a faction of an ohm.

The impedance of inductors and capacitors follows the form of

$$X_L = 2\pi f L \tag{1.1}$$

$$X_C = \frac{1}{2\pi f C} \tag{1.2}$$

where f denotes frequency. Inductors increase impedance with increasing frequency while capacitors decrease impedance with increasing frequency. These equations, combined with the ESR, form the final impedance equation:

$$Z_C = \sqrt{R^2 + (X_C - X_L)^2}. \tag{1.3}$$

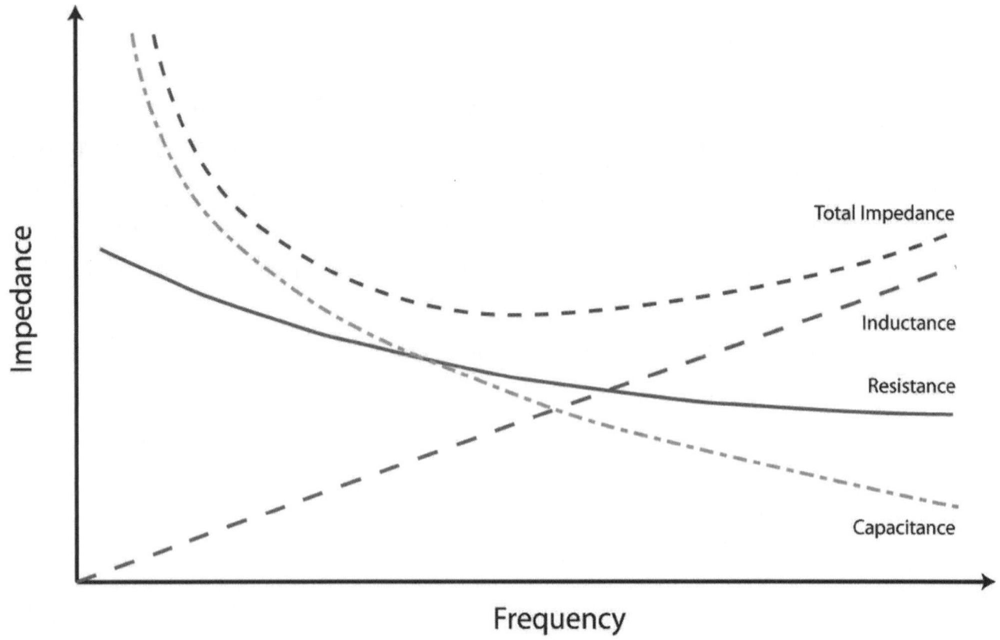

FIGURE 1.1: Total impedance characteristic of a bypass capacitor

Fig. 1.1 shows the impedance of a capacitor with parasitic inductance and resistance over a large frequency range. The total characteristic impedance is dominated by the capacitance at low frequency and by the inductance at high frequency. If the capacitance is increased, the impedance curve moves down and left. With a fixed package, the total impedance can be decreased by choosing higher capacitances. Likewise, the total impedance can be decreased by using multiple capacitors in parallel since inductance decreases in parallel. This is the same reason why logic device packages have multiple power and ground pins. The inductance of the power and ground wire leads is decreased by having many of them in parallel.

The goal of a power distribution system is to have low impedance over all frequencies. Each capacitor will have minimum impedance at a specific frequency; therefore an array of capacitors must be used to target different frequencies. A capacitor is needed in every decade of the capacitor value range. Also, smaller capacitors have less impact on the overall impedance so more of them are needed. Typically, the largest capacitor needed is in the range of 100 μF to 1000 μF. For each logic device, one capacitor at this value is needed. For every decade lower than this, twice as many capacitors are needed. This means two capacitors are needed at 10.0–47.0 μF range, four at 1.0–4.7 μF range, eight at 0.1–0.47 μF range, sixteen at 0.01–0.047 μF range, etc. Also, within each range, the number of capacitors should be split at the upper end and at the lower end. This means for the sixteen capacitors needed in the 0.01–0.047 μF range, eight must be 0.01 μF and eight must be 0.047 μF.

TABLE 1.1: Capacitors Needed for a Power Distribution System

CAPACITOR RANGE (μF)	RATIO OF TOTAL CAPACITORS (%)	CAPACITOR PACKAGE
100–1000	3.2	Tantalum/Electrolytic
10–47	6.5	Tantalum/Electrolytic
1.0–4.7	12.9	0805
0.1–0.47	25.8	0603
0.01–0.047	51.6	0402

With these ratios, a total of 31 capacitors are needed. This should be the minimum number of capacitors used for any high-speed design. As stated above, there should be one capacitor for each power pin on each logic device, so if more are needed, capacitors should be added while maintaining about the same ratio. Note that the quantity of the smallest value of capacitors represents about half the total number of needed capacitors. Specifically, 16 of the total 31 capacitors represent about 51.6% of the total. Eight capacitors represent 25.8%. Therefore, the number of capacitors for any design can be weighted with these ratios. All the ratios are listed in Table 1.1.

Smaller package should be used as capacitance decreases. The largest capacitor will need to be a tantalum or low-impedance electrolytic. The smallest range should use a 0402 package. This will minimize the parasitic inductance.

The effectiveness of this power distribution system should be simulated before the design of the circuit board to measure its effectiveness. The tantalum package typically has a wide frequency range, so sometimes the capacitors in the 10–47 μF range may not have a large impact on the overall impedance. The package datasheet will have the values of parasitic inductance and parasitic resistance to use in the simulation to determine their impact. A simple lumped-element SPICE simulation will be adequate for a preliminary evaluation.

Example 1.1. In my design, I have two high-speed logic devices with 20 power/ground pins on one and 30 power/ground pins on the other. This means I will need a total of 50 capacitors.

For the best filtering, I must choose two capacitor levels from each range. For the highest capacitors, I will choose one 470 μF and one 100 μF capacitor (Option A). For a reduced cost of materials for my PCB, I could choose one capacitor for each range (Option B), but the noise filtering will not be quite as good. I will simulate both options to qualitatively decide if the reduced cost option would be acceptable. The actual number of capacitors I need in each range is listed in Table 1.2.

TABLE 1.2: Calculating Needed Capacitors for a 50 Capacitor Array

CAPACITOR RANGE (μF)	CALCULATED NUMBER	ACTUAL NUMBER USED
100–1000	3.2% of 50 = 1.6	2
10–47	6.5% of 50 = 3.25	3
1.0–4.7	12.9% of 50 = 6.45	6
0.1–0.47	25.8% of 50 = 12.9	13
0.01–0.047	51.6% of 50 = 25.8	26

I decide to purchase the capacitors from multiple vendors, and I reference the datasheets to find the parasitic inductance and resistance to use in my simulation. Table 1.3 lists the parasitics for each capacitor and the quantity needed for each option.

I will use PSPICE for my circuit simulation. I want to measure the impedance of the capacitor array over a wide range of frequencies. I can assume that below 10 kHz, the power supply does not need filtering, so I will plot the impedance from 10 kHz to 1 GHz. I would like

TABLE 1.3: Capacitor Parasitics

CAPACITOR VALUE	PARASITIC INDUCTANCE (pH)	PARASITIC RESISTANCE (Ω)	QUANTITY OPTION A	QUANTITY OPTION B
470 μF Electrolytic	2000	0.07	1	2
100 μF Tantalum	2000	0.07	1	0
47 μF Tantalum	2000	0.07	1	0
10 μF Tantalum	2000	0.07	2	3
4.7 μF X7R 0805	600	0.12	3	0
1.0 μF X7R 0805	600	0.29	3	6
0.47 μF NP0 0603	500	0.07	6	0
0.1 μF NP0 0603	500	0.12	7	13
0.047 μF NP0 0402	400	0.13	13	0
0.01 μF NP0 0402	400	0.13	13	26

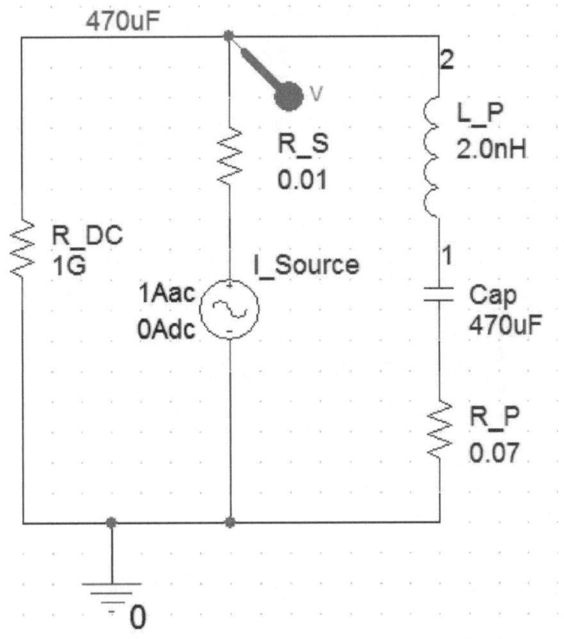

FIGURE 1.2: Simple bypass capacitor simulation circuit

to see very low, flat impedance over that range. Low impedance between power and ground at those frequencies means the noise on the power rail will be shorted to ground.

In my PSPICE simulation, I will use an ac current source with a very small series resistance. The source will be set to 1 A ac current and 0 A dc current. The capacitor model is placed between power and ground with an inductor, capacitor and resistor in series. Since the current source is ideal, there must be a dc path to ground in order for the circuit to simulate. Since the capacitors block dc current, a large resistor (~1 G Ω) should be placed between the power and ground. An example of this circuit is shown in Fig. 1.2 with only one bypass capacitor.

The capacitor model is repeated for each capacitor needed in the array. The final circuit for simulation is shown in Fig. 1.3.

The impedance is measured by dividing the voltage at the power rail by the current. The plot of impedance over the frequency range is best shown in log/log format as in Fig. 1.4. This plot shows three different capacitor arrays. The circuit with only one 470 μF capacitor has relatively high impedance which only filters noise up to about 3 MHz. Above this frequency, very little filtering will occur. The other capacitor arrays have very low impedance over the entire frequency range. At 1 GHz, the impedance is about equal to that of the 470 μF capacitor at its best. At 3 MHz, the capacitor arrays provide about 10 times better filtering. The differences between the high-quality capacitor array and the cheaper capacitor array are not very significant. I would probably use the cheaper capacitor array since it not only costs less, but each size of the

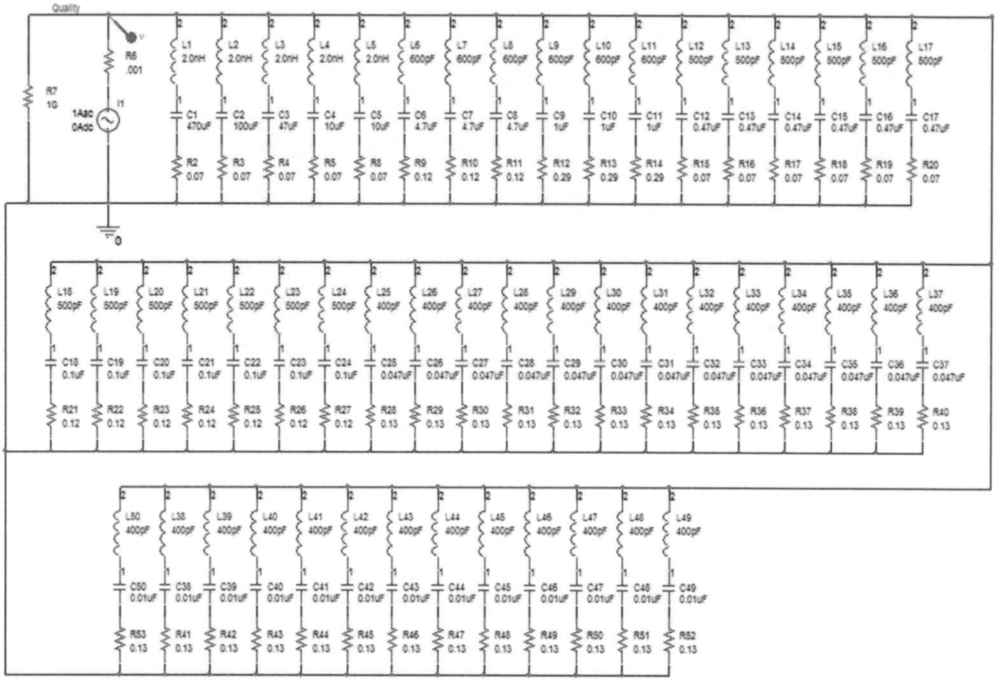

FIGURE 1.3: Total bypass capacitor simulation circuit

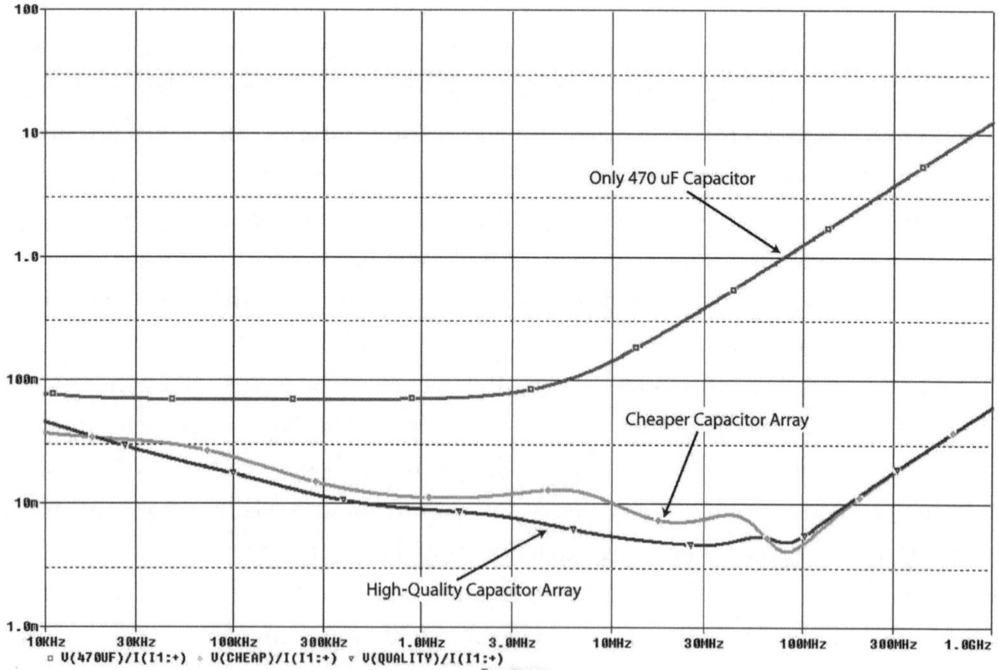

FIGURE 1.4: Impedance plot of multiple bypass capacitor arrays

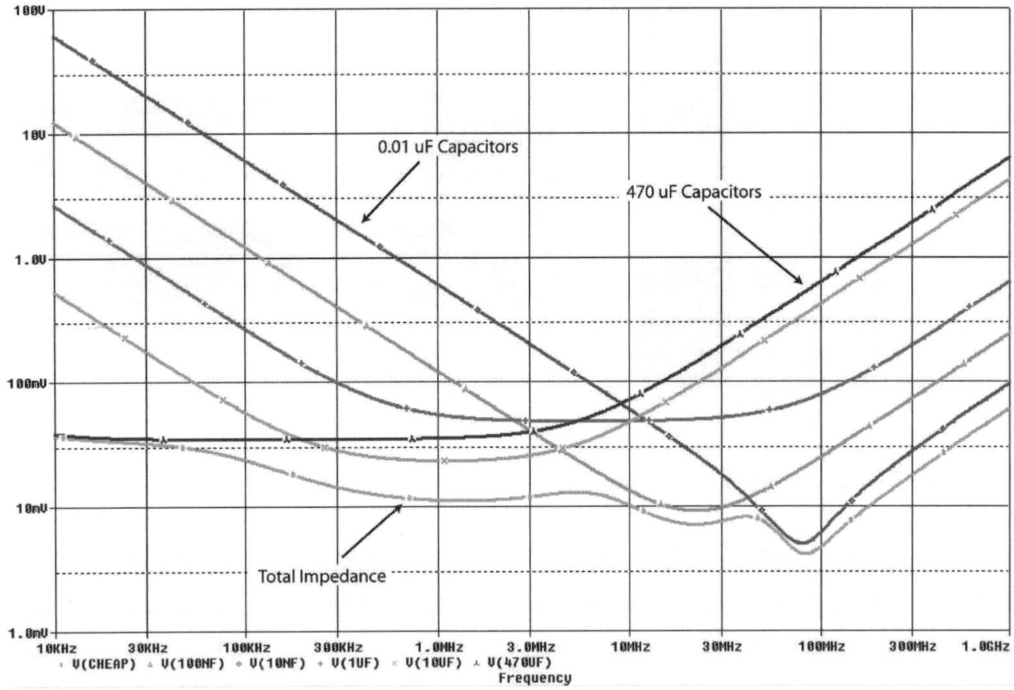

FIGURE 1.5: Effect of individual bypass capacitor elements on the total impedance

capacitor has only one capacitor value. For example, the only capacitor value at the 0805 size is 1.0 μF. Since capacitors of this size and less have very small or sometimes no text on them, it can be very hard to keep track of which capacitors are of which value.

The last plot in Fig. 1.5 shows the contribution of each capacitor range on the overall impedance. Each capacitor range filters most effectively at a specific frequency. The smallest capacitors filter at the highest frequency.

Layout Considerations for Bypass Capacitors

While capacitors have a parasitic inductance associated with the leads, this is not the only inductance when the capacitor is mounted onto a PCB. The current will flow from one plane, through the via, through any trace to the solder pad, through the solder, and into the capacitor. This path is repeated on the other side of the capacitor. This inductance can be two to four times as large as the lead inductance of a surface-mount capacitor. This forms a current loop which has some inductance relative to the size of the loop. With a fixed amount of current, smaller loops will have smaller inductance.

Minimizing the loop can be done with a few different methods. The first is to minimize the length of the trace between the via and the solder pad. If possible, the via should touch the edge of the solder pad. If not, the trace to the via should be as wide as the solder pad (Fig. 1.6(a)).

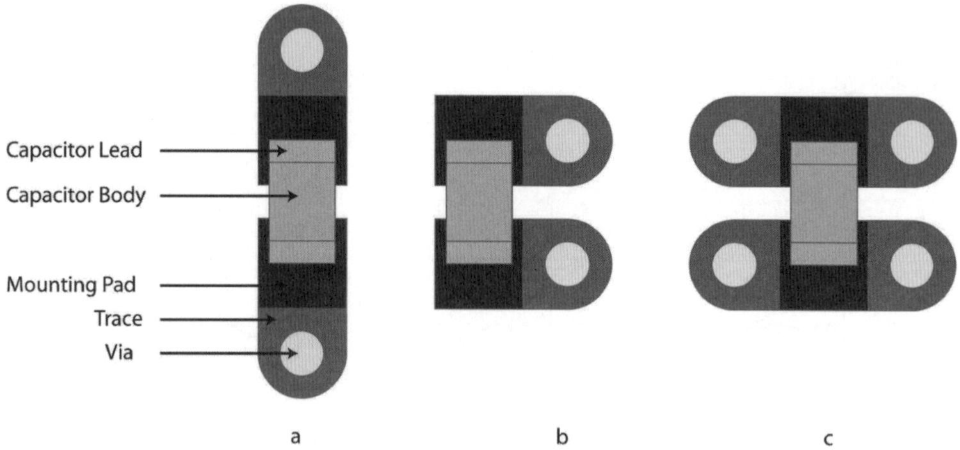

FIGURE 1.6: Bypass capacitor pad arrangements

The vias should also be perpendicular to the capacitor instead of in-line with it (Fig. 1.6(b)). Space permitting, multiple vias could be used to reduce the amount of current through each via which minimizes the inductance. Fig. 1.6(c) shows two vias on each side of the capacitor; however, three vias per side is also possible above and below the solder pad. In any case, each capacitor should have its own vias, and multiple capacitors should not share vias. The layouts in Fig. 1.6 are larger than necessary for the capacitor size shown. The mount pads should be just large enough to reliably solder the capacitor without bridging solder across to the other pad. If the capacitors are soldered by hand, a larger mounting pad may be necessary.

While the orientation of the capacitors matters, so does the relative location to the logic device. The smallest capacitor values should be as close to the power and ground pins of the device as possible. As a rule of thumb, the smallest capacitors should not be farther than about an inch away. They can be mounted on either the top or the bottom of the PCB as long as they are within this distance to the power/ground pins (not the center of the chip). If they are mounted farther away than this, their response time to changes in power demand is not fast enough to make them useful. Capacitors larger than 1.0 μF are not as closely constrained by distance, but they should be relatively close to the logic devices.

The bypass capacitor network should provide power supply filtering of noise up to 500 MHz. Above this, the inductance of the leads of the logic device package will limit the effectiveness of adding smaller capacitors. At this point, the only board-level filtering that can occur is from the embedded capacitance of the power and ground planes. The next step in attaining higher frequency noise filtering is adding small capacitors within the mounted packages. This is effective into the low gigahertz range. For circuits which operate higher than this, on-die filtering is required. See Fig. 1.7 for the total power distribution system including the PCB, package, and die.

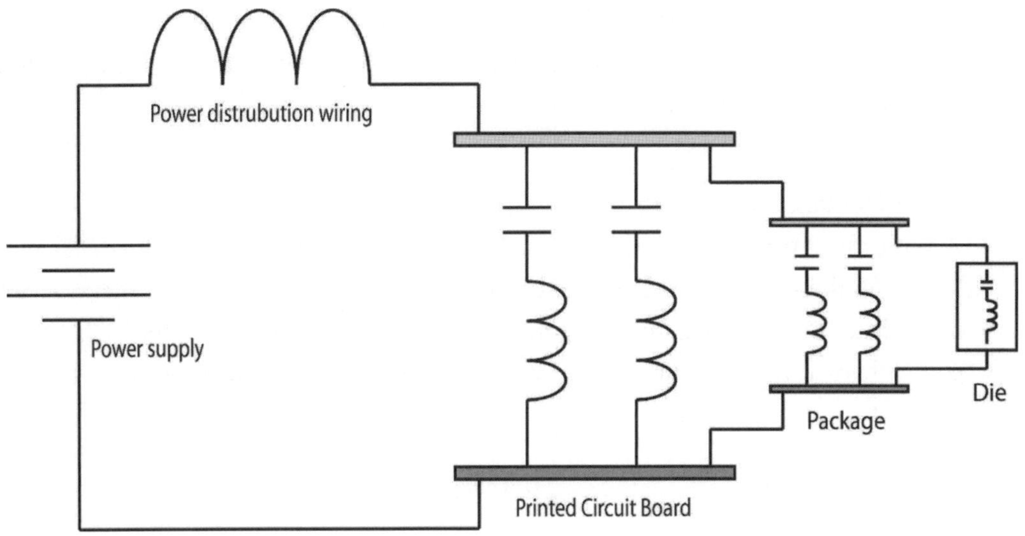

FIGURE 1.7: Total power distribution system

1.3 LAYER STACKING

A high-speed digital system must have at least one power plane and one ground plane. If a single trace is used to route power, the inductance will be high. Using a wider trace will lower the inductance. The widest trace possible is the entire width of the circuit board. Therefore, using an entire plane will have the lowest possible inductance. Power and ground planes also make routing signals significantly easier. Often multiple power and ground planes are necessary.

Layer Basics

A two-layer printed circuit board starts with material referred to as *core* with a plane of copper on either side. The copper is etched away using a chemical solvent. If a multilayer PCB is being made, then multiples of these can be glued together using a sheet of epoxy material called *prepreg*. The sheet is aligned with the cores and then heated and pressed. The prepreg should have the same dielectric constant as that of the core material, but not necessarily the same thickness. The prepreg and core layers will alternate. With a four-layer board, sometimes two cores are used with copper on either side and then glued together with prepreg. Sometimes one core is used, and then prepreg is placed on either side with bare sheets of copper on the outside of that.

After the boards are glued together, the vias/holes are drilled. These holes are then plated with metal to electrically connect the layers and provide a reliable solder connection for any through-hole packages or connectors. The board is then tinned, coated with a solder mask to prevent oxidation of the copper traces, and silk-screened on one or both sides. Sometimes gold plating is applied to reduce oxidation of exposed solder pads and to ensure a highly reliable connection. If the logic devices will not immediately be soldered onto the PCB, such as in prototyping, gold plating is recommended.

Embedded PCB Capacitance

On a PCB, typically entire metal planes are used for both power and ground. A small capacitor is formed between these layers since a capacitor by definition is two planes of metal separated by a nonconducting material. In this case, the nonconducting material is the PCB substrate. The planes are charged to different voltages which creates an electric field between them. The specific value of this capacitance is

$$C = \frac{0.225\varepsilon_r A}{d} \tag{1.4}$$

where

- ε_r is the relative electric permeability of the PCB substrate (4.5 for FR-4);
- A is the area of the planes (usually the size of the PCB) in in.^{2};
- d is the distance between the layers in inches;
- C is the capacitance of the planes in picofarads.

A circuit board with 0.01 in. separation (10 mil) between the ground and power layers will have a capacitance of about 100 pF in.$^{-2}$. If the same board is 5 in.^{2}, it will have a capacitance of 2531 pF or 0.0025 μF. This will provide high-frequency noise filtering which is above what the on-board capacitors can provide (greater than 500 MHz). Special PCB fabrication techniques can reduce the distance between the power and ground planes to as low as 2 mil providing significantly higher capacitance.

The power and ground planes also have an associated inductance. As current flows through these planes it spreads out over the plane and causes *spreading inductance* specified in henries per square (a unitless dimension). With a fixed area of the planes, the spreading inductance of the power and ground planes is a function of the distance between the planes. Closer spacing will result in lower spreading inductance. However, decreasing the distance between planes also lowers the capacitance between the planes.

This interplane capacitance provides extra filtering from about 50 MHz to above the high-frequency limit of what bypass capacitors can provide (about 500 MHz). When deciding the layer stacking, a high priority should be placed on keeping the power and ground planes adjacent.

Layer Order

The ordering of PCB layers needed for a high-speed design is fundamental in maintaining quality signals across the board. The power and ground planes not only provide a low-impedance current path, but also a way of shielding the high-speed traces from external noise. Also, the metal planes reduce the amount of noise injected into other parts of the circuit board. As a general rule of thumb, a four-layer PCB will produce 15 dB less noise than a two-layer board. Anytime a circuit is operating above 15 MHz, at least a four-layer board should be used.

There are five objectives when designing a multilayer board. In order of importance they are as follows:

1. Signal layers should be adjacent to a power/ground plane.
2. Signal layers should be tightly coupled to their adjacent power/ground plane.
3. Power and ground planes should be closely coupled together.
4. High-speed signals should be routed on buried layers located between power/ground planes.
5. Multiple ground planes should be used wherever possible.

To achieve all of the above conditions, a minimum of eight layers are required. If less than eight are needed, then a compromise can be made. The typical four-layer layout is the signal layers on the top and bottom with the power and ground layers in the middle shown in Fig. 1.8(a). This will satisfy objective 1. If the layers are equally spaced, then the separation between the layers will be large. To achieve objective 2, distance between the signal layers and

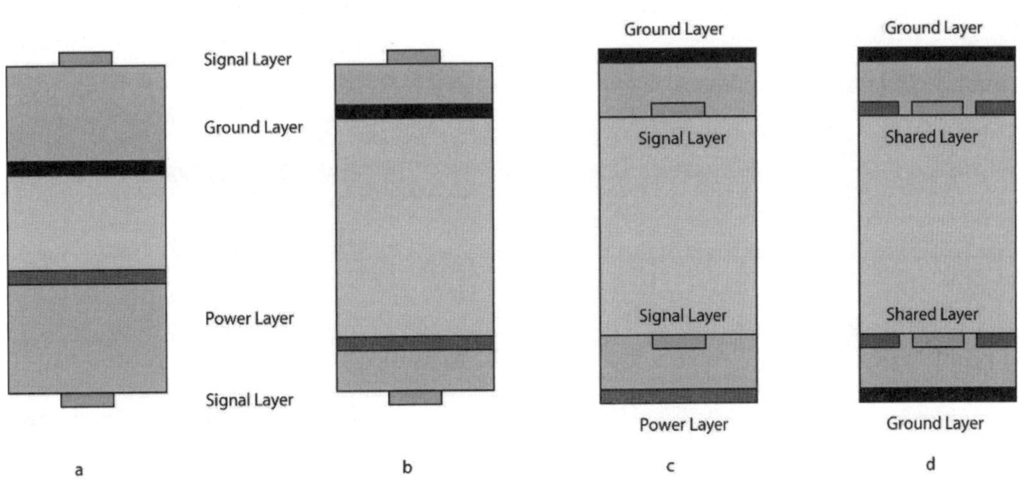

FIGURE 1.8: Four-layer stacking options

the power/ground layers can be reduced. The prepreg between layers 1 and 2 should be less than 0.010 in. while the core between layers 2 and 3 should be more than 0.040 in. as shown in Fig. 1.8(b). This will reduce the noise generated by the signal layers within each layer (called crosstalk). Objectives 3, 4, and 5 will be unattainable at this point.

A nonstandard layering shown in Fig. 1.8(c) has the ground and power planes on the top and bottom layers with the signal layers internal. The major advantage of this is that the outer planes act as a noise shield for the signals. There are two circumstances in which this layout should be used: if the board will be used in a very noisy environment without a grounded metal chassis, or the noise emission of the board must be very low. Packages must still be mounted on the external layers, so the plane will not be a uniform layer of metal which will reduce the signal quality. Also, burying signals will make them inaccessible if any rework is needed. Objectives 1 and 2 will be satisfied with objective 4 partially satisfied. The reduction in signal quality from having nonuniform ground and power planes can be significant, so this option should be reserved for special cases.

A ground plane will shield the signals much better than a power plane. Therefore, both outside layers should be ground planes. This means that one of the signal layers must become the new power plane, or the signal layers must share the power plane shown in Fig. 1.8(d). While maximizing shielding, it has the drawbacks of also increasing the impedance of the power distribution system and decreasing the signal quality.

Six-layer boards provide extra flexibility and attain more objectives. The addition of two layers is for either more signal layers or more power and ground layers. If more signal layers are required, then the layout in Fig. 1.9(a) is preferred. High-speed signals should be routed on

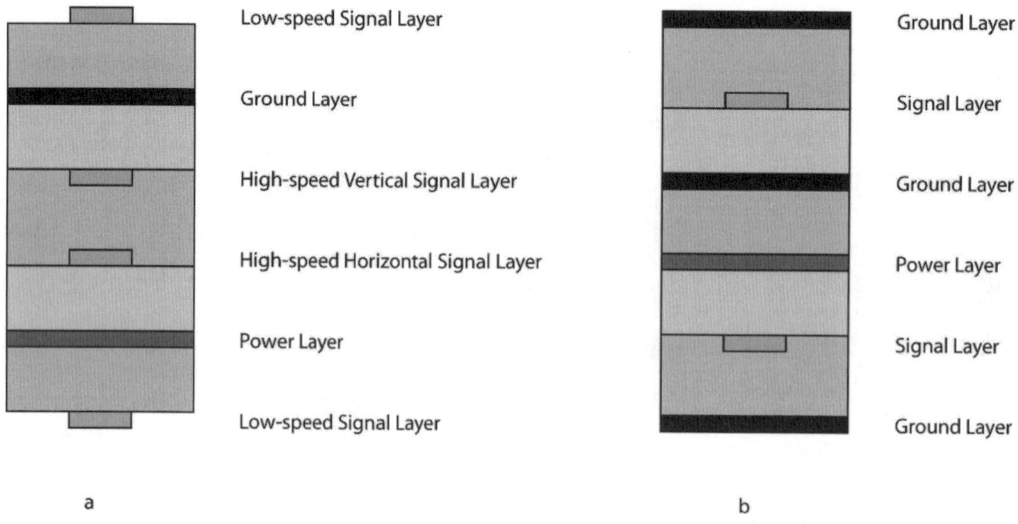

FIGURE 1.9: Six-layer stacking options

the internal layers to provide maximum shielding. Lower speed signals should be routed on the top and bottom layers. In this option, the ground and power planes have a significant distance between them which minimizes their embedded capacitance. Since the internal layers do not have a metal shielding plane between them, the signal should be routed orthogonally on these layers to prevent noise from coupling between them. One layer should have all vertical traces, and the other should have all horizontal traces. The only exception is when the high-speed traces are differential pairs. In this case, the signals should be routed directly on top of each other.

The alternate six-layer board will use three ground planes, one power plane and two signal planes as shown in Fig. 1.9(b). This is the optimal layer stacking for a high-speed system. All five objectives are met with this board. While only two layers are for signals, this board has the best noise shielding while maintaining a good embedded capacitance between the power and ground planes. The external ground layers will not be solid planes since the devices must be mounted, but with the extra planes in the center of the board this should not be a problem. The only difficulty with this stacking is that the signal layers are inaccessible for any rework or probing.

The next best board is an eight-layer board. While a six-layer board will satisfy all the objectives, usually the eight-layer board shown in Fig. 1.10(a) is preferred. The high-speed signals will be restricted to the center layers while all other signals and test points will be on the

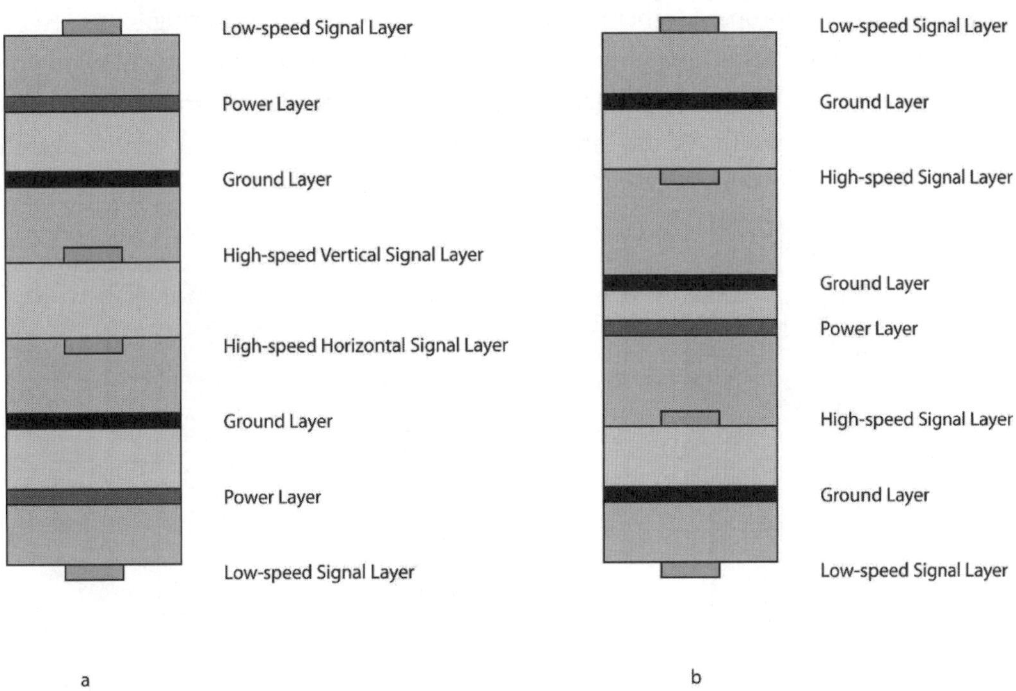

FIGURE 1.10: Eight-layer stacking options

external layers. The main advantage of the eight-layer board is not for its extra signal layers, but better noise reduction. Therefore, an eight-layer board is an improved version of the same six-layer board. All signals should be accessible either directly on the external layers or through carefully designed test points. The ground and power layers should be as closely spaced as possible to provide a good embedded capacitance. While the stacking shown in Fig. 1.10(b) is also acceptable, it should not be used when two power supply voltages are required as is typical of many high-speed systems. In no circumstance should an eight-layer board have six signal layers. If six signal layers are needed, then a minimum of ten layers should be used. Boards with more than eight layers should follow the same type of shielding as seen in eight-layer boards.

Stacking Stripes

An aid in ensuring a quality PCB is *stacking stripes*. They are traces about 50 mil wide on each layer. These traces should straddle the edge of the PCB where it will be cut from the panel. This means that copper traces will be visible on the edge of the PCB. On the top layer, this trace is 50 mil long. Each successive layer's stripe is 50 mil longer than the previous one. When the PCB is returned from the manufacturer, a quick inspection will determine if the layers were produced in the correct order. A stair-step pattern should be obvious as seen in Fig. 1.11. These traces must not contact any other metal in the design including power and ground planes. Without these stacking stripes, problems with the layer order are very difficult to diagnose. These problems can arise from either improper Gerber generation or incorrect manufacturing.

A second feature of stacking stripes is a small section of trace about 5 mil wide on each layer called *shape traces*. Measuring the actual etched trace width will determine the accuracy of the manufactured trace widths of the internal layers. Sometimes the traces can be overetched

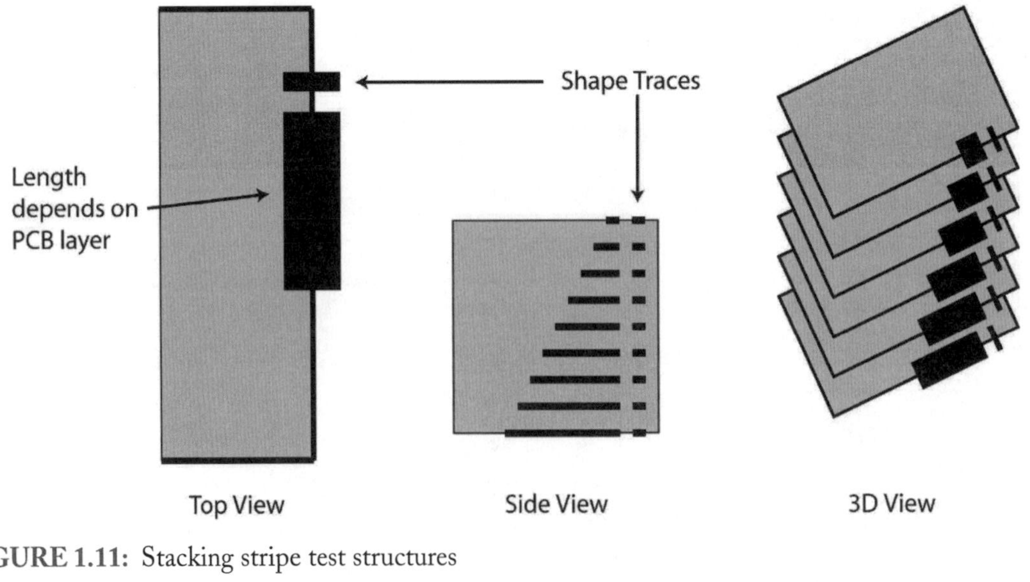

FIGURE 1.11: Stacking stripe test structures

or underetched. Since the trace width dictates the characteristic impedance (as discussed in the next chapter), knowing the accuracy of the trace will help diagnose any signal quality problems.

Lastly, using stacking stripes will enable easy measurement of the thickness of dielectric and copper layers. The thickness between metal layers dictates many parameters essential for quality signals. Sometimes PCB manufacturers will object to having metal at the edge of a board; however, they can usually be convinced. If not, there are many other PCB manufacturers.

1.4 VIAS

A "via" is a physical hole in a printed circuit board. Vias typically serve two purposes: to provide a path for signals between layers, and to provide a place to mount through-hole components. The size of vias is determined early in the development of a circuit board layout since their parasitic effects impact the power distribution system and signal quality.

Vias are roadblocks in printed circuit boards since they usually penetrate all levels of the circuit board. Signal traces must be routed around them, and return current from the ground plane must flow around them. Minimizing the size of vias will allow more room on the PCB. It will also minimize the unwanted electrical effects as well. Ideally all vias should be as small as possible, but the cost of drilling small vias increases with decreasing size. Smaller vias require small drill bits which are more prone to breaking. Also small drill bits cannot penetrate a thick board without drifting off center. These vias must be drilled in smaller batches which adds to the manufacturing time and increases the cost. Ultimately, the manufacturer will determine the price based on the size of the hole and the thickness of the board. The minimum hole size is usually one fifth of the thickness of the board.

Vias do not have to penetrate the entire thickness of the board. A *blind via* only penetrates a certain depth of the board. An *embedded via*, also called a *buried via*, is an internal via which does not reach an external surface of a board. If a via does not penetrate the entire thickness of the board, it will not be a roadblock on those layers. The parasitic effects of vias can be minimized by limiting its depth.

When placing a via in a PCB design this is typically the drilled hole size, but this may not be the final size of the hole. Often vias are plated with metal on the inside so all electrical layers will be connected to it. This is usually done on all vias except those specifically designated for mechanical connectors which will not be carrying current. The plating will decrease the hole size by a few mils. Therefore, if a through-hole lead is to be placed inside a via, the via must be drilled large enough to allow for this plating reducing the size of the hole. The difference between the drilled hole size and the final hole size is called the *plating allowance*. Fig. 1.12 shows the relationship between drilled size and final size. Sometimes a via is so small that there is no hole after plating.

The second consideration for via size is the error in drilling size. Even though a drill bit may be a specific size, the hole may not be exactly that size. Often manufacturers will give an

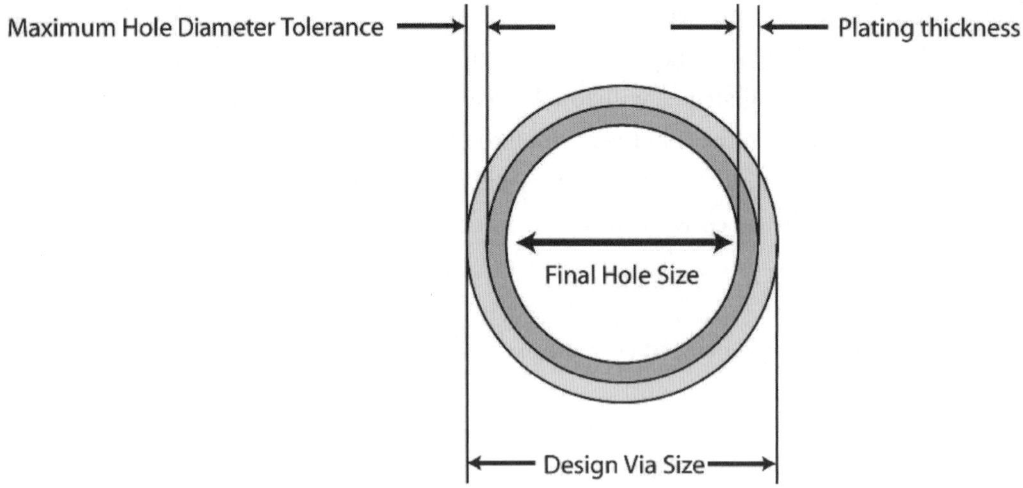

FIGURE 1.12: Final via size versus drilled via size

error associated with each drill size. A 30 mil drill bit might have a 1 or 2 mil error either too small or too big. If the drilled hole is on the small side, it may be too small for the designated wire lead to fit through. Therefore, the hole must be designed slightly larger to account for the possibility that the hole might be drilled smaller than intended.

The size of the drilled hole is determined by

$$\text{DRILL} = \text{FINAL} + \text{PA} + \text{HD} \tag{1.5}$$

where

DRILL	is the size of the hole to drill;
FINAL	is the final hole size needed (from connector datasheet);
PA	is the plating allowance (from PCB manufacturer);
HD	is the hole diameter tolerance (from PCB manufacturer).

Example 1.2. I am about to start adding vias to my PCB for a special socket which uses through-hole leads. The datasheet indicates the maximum size of the leads as 20 mil. I will add another 5 mil to allow for easy insertion of the leads. Therefore, the minimum size my vias need to be is 25 mil. I call the PCB manufacturer I plan on sending my design to and I discover that their 25 mil drill bit has a hole diameter tolerance of ±3 mil. Their plating thickness is 2 mil, which means it will add 2 mil on either side of the hole. Therefore, the plating allowance is 4 mil. I use the equation to find the final drill hole size

$$\text{DRILL} = 25 + 4 + 3 = 32\,\text{mil}. \tag{1.6}$$

In my CAD program, I use a via size of 32 mil.

Sometimes a via should not connect to all layers on a printed circuit board. For example, the ground plane should not be connected to the power plane. By default all vias will have a clearance ring, or keep-out ring, around them on the solid metal planes. To connect the via to that plane, a trace or strap must be placed across the via to extend beyond this ring. The hole alignment with the clearance ring may have some associated error. Sometimes the drill may not be perfectly aligned with the board which would result in the hole being drilled slightly off center. The *hole alignment allowance* is given in mils and refers to how far off the target in any direction the via may be drilled. If the via is drilled so that it extends beyond the keep-out area, when the via is plated it will contact all layers, which creates a short between power and ground. To prevent this, use a keep-out area at least twice the hole alignment allowance.

The clearance area on the power and ground planes can cause problems. Often vias for a connector or through-hole logic device are laid in a long row or grid pattern. Usually enough space is left between the vias for one or more signal traces to pass, but the keep-out areas may overlap. This would appear on the metal plane as a large hole as seen in Fig. 1.13. Current flowing on these planes will have to flow around the hole which can cause an increase in impedance and noise. This is also called a *slot*.

For current to flow, a loop of metal must be formed for it to flow through. The current will flow from one logic device to another logic device through a signal trace, but current will also flow in the opposite direction through the ground plane. For low-speed signals, the current will follow the path of least resistance. On a metal ground plane, this means the current will spread out over the plane. For high-speed signals, the current will follow the path of least inductance. On a metal ground plane, the path of least inductance is directly beneath the signal

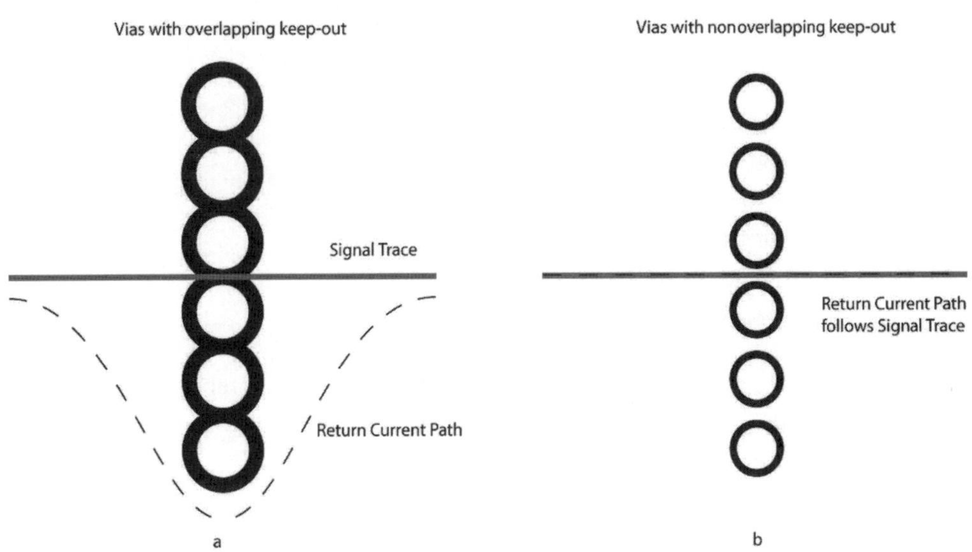

FIGURE 1.13: Ground plane slot created by vias

trace. If the signal cannot flow directly beneath the signal trace, the inductance will increase significantly. Therefore, anytime a signal trace passes over a hole in the ground plane, the current cannot flow underneath the trace. The return current must flow around the hole creating a large loop of current which creates a large inductance. This inductance will increase the rise/fall times on the signal being transmitted.

A few methods of avoiding unnecessary inductance caused by ground plane slots are minimizing the keep-out area of the vias, spacing the vias far apart so the keep-out areas do not overlap, and finally avoiding routing high-speed signals between vias. The minimum keep-out diameter is dictated by the hole alignment allowance, so this may not be a possible solution. The connector or package will have a defined spacing between the vias, so increasing the distance may not be possible either. By not routing high-speed signals between vias, the problem will be solved regardless of the above limitations.

Via Models

A significant amount of current flows through vias especially from the bypass capacitors. This current switches at high frequency, so the parasitic effects of the vias may affect the circuit. Two models which can be used are either a series inductance, or series inductance with capacitors to ground on either side (a pi model).

A series inductance works well as long as the rise time of the signal passing through it is at least three times larger than the total delay through the via. The delay through the via is dependent on its inductance and capacitance by the equation

$$t_{pd} = \sqrt{L_v C_v}. \tag{1.7}$$

The parasitic inductance of a via is based on its length and diameter. The equation to calculate its inductance is

$$L_v = 5.08\, h \left[\ln\left(\frac{4h}{d}\right) + 1 \right] \tag{1.8}$$

where

L_v is the inductance of the via in nH;
h is the height of the via (usually thickness of the PCB) in inches;
d is the diameter of the via in inches.

The parasitic capacitance of a via is based on its length and the diameter of the pad surrounding the via. The equation to calculate its inductance is

$$C_v = \frac{1.41 \varepsilon_r h d_p}{d_c - d_p} \tag{1.9}$$

where

C_v is the parasitic capacitance of the via in pF;
h is the height of the via (usually thickness of the PCB) in inches;
ε_r is the dielectric constant of the insulating material;
d_p is the diameter of the pad surrounding the via in inches;
d_c is the diameter of the clearance hole in inches.

Note that the diameter of the clearance area on the ground plane will have a significant impact on capacitance. A large clearance area will result in a small capacitance; however, large clearance areas can create undesirable ground slots.

Example 1.3. I am planning on using 10 mil vias. My board is going to be the standard 63 mil thick using FR-4. The pad diameter will be 15 mil, and the clearance diameter will be 20 mil. Therefore,

$$h = 0.063 \tag{1.10}$$

$$d = 0.010 \tag{1.11}$$

$$L_v = (5.08)(0.063)\left[\ln\left(\frac{4 \times 0.063}{0.010}\right) + 1\right] \tag{1.12}$$

$$L_v = (0.32)\left[\ln(25.2) + 1\right] \tag{1.13}$$

$$L_v = (0.32)(4.22) \tag{1.14}$$

$$L_v = 1.35\,\text{nH} \tag{1.15}$$

$$d_p = 0.015 \tag{1.16}$$

$$d_c = 0.020 \tag{1.17}$$

$$\varepsilon_r = 4.5 \tag{1.18}$$

$$C_v = \frac{(1.41)(4.5)(0.063)(0.015)}{0.020 - 0.015} \tag{1.19}$$

$$C_v = \frac{0.006}{0.005} \tag{1.20}$$

$$C_v = 1.2\,\text{pF}. \tag{1.21}$$

Next I determine which model I need to use. My signals will be operating with a 1.5 ns rise time:

$$t_{pd} = \sqrt{(1.35)\,(1000)\,(1.2)} = 40.2\,\text{ps}. \tag{1.22}$$

Since my rise time of 1.5 ns is larger than three times my delay through the via, I can use a single series inductor. So I add an additional 1.35 nH inductor on each terminal of my bypass capacitors and repeat my simulation. If my rise time was a little bit smaller, I would use the pi model and put a capacitor on either side of the inductor. These capacitors would each have a value of $C_v/2$. This would make my simulation much more complex, but also much more accurate.

Often a signal is not being routed through the entire length of a via. The only case where it would is when routing a signal from the top layer to the bottom layer. As the number of layers increases, the likelihood of this happening decreases. The part of the via that signal is not passing through is called a *stub*. This stub can reduce the quality of the signal passing through the via. Sometimes blind vias or embedded vias are used to alleviate this problem; however, they are expensive to manufacture. Another method is called *back-drilling* which uses a drill bit slightly larger than the original used to create the via. The back-drilling bit is aligned over the via and then penetrates one side of the board partway through. This removes the metal where signal is not being routed. Removing the extra metal reduces the height of the via, which reduces the inductance and capacitance of the via.

As the rise time of the signal approaches the delay of the via, pi model may not realistically predict the via. A more accurate model of the via is necessary, and using a three-dimensional (3D) field solver will improve the simulation. Even though accurate modeling will help predict the behavior of the signal, a better solution would be to decrease the size of the via. If the via is so large that the pi model is not good enough, a digital signal will not perform well passing through it.

CHAPTER 2

Ideal Transmission Lines

The purpose of this chapter is to describe how a printed circuit board (PCB) trace acts like a transmission line. This chapter assumes that the reader is familiar with analog components to analyze simple circuits, basic PCB design, digital circuits, and differential signaling.

2.1 LEARNING OBJECTIVES

After reading this chapter, you will be able to perform the following tasks:

- Choose the best PCB design parameters such a trace width, height, and routed path to ensure the most stable characteristic impedance.
- Determine the correct termination to minimize reflections.
- Draw a bounce diagram for reflections within a transmission line.
- Predict the delay caused by a given PCB trace.
- Minimize driver power consumption using ac terminations.
- Compensate for discontinuities along a PCB trace.

2.2 CHARACTERISTIC IMPEDANCE

An ideal digital signal has instantaneous transitions from zero to one and from one to zero. This ideal signal will travel along a wire instantaneously and will be received at the other end with no distortion. But once students assemble their first circuit, the ideal world is left behind, and they are hit with reality. In reality, all digital signals have a finite rise or fall time. The signal travels down a wire with loss and noise at some rate less than the speed of light. The signal is received in by another device which may or may not be able to interpret the signal afterward. This chapter will describe the reality of sending high-speed signals across a PCB, and how to make the signal approach the ideal waveform.

The circuit in Fig. 2.1 is a very basic model of a digital circuit. The voltage supply will be a unit step from a low voltage to a high voltage. The supply has a series impedance associated with it. For simplicity, assume that this impedance only has a real component (a resistor). The load is also an impedance with only a real component. Assume that all wires have no resistance.

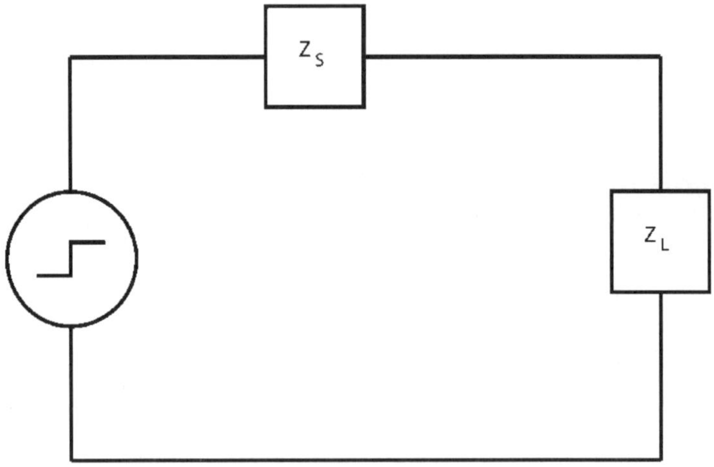

FIGURE 2.1: Simple digital circuit

When the unit step is applied, a proportional unit step voltage is seen across the load at the same instant.

One well-known fact in physics is that nothing can travel faster than the speed of light. The speed of light is 186,000 miles per second. In the circuit above, assume that the distance between the series impedance and the load impedance is 186,000 miles. When the unit step is applied, the proportional unit step is not seen at the load until at least one second later. This means that the voltage along the wire can be different at different locations along the wire. The voltage travels like a wave down the wire as seen in Fig. 2.2. The voltage is shown along the entire length of the wire at four different times.

Suppose that the same circuit has infinitely long wires connecting the source impedance and the load impedance. Effectively, the load will have no impact since the voltage will never reach it. This can be modeled as an infinite load, or an open circuit. This does not mean that there will be no current flow along those wires. Since there are two wires in parallel, it will behave as a very long capacitor. Before the unit step, this capacitor will have no charge on it. After the unit step, the capacitor will draw current until it is fully charged. The capacitance is dependent on the distance between the two wires, and the surface area of the wires. Since the wires are infinitely long, it will have an infinite capacitance. More realistically, the wires will act like an infinite number of parallel capacitors. The capacitors close to the voltage source will charge first.

The capacitance between these two wires will store energy in the form of an electric field. Any change in voltage will be opposed by this electric field by supplying or sinking a current. This follows the equation for a capacitor:

$$i = C\left(\frac{\mathrm{d}V}{\mathrm{d}t}\right). \tag{2.1}$$

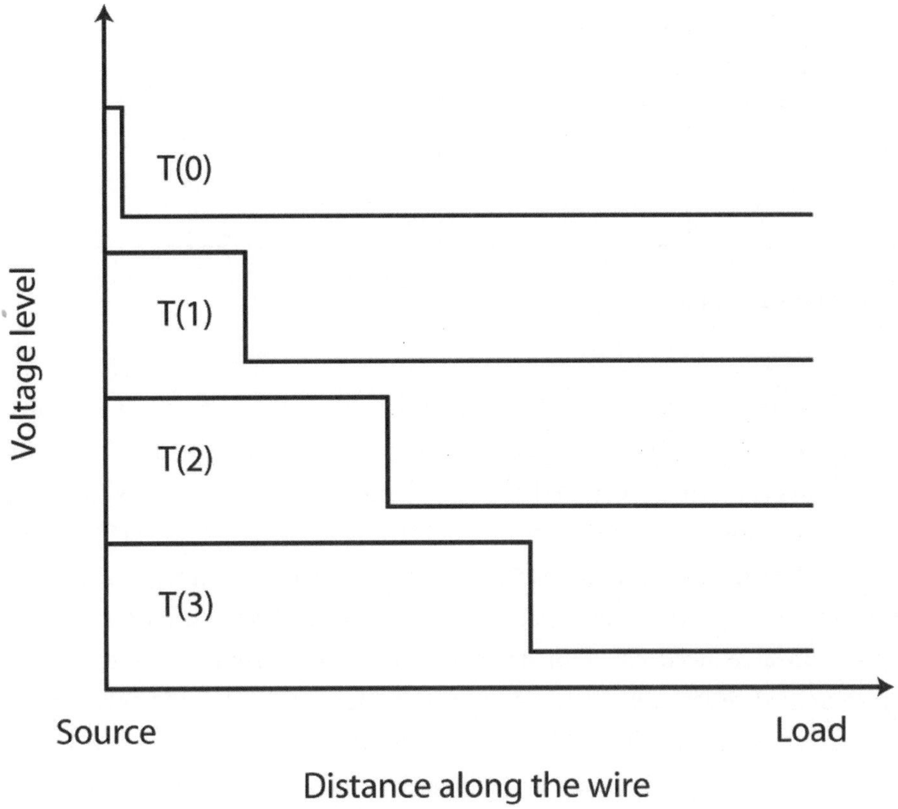

FIGURE 2.2: Wave motion of voltage on a transmission line

If the capacitance is infinite then the capacitor will draw an infinite amount of current. Also, if the voltage changes instantly, the capacitor will draw an infinite amount of current. This high current draw will make the parasitic inductance in the wires apparent.

Any changing current through a wire will create a magnetic field relative to the parasitic inductance of the wire. The inductance will store energy in the form of a magnetic field around the wire. Any change in current will be opposed by the magnetic field by changing the voltage across the inductor. This follows the equation for an inductor:

$$v = L\left(\frac{\mathrm{d}I}{\mathrm{d}t}\right). \tag{2.2}$$

This inductance will prevent the current from ever reaching an infinite magnitude. This inductance is modeled between parallel capacitors as seen in Fig. 2.3.

The end result of the series of inductors and capacitors is a constant current of less-than-infinite magnitude from the voltage source. The wire will draw a constant current from the source for an unlimited amount of time. In this way, the infinite wire will act like a simple

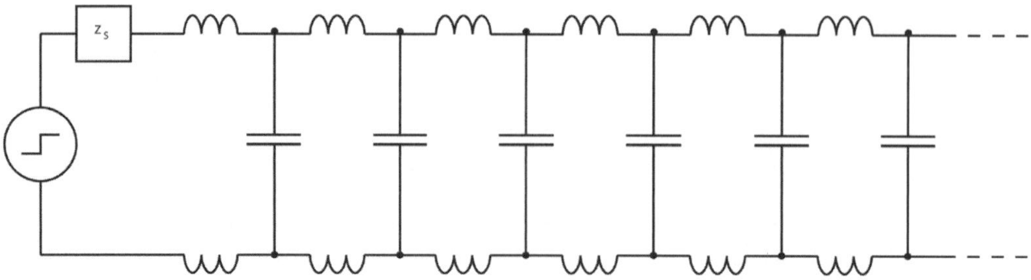

FIGURE 2.3: Distributed model of an infinite wire

resistive load from the perspective of the voltage source. The resistance of this set of wires is called the *characteristic impedance* measured in ohms. Coaxial cable is often rated in 50 Ω, or 75 Ω, because of its characteristic impedance. This impedance is determined by the geometries and distance of the wires. The set of wires is collectively known as a *transmission line.*

The characteristic impedance is set by the geometry of the two wires. If the separation of the conductors is increased, the capacitance is decreased and the inductance is increased. This will result in a reduced constant current being drawn from the voltage source, which acts like an increased resistance. If the separation is decreased, the opposite effect will occur and the characteristic impedance will decrease. While in this example the inductance and capacitance is a series of finite elements, the actual transmission line is measured in instantaneous capacitance and inductance. For a specific geometry, there will be a capacitance per unit length and inductance per unit length. The equation to determine the characteristic impedance is

$$Z_0 = \sqrt{\frac{L/l}{C/l}} \tag{2.3}$$

where

Z_0 is the characteristic impedance in ohms;
C/l is the capacitance per unit length;
L/l is the inductance per unit length.

The unit length measurements will cancel out leaving the equation

$$Z_0 = \sqrt{\frac{L}{C}}. \tag{2.4}$$

Note that for this equation the characteristic impedance is not dependent on the length of the transmission line.

Practical values for characteristic impedance on PCBs are usually either 50 Ω or 75 Ω. If the signals will be entering/leaving the PCB by means of cables or connectors, those connectors should have the same impedance. Typical cables have impedances from as low as a few ohms to 300 Ω. If possible, the PCB traces should match the cable impedance. The choice of cables should be decided early in the design phase to make PCB design easier.

Measuring Characteristic Impedance

Measuring the characteristic impedance of a transmission line is a relative easy process depending on the test equipment on hand. For a rough estimate, a network analyzer can be used. The network analyzer will measure the inductance and the capacitance at a set frequency. This frequency should be at least in the MHz range for transmission line tests. The equation above can then be used to calculate the characteristic impedance. The network analyzer will have two connectors: the first should be connected to one end of the trace and the other should be connected to the closest ground point available to that end. The other end of the trace should be left open when performing the capacitance test. When performing the inductance test, the other end should be shorted to the closest ground point. If there is no nearby ground point, it can significantly affect the measurement.

Ideally, a *test coupon* should be made along with the PCB to perform these measurements. A test coupon uses the same layering as the PCB and has a straight, long trace with a nearby via to the ground plane at both ends. This enables a precise measurement of the characteristic impedance. Using a test coupon with a network analyzer can have up to a 5% error.

A better tool to measure the characteristic impedance is a time domain reflectometer (TDR). This tool is designed specifically to measure transmission lines and has a very high accuracy. A simplified TDR is a pulse generator connected through a signal splitter to an oscilloscope. The other end of the splitter is connected to the signal trace under test. The pulse generator transmits a fast-rising pulse through the signal trace and the oscilloscope measures the exact response of the pulse. Some oscilloscopes have built-in TDR measurement tools, and some have add-on modules to perform this task. If a TDR is not available, then one can be put together using a pulse generator and oscilloscope; however, care must be taken when splitting the signal. To manually determine the characteristic impedance, the amplitude of the reflected signal must be measured on the oscilloscope. Reflections will be covered later in this chapter. Measuring the characteristic impedance is only one of the basic uses of a TDR.

Designing for Characteristic Impedance

While the above equations are useful for determining the characteristic impedance after a PCB is fabricated, designing a PCB from scratch uses a different procedure. A few simple equations

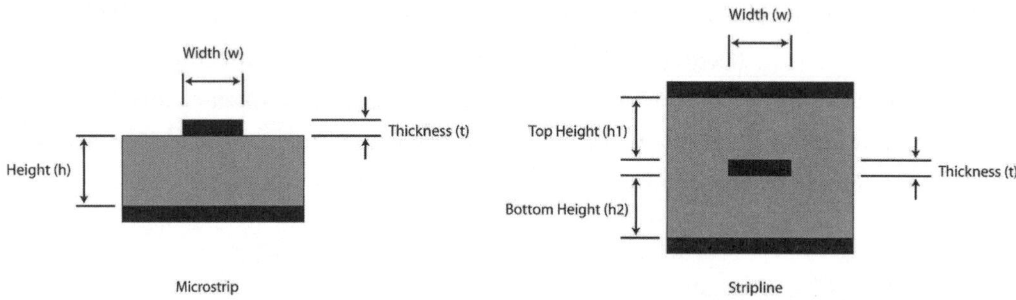

FIGURE 2.4: Microstrip and stripline geometries

exist for an estimate of the geometry of the traces; however, these equations can have significant error.

Two types of printed circuit board traces exist on a PCB: microstrip and stripline. Microstrip traces are routed on the outside layers with the next layer a ground plane. Stripline traces are embedded in an internal layer with a ground plane on either side. Sometimes two signal layers are embedded between ground planes. This is a special case of stripline. The equations below do not apply to this case. The relevant variables are shown in Fig. 2.4.

For microstrip traces, the characteristic impedance is based on the following equation:

$$Z_0 = \frac{87}{\sqrt{\varepsilon_r + 1.41}} \ln\left(\frac{5.98h}{0.8w + t}\right). \tag{2.5}$$

This equation only holds true under the following conditions:

$$0.1 < \frac{w}{h} < 3.0. \tag{2.6}$$

$$1 < \varepsilon_r < 15. \tag{2.7}$$

These conditions are usually met with most standard PCB designs. The capacitance and inductance can also be found using the following equations:

$$1 < \varepsilon_r < 15 \tag{2.8}$$

$$C_0 = \frac{0.67\,(\varepsilon_r + 1.41)}{\ln\left(\dfrac{5.98h}{0.8w + t}\right)} \tag{2.9}$$

$$L_0 = C_0 Z_0{}^2. \tag{2.10}$$

The stripline trace does not need to be symmetric between the two planes. If it is asymmetric, then assume that the top height is smaller than the bottom height:

$$Z_0 = \frac{80}{\sqrt{\varepsilon_r}} \ln\left(\frac{1.9\,(2h_1 + t)}{0.8w + t}\right)\left(1 - \frac{h_1}{4h_2}\right). \tag{2.11}$$

This equation only holds true under the following conditions:

$$h_1 < h_2 \tag{2.12}$$

$$0.1 < \frac{w}{h_1} < 2.0 \tag{2.13}$$

$$\frac{t}{h_1} < 0.25 \tag{2.14}$$

$$1 < \varepsilon_r < 15. \tag{2.15}$$

These conditions are usually met with most standard PCB designs. The capacitance and inductance for stripline can be found using the following equations:

$$C_0 = \frac{1.06\varepsilon_r}{\ln\left(\frac{1.9\,(2h_1 + t)}{0.8w + t}\right)\left(1 - \frac{h_1}{4h_2}\right)} \tag{2.16}$$

$$L_0 = C_0 Z_0^2. \tag{2.17}$$

These equations are good enough for first estimates of trace geometries. Field solvers can be used to find precise values. Many CAD tools have a type of field solver built-in or as an option to perform "what-if" simulations. They usually incorporate the layer stacking, via dimensions, and other parameters to determine the best layout for the signals. The results of these simulations will be good enough to start a design.

Postprocessing tools can be used after a board has been designed to determine the exact characteristic impedance for each line. It should be able to give a model of every transmission line showing each discontinuity in the line and voltage waveforms at any place on the line. In addition to this, the postprocessing tools can give recommendations for how the board can be modified to achieve better signal quality, such as a better path for signal routing or where components should be added/removed. Modern simulation tools are very powerful for analyzing a PCB before it is sent for manufacturing. For any high-speed design, postprocessing simulation tools should be used.

Example 2.1. I plan on making a four-layer PCB with my signals on the top and bottom traces. The width of my traces is 15 mil, with a 10 mil separation in my layers. The thickness of my traces is 1.37 mil. I will be using FR-4, so I will use 4.5 for my dielectric constant. I first compute my characteristic impedance:

$$Z_0 = \frac{87}{\sqrt{4.5 + 1.41}} \ln\left(\frac{(5.98)(10)}{(0.8)(15) + 1.37}\right) \tag{2.18}$$

$$Z_0 = \frac{87}{\sqrt{5.91}} \ln\left(\frac{59.8}{13.37}\right) \tag{2.19}$$

$$Z_0 = (35.79)(1.498) \tag{2.20}$$

$$Z_0 = 53.61\,\Omega. \tag{2.21}$$

Next I predict the capacitance of my traces:

$$C_0 = \frac{0.67(4.5 + 1.41)}{\ln\left(\frac{(5.98)(10)}{(0.8)(15) + 1.37}\right)} \tag{2.22}$$

$$C_0 = \frac{0.67(5.91)}{\ln\left(\frac{59.8}{13.37}\right)} \tag{2.23}$$

$$C_0 = \frac{3.96}{1.498} \tag{2.24}$$

$$C_0 = 2.64\,\text{pF in.}^{-1}. \tag{2.25}$$

Finally the inductance is

$$L_0 = (2.64)(53.61)^2 \tag{2.26}$$

$$L_0 = 7587\,\text{pH in.}^{-1} \tag{2.27}$$

$$L_0 = 7.587\,\text{nH in.}^{-1}. \tag{2.28}$$

2.3 PROPAGATION VELOCITY

The voltage travels down the transmission line like a wave with a velocity of the speed of light in a vacuum. Most wires do not operate in vacuum, so the actual velocity of propagation is less than the speed of light. The *propagation velocity,* also known as the *propagation delay* or *velocity factor*, is the actual speed of the signal traveling down the trace. The propagation velocity is determined by the properties of the material surrounding the wires. The important properties are how well an electric field can permeate the material (called the *dielectric constant*), and how well a magnetic field can permeate the material,

$$v_p = \frac{c}{\sqrt{\varepsilon_r \mu_r}} \tag{2.29}$$

where

v_p is the propagation velocity in meters per second;
c is the speed of light in meters per second;
ε_r is the dielectric constant of the surrounding material;
μ_r is the magnetic permeability of the surrounding material.

Usually the magnetic permeability is equal to 1, and therefore has no impact on the equation. This equation applies only with a homogenous insulating material. Sometimes the material is not homogenous, such as when a printed circuit board trace is on the external surface with FR-4 on one side and air on the other. In this case, the dielectric material will not slow the propagation velocity as much as if it were surrounded by it such as with stripline.

The dielectric constant of PCBs which use FR-4 as the insulating material is about 4.5–4.8. This means that the velocity of a voltage wave traveling through a circuit board trace is a little slower than half the speed of light.

A second equation also describes the propagation velocity:

$$v_p = \frac{1}{\sqrt{LC}}. \tag{2.30}$$

In this equation, the velocity is described using the inductance and capacitance. If the two equations are combined,

$$\frac{c}{\sqrt{\varepsilon_r \mu_r}} = \frac{1}{\sqrt{LC}}. \tag{2.31}$$

This equation shows how the inductance and capacitance are interrelated for a given material. Since the speed of light, the dielectric constant, and the magnetic permeability are all constants, changing the inductance of a transmission line will result in a relative change to the capacitance. Changing the propagation velocity is impossible without changing material. Again note that this is only true for a homogenous material surrounding the transmission lines. In a nonhomogenous material the inductance and capacitance might be changed independently or at least at a different rate.

The speed of light in units relative to the size of PCBs is 11.8 in. ns^{-1}. A board with FR-4 insulating material will have signals travel at about 5 in. ns^{-1}. High-speed designs can have clock periods in the picosecond range. This means that a device can send tens to hundreds of pulses down a signal trace before they reach the other end of the circuit board.

For FR-4, a signal will have about 160 ps in.$^{-1}$ delay. Therefore, close placement of devices which transmit/receive signals at high speed is necessary to minimize delay. This also means that having a synchronized clock between all logic devices on a high-speed PCB is very difficult. Other methods such as source-synchronous clocks or regenerative clock methods (delay locked loops) must be used.

Example 2.2. I have just read the previous material and want to check an old circuit board that I have. The only measurement tool I have is an RLC meter. So I pick a trace on the circuit board that is mostly straight. The next step is to find a nearby via connected to ground. I use the RLC meter and probe nearby vias to find one that has a zero resistance to ground. I then switch my RLC meter to capacitance mode and measure the capacitance of the trace by probing those two spots. I measure the capacitance as 36.6 pF. Next I find a via connected to ground near the other end of the trace. I then get out my soldering iron and very carefully solder a small piece of wire from the trace to that ground via. I switch my RLC meter into inductance mode and measure the inductance of the trace by probing the same two spots. I measure the inductance as 92.3 nH. I must convert nanohenries into picohenries to cancel picofarads in my characteristic impedance calculation. Since the length of the trace is not important, I can use these numbers into the equation for characteristic impedance:

$$Z_C = \sqrt{\frac{L}{C}} = \sqrt{\frac{92.3 \text{ nH} \times 1000}{36.6 \text{ pF}}} = \sqrt{2521.9} = 50.21\ \Omega. \tag{2.32}$$

So my characteristic impedance is about 50 Ω. Next I want to find how fast my signal can propagate down this trace. The length of the trace cancels out in the characteristic impedance equation, but not for the propagation velocity. To find the inductance per unit (in this case inches) I measure the length of the trace to be 10 in. Currently my inductance is in inductance per 10 in., so I divide it by 10 to get inductance per inch. The same is done with capacitance. So I use the next equation

$$v_p = \frac{1}{\sqrt{LC}} = \frac{1}{\sqrt{(9.23 \text{ nH inch}^{-1} \times 1000)(3.66 \text{ pF inch}^{-1})}} = \frac{1}{\sqrt{33{,}781}}$$
$$= 0.00544 \text{ inch ps}^{-1}. \tag{2.33}$$

So my signals will travel 0.00544 in., or 5.44 mil, every picosecond. I can reasonably determine what type of insulating material was used to create my PCB. Assuming a magnetic permeability of 1,

$$v_p = \frac{c}{\sqrt{\varepsilon_r}} \tag{2.34}$$

$$\sqrt{\varepsilon_r} = \frac{c}{v_p} \tag{2.35}$$

$$\varepsilon_r = \left(\frac{c}{v_p}\right)^2 = \left(\frac{0.0118 \text{ in. ps}^{-1}}{0.00544 \text{ in. ps}^{-1}}\right)^2 = (2.1691)^2 = 4.7. \tag{2.36}$$

The dielectric constant of 4.7 is about that of FR-4, which is the standard material used in PCB construction.

2.4 REFLECTIONS

An infinite transmission line is not physically possible. Therefore, a voltage wave will encounter the end of the transmission line at some point in time. This means that after a certain amount of time, the distributed capacitance along the transmission line will fully charge and stop behaving like a transmission line. This stable point will be reached after a finite time for any finite length transmission line.

If the end of the transmission line is an open circuit, then current will stop flowing once the capacitance is fully charged. On the other hand, if the end of the transmission line is a short circuit, there is no voltage drop across the wires at that point. At the stable point of the transmission line, both wires should have the same voltage across the entire length. This means that at some point in time the capacitors will start to charge from the transmitted voltage wave, and then discharge some time after that. When the voltage wave is first transmitted, the load at the other end of the transmission line does not affect the wave. The only factor is the characteristic impedance at first. When the incident voltage wave reaches the end of the line, something else happens to reduce the voltage across the capacitors. Logically, the capacitance closest to the load will discharge first, and the capacitor farthest from the load will discharge last. This can be modeled as another voltage wave traveling in the opposite direction back toward the voltage source. This is called a reflection.

A transmission line acts like any type of medium with a wave traveling through it. This can be sound or light traveling through air. Sound travels through air at a certain rate depending on a number of factors such as density and humidity. When sound hits a house's wall, it will bounce off it and travel back to the origin of the sound. However, if the sound was completely reflected, houses would be silent inside when a large truck passes outside. Therefore, some of the sound passes into the wall and then into the air inside the house. The sound will be quieter on the inside, so most of the sound energy is reflected, and a smaller amount is transmitted.

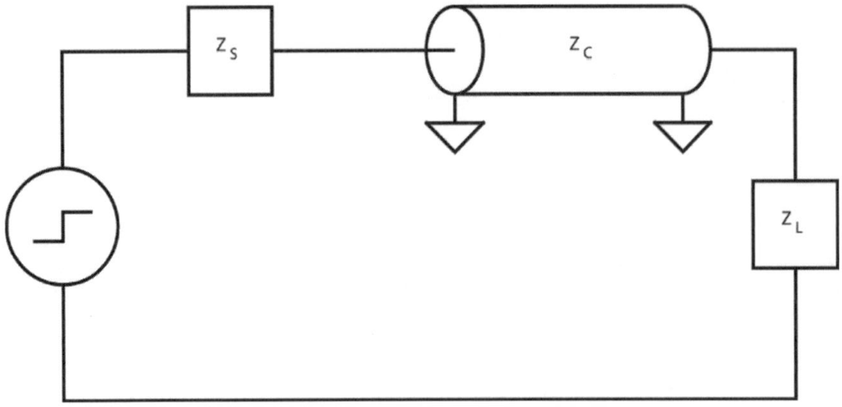

FIGURE 2.5: Digital circuit with a transmission line

This is because the density of the wall is different from the density of the air. This is similar to a load on a transmission line. Instead of density, the difference is in impedance. As with all waves, they will reflect when there is a change in the medium they travel through.

The amount of reflection can be predicted by knowing the difference in impedance. The characteristic impedance of a transmission line is based on the geometry of the wires. On a PCB, if the signal trace is a straight uniform wire, then the characteristic impedance will be the same everywhere on the trace. The only differences in impedance will be at the ends of the trace. The transmission line can be modeled using the original circuit with the transmission line as a series impedance as seen in Fig. 2.5.

A voltage wave must first be input to one side of the trace for a wave to propagate down it. The voltage step applied to one end of the trace may not have the same amplitude as the wave that travels down the line. This fraction of the incident voltage is called the *input acceptance function*. It is a function of the source impedance and the characteristic impedance of the line,

$$A = \frac{Z_0}{Z_S + Z_0} \tag{2.37}$$

$$v_C = A v_S \tag{2.38}$$

where

A is the input acceptance function;
Z_0 is the characteristic impedance of the transmission line;
Z_S is the impedance of the source;
v_C is the voltage inside the transmission line;
v_S is the voltage incident from the source.

This equation follows the same form as a simple voltage divider. The voltage amplitude inside the transmission line will be uniform until it reaches an impedance discontinuity. When the voltage wave reaches the discontinuity, part of the wave will transmit to the new impedance and part will reflect back through the transmission line. The amount transmitted is

$$T = \frac{2\,Z_L}{Z_L + Z_0} \tag{2.39}$$

where

T is the transfer coefficient;
Z_0 is the characteristic impedance of the transmission line;
Z_L is the impedance of the load.

The transfer coefficient can range from 0 to 2. If the characteristic impedance is low relative to the load impedance, more signal will transfer to the load. If the load impedance is low relative to the characteristic impedance, very little signal will transfer. The amount of signal reflected is

$$R = \frac{Z_L - Z_0}{Z_L + Z_0} \tag{2.40}$$

where

R is the reflection coefficient.

Note that the reflected wave can be positive or negative depending on the relative size of the impedances. If the load impedance is zero (meaning there is a short) then the reflection coefficient will be -1, which means that the signal will completely reflect, but will be inverted. If a 5 V pulse is transmitted, then a -5 V pulse is reflected. The sum of these two pulses will result in 0 V on the transmission line, which is what to expect when the lines are shorted together. The transfer coefficient will be zero in this case.

If the load impedance is infinite, meaning an open circuit, the reflection coefficient will be 1, so the signal will also be completely reflected, but not inverted. The transfer coefficient will be two in this case. The only way for the reflection coefficient to be zero is if the impedances are the same.

A reflected wave will continue back toward the source until it reaches another impedance discontinuity. If it does, it will follow the same equations for transmission and reflection of the wave. If the source and load impedances are different from the transmission line, the wave can continue to reflect back and forth across the transmission line.

The voltage on the transmission line will remain the same unless a voltage wave travels through it. As a voltage wave passes each point on the line, it will add its voltage to the current voltage at that location. Therefore, any reflected wave will add its amplitude when it passes each point on the line. If the wave repeatedly reflects down the transmission line, the equation to determine the current voltage at any location becomes very large.

Bounce Diagrams

The simplest way to keep track of the voltage at any point along a transmission line is through the use of *bounce diagrams*. Fig. 2.6 shows a bounce diagram for the circuit above. The horizontal axis represents the length of the transmission line. The vertical axis represents the starting time when the wave first enters the transmission line. The plot shows how the wave travels back and forth across the transmission line over time. The normalized amplitude of the wave is shown on the plot for each reflection.

The source impedance is at location $X(0)$, and the load impedance is at location $X(4)$. The incident wave will have a normalized amplitude of A starting at $X(0)$. Once the wave reaches the load at $X(4)$, the reflected wave will have an amplitude of AR_L and travel back toward the source. After the wave makes a full round trip across the transmission line, it will reflect again with amplitude $AR_L R_S$. The wave will continue to reflect with the amplitude being modified after each trip across the line. The time of travel across the transmission line is from $T(0)$ to

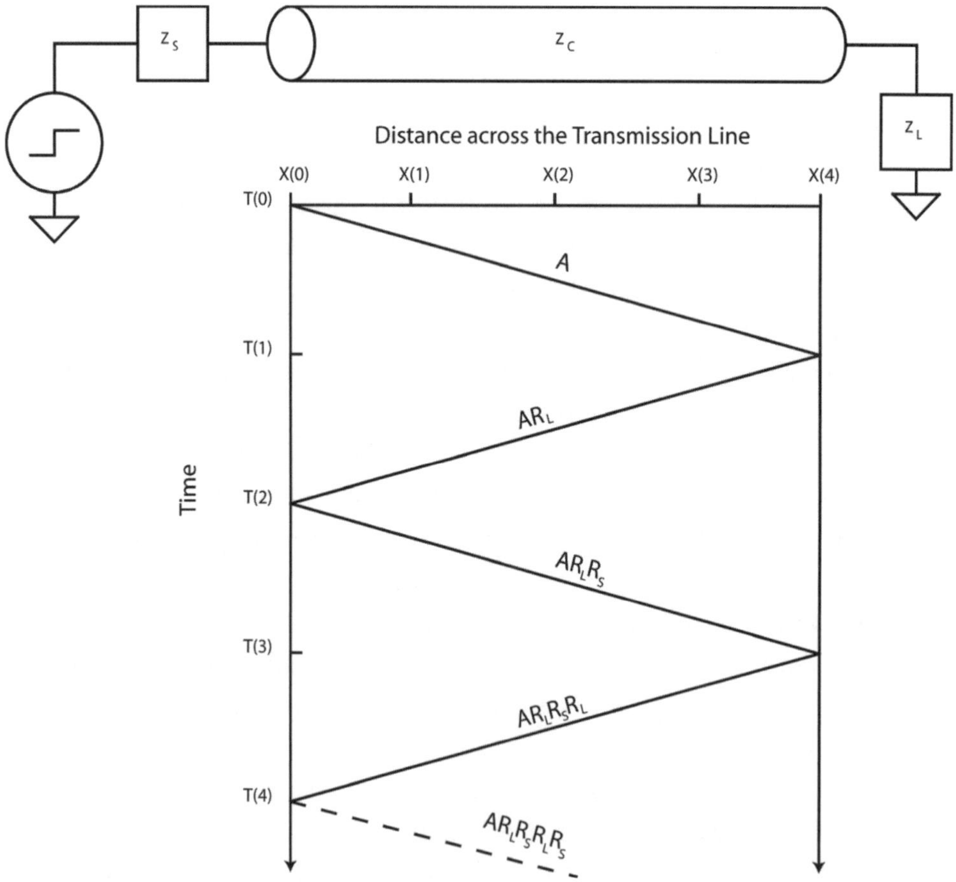

FIGURE 2.6: Generic bounce diagram

$T(1)$. The velocity of the wave can be written as

$$v_p = \frac{T(1) - T(0)}{X(4) - X(0)}. \tag{2.41}$$

Using this bounce diagram, a plot of the voltage amplitude can be created for any point along the transmission line. A line is traced down the bounce diagram at the given location starting at $T(0)$. The voltage remains the same until the line representing the wave is reached. The voltage amplitude of the wave is then added to the current voltage at that point in time.

Example 2.3. Given the circuit above, I want to transmit a 10 V signal to a 25 Ω load. My voltage source has 75 Ω impedance. I have measured the characteristic impedance of the transmission line between the source and load to be 50 Ω. I want to know what the voltage response will be at the center of the transmission line. I will first construct a bounce diagram. I compute my incident acceptance function

$$A = \frac{50}{75 + 50} = \frac{2}{5} \tag{2.42}$$

$$Av_S = \frac{2}{5} \times 10 = 4 \text{ V}. \tag{2.43}$$

The wave will have a 4 V amplitude when it first travels across the transmission line. When it reaches the far end, part of the 4 V will transmit into the load and part will reflect back:

$$R_L = \frac{25 - 50}{25 + 50} = -\frac{25}{75} = -\frac{1}{3} \tag{2.44}$$

$$(v_S A) R_L = 4 \times \left(-\frac{1}{3}\right) = -\frac{4}{3} = -1.33 \text{ V}. \tag{2.45}$$

Therefore one-third of the wave, or −1.33 V, will invert and reflect back to the source. When the wave reaches the source, it will reflect again;

$$R_S = \frac{75 - 50}{75 + 50} = \frac{25}{125} = \frac{1}{5} \tag{2.46}$$

$$(v_S A R_L) R_S = -1.33 \times \left(\frac{1}{5}\right) = -0.2667 \text{ V}. \tag{2.47}$$

The reflected wave will have a fifth of the voltage, or −0.2667 V. I can continue to use these reflection coefficients to calculate the amplitude of the remaining reflections,

$$(v_S A R_L R_S) R_L = -0.2667 \times \left(-\frac{1}{3}\right) = 0.089 \text{ V} \tag{2.48}$$

$$(v_S A R_L R_S R_L) R_S = 0.089 \times \left(\frac{1}{5}\right) = 0.0178 \text{ V}. \tag{2.49}$$

The final bounce diagram will look like Fig. 2.7.

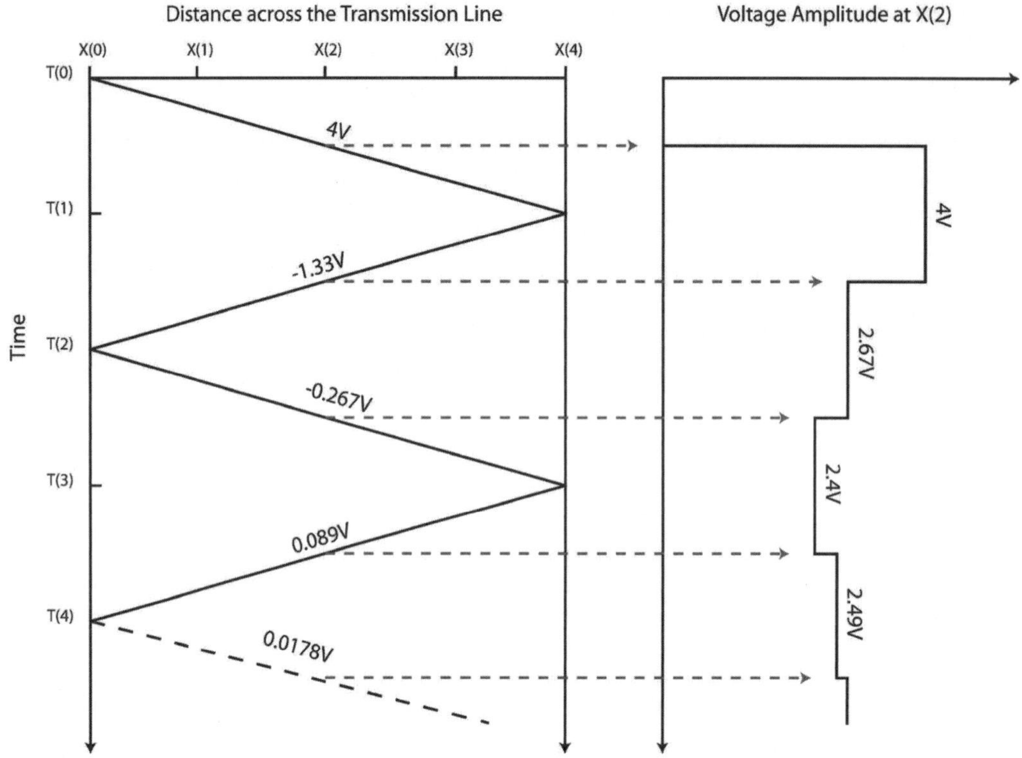

FIGURE 2.7: Example bounce diagram

I will first draw the voltage waveform for the point in the center of the transmission line. This corresponds to point $X(2)$. Since the wave travels at a constant speed, it will reach point $X(2)$ exactly halfway between $T(0)$ and $T(1)$. The voltage at this point will jump to 4 V. The next crossing will be halfway to $T(2)$. The voltage wave will be -1.33 V this time. This voltage will be added to the current voltage of 4 V, so the new voltage is 2.67 V. The next wave will reduce the voltage by 0.267 V leaving 2.4 V. In the next passing it will be 2.49 V, and then 2.5 V. Each time the voltage wave passes point $X(2)$, it has a smaller amplitude than the previous time.

I can check my work by looking at the steady-state voltage. If I remove the transmission line, the voltage across the load can be solved using the voltage divider formula

$$v_L = \frac{Z_L}{Z_S + Z_L} v_S \tag{2.50}$$

$$v_L = \frac{25}{25 + 75} \times 10 = \frac{1}{4} \times 10 = 2.5 \text{ V}. \tag{2.51}$$

At steady state, the voltage across the load should be 2.5 V. From the voltage waveform, this is where the voltage eventually stabilizes.

One thing to note here is that if the voltage waveform was plotted at a point very near the load, the amount of time 4 V would be present on the waveform would be very small. Almost instantly it would drop down to 2.67 V. This means that the load would never actually see this 4 V pulse. The voltage would instantly become 2.67 V.

The voltage across a transmission line will eventually stabilize at a certain level; however, while the voltage is bouncing back and forth, the receiver will be measuring this voltage level. A large change can occur at the receiver, which may be interpreted as a logic transition. In the previous example, the voltage bounces between high and low voltages, which the receiver may interpret as multiple 0 to 1 transitions. This overshoot and ringing response is seen in the previous example.

Often with CMOS logic devices, the input and output resistance is very high. This means that the signal will reflect entirely at the load and almost entirely at the source. Compared to the impedance of the transmission line, it can be considered an open circuit. This will result in significant reflections. Because of the significant difference in the source and the transmission line, very little signal will be injected into the transmission line. If the voltage waveform was plotted at the load, it would look like a stair-step pattern rising slowly to the steady-state voltage. Therefore, a mismatch in transmission line impedance can significantly slow down the effective rise time of the transmitted pulse.

The input impedance to CMOS devices can be modeled as a capacitor. This capacitance will add delay to the signal. If the transmission line is modeled as a resistor with the capacitor, the load will act like an RC low-pass filter. Specifically, the equation for the load is

$$v_L(t) = v_s\left(1 - e^{\left(-\frac{t-t_{pd}}{\tau}\right)}\right) \quad \text{when } t > t_{pd} \tag{2.52}$$

where

$v_L(t)$	is the voltage seen by the receiver;
v_s	is the voltage in the transmission line;
t_{pd}	is the delay of the transmission line;
$\tau = Z_0 C$	is the time constant;
Z_0	is the characteristic impedance of the transmission line;
C	is the capacitive load.

If t_d is the time when $v_L(t = t_d) = 0.9v_s$, then

$$t_d = t_{pd} + 2.3\tau. \tag{2.53}$$

This equation means that the receiver will have an extra $2.3\,Z_0 C$ delay before the signal is detected.

Delay can also be introduced from capacitive and inductive discontinuities present along the transmission line. Usually capacitance will be shunted to ground, and inductance will be in series with the transmission line. Vias can appear as either a capacitive or inductive discontinuity. Bends in the trace can change the characteristic impedance of the wire because the width cross-section can increase around the bend. Each bend produces a capacitive discontinuity. Sharper bends produce a larger capacitance. Bends usually do not have enough discontinuity to effect signal quality until gigahertz speeds. The delay for both types of discontinuity follows the same equation

$$v_t(t) = v_s\left(1 - e^{-\frac{t}{\tau}}\right). \tag{2.54}$$

Here $v_t(t)$ reaches $0.9v_s$ at $t = 2.3\tau$. For capacitive discontinuities,

$$\tau = \frac{Z_0 C}{2}, \tag{2.55}$$

which adds a delay of $1.15CZ_0$. For inductive discontinuities,

$$\tau = \frac{L}{2Z_0}, \tag{2.56}$$

which adds a delay of $1.15\frac{L}{Z_0}$.

Not only is the signal delayed, but part of the signal will reflect at the discontinuity. A capacitive discontinuity will reflect a negative voltage while an inductive discontinuity will reflect a positive voltage. The amount of signal reflected is

$$R = \frac{\tau}{t_r} \tag{2.57}$$

where

R is the reflection coefficient;
τ is the time constant (for either capacitive or inductive discontinuity);
t_r is the rise time of the voltage wave.

Example 2.4. The transmission line system shown in Fig. 2.8 has an input voltage step from 0 V to 64 V. The voltage waveform V_A is shown in Fig. 2.9, measured at the source resistance.

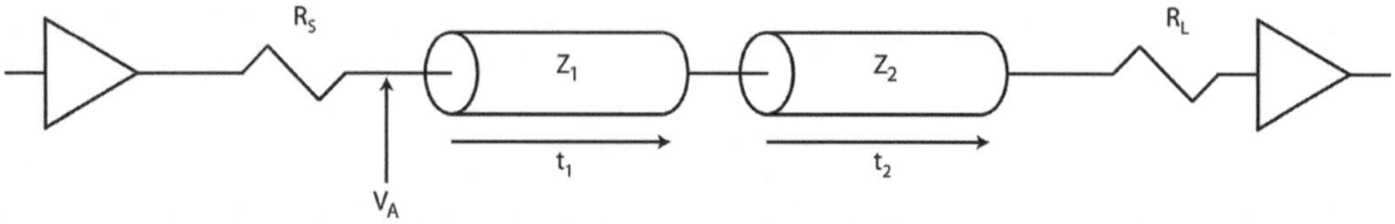

FIGURE 2.8: Two transmission lines in series

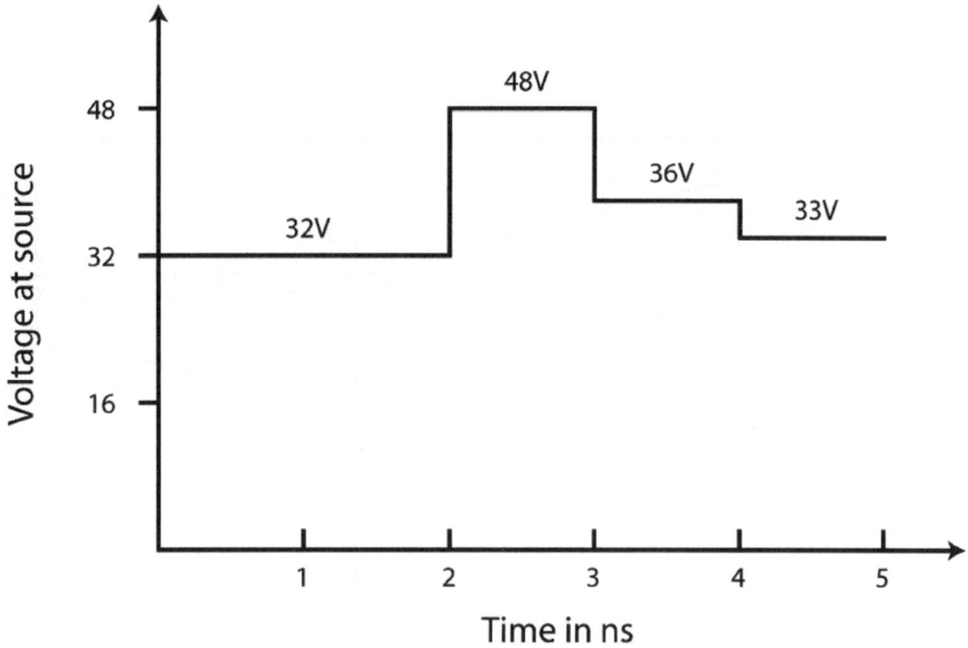

FIGURE 2.9: Waveform measured at source

There are two transmission lines with different characteristic impedances labeled Z_1 and Z_2. The time of travel across impedances Z_1 and Z_2 is t_1 and t_2 respectively. Given the source impedance of 50 Ω, I can find the values of Z_1, Z_2, t_1, t_2, and R_L.

The first reflection reaches the source 2 ns after it is sent which is the round trip time across Z_1. Therefore one-way trip across (t_1) is 1 ns. The second reflection reaches the source after 3 ns. Subtracting the trip across Z_1, the round trip time across Z_2 is 1 ns, which means the one-way trip (t_2) is 0.5 ns. I can next use the input acceptance function to find the value of Z_1:

$$v_C = \frac{Z_1}{Z_1 + R_0} v_S \tag{2.58}$$

$$v_C Z_1 + v_C R_0 = Z_1 v_S \tag{2.59}$$

$$v_C R_0 = Z_1 v_S - v_C Z_1 \tag{2.60}$$

$$\frac{v_C R_0}{v_S - v_C} = Z_1 \tag{2.61}$$

$$\frac{32 \times 50}{32 - 64} = Z_1 = \frac{1600}{-32} = 50\,\Omega. \tag{2.62}$$

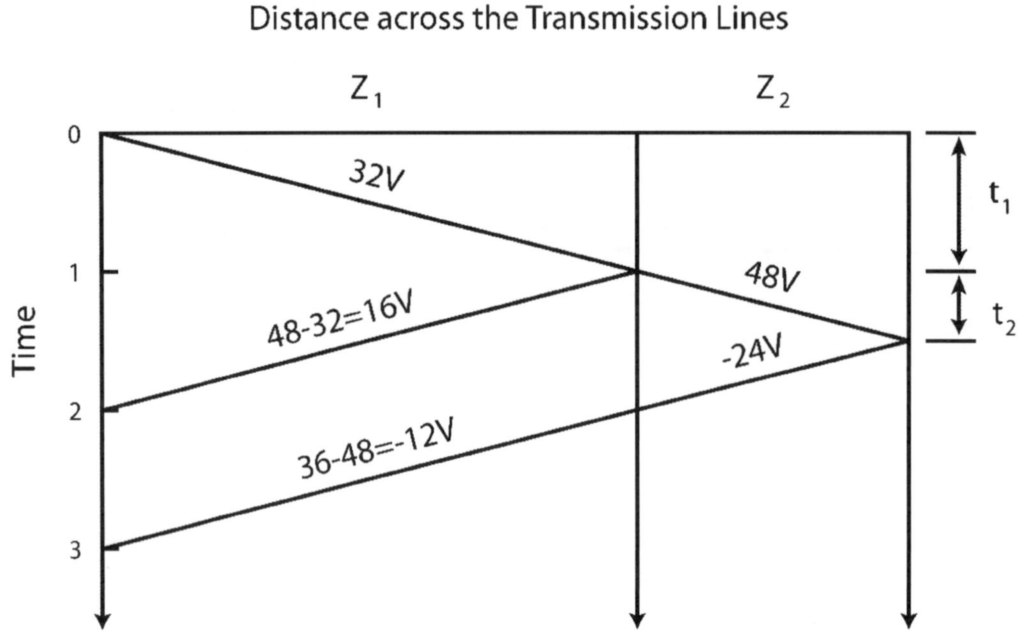

FIGURE 2.10: Bounce diagram for two transmission lines

Since the value of Z_1 and R_S are the same, there will be no reflections at the source. This will make solving the remaining values much easier. The value of Z_2 and R_L can easily be found by using a bounce diagram like Fig. 2.10.

The first returning voltage is 48 V. The amplitude of the returning voltage wave will be the difference between the first returning voltage and the first transmitted voltage (32 V). Therefore, the amplitude of the returning wave is 16 V. This means that when the 32 V wave meets Z_2, half is reflected back. So the reflection coefficient is +0.5. The Z_2 impedance can be found using the reflection coefficient

$$R_{1-2} = \frac{Z_2 - Z_1}{Z_2 + Z_1} = \frac{1}{2} = \frac{Z_2 - 50}{Z_2 + 50} \tag{2.63}$$

$$Z_2 + 50 = 2(Z_2 - 50) \tag{2.64}$$

$$Z_2 + 50 = 2Z_2 - 100 \tag{2.65}$$

$$Z_2 = 150\,\Omega. \tag{2.66}$$

The last step of finding the value of R_L will be a little more complicated. First, the transmitted voltage must be found going into Z_2:

$$T_{1-2} = \frac{2Z_2}{Z_1 + Z_2} = \frac{2 \times 150}{50 + 150} = \frac{300}{200} = 1.5. \tag{2.67}$$

The amplitude of the voltage wave transmitted is 1.5 times the first voltage wave resulting in 48 V. Also, the voltage wave transmitted in the other direction must be found:

$$T_{2-1} = \frac{2Z_1}{Z_1 + Z_2} = \frac{2 \times 50}{50 + 150} = \frac{100}{200} = \frac{1}{2}. \tag{2.68}$$

The amplitude of the returning voltage wave at 3 ns will be the difference of the measured voltage at 2 ns and 3 ns. Therefore, the returning voltage wave will be -12 V. Since -12 V was transmitted from the wave in Z_2, the voltage wave in Z_2 must be $-12/T_{2-1}$. This means that the voltage wave in Z_2 must be -24 V. Since 48 V is transmitted into Z_2 and 24 V is reflected back, the reflection coefficient is -0.5:

$$R_{2-L} = \frac{R_L - Z_2}{R_L + Z_2} = -\frac{1}{2} = \frac{R_L - 150}{R_L + 150} \tag{2.69}$$

$$-R_L - 150 = 2(R_L - 150) \tag{2.70}$$

$$-R_L - 150 = 2R_L - 300 \tag{2.71}$$

$$3R_L = 150 \tag{2.72}$$

$$R_L = 50\,\Omega. \tag{2.73}$$

2.5 IMPEDANCE COMPENSATION

To ensure no reflections in a transmission line, the impedance of the entire path from the transmitter to receiver must be constant. Since the source and load impedance can vary with the device technology, designing transmission line impedance to match it can be impossible. Impedance can change along the length of the transmission line as well. This section discusses methods of ensuring the most balanced transmission line by modifying the trace or adding components.

The goal of a good PCB design is to ensure the source and/or load impedance is the same as the characteristic impedance. Since the impedance of transmission lines is purely resistive, a resistive network can be added at the load and/or source. This resistor can either add or subtract from the impedance.

Load Termination

For very high impedance loads, a resistor in parallel to the load will lower the impedance to match the transmission line. For example, if a 50 Ω resistor is placed in parallel with a 1 MΩ resistor, the resulting resistance will be very slightly smaller than 50 Ω. If a 50 Ω transmission line is used, a 50 Ω resistor is connected to the load with the other end connected to ground. This resistor should be placed as close to the load as possible. A well-balanced load impedance will allow the full amplitude of the voltage on the transmission line to transmit to the receiver.

For low-impedance loads, a resistor in series to the load raises the impedance. The value of this resistor should be the difference between the transmission line and the load. This should also be placed as close as possible to the load.

One major problem with terminating resistors is the physical location. Resistors are large relative to the size of the pins of the packages containing logic devices. If many traces are routed close together to nearby pins, mounting nearby resistors can be very difficult. The priority for routing should be on the terminating resistor. It should always be at the very end of the transmission line. If the resistor is not placed at the very end of the transmission line, the resistor forms a different type of resistive divider. The transmission line can be modeled as two different transmission lines with a resistive discontinuity in the center. The short transmission line which connects the load and resistor is a stub. When the larger transmission line encounters the terminating resistor, it detects two resistors in parallel both with the same resistance. This can cause a significant reflection. This effect is minimized with shorter stubs. A stub can be modeled as a capacitive discontinuity. If the signal delay across the stub is 20% of the signal rise time, the advantages of using a terminating resistor are eliminated.

Some modern devices have the terminating resistor placed inside the package. This is a good solution for PCB designers since they do not have to worry about finding places for all the resistors. The transmission line stub will not be a problem since the trace will only be routed to the pin. Sometimes this internal resistance is user defined by external reference resistors. The only concern is the lead inductance of the package to the PCB and to the die. Ball grid array (BGA) or flip-chip packages have very small leads and therefore minimize inductance.

Source Termination

A source termination is not critical if there are no reflections returning on the transmission line; however, if there are reflections from the load, the source must be terminated to prevent repeated reflections within the transmission line. In some cases, a load termination is not possible, so a source termination can be used knowing that there will be one reflection from the load, but it will not reflect back again.

A source termination is a little more complex than a load termination because it will modify the signal injected to the transmission line. The impedance of the source when measured from the transmission line should be equal to the transmission line. The impedance can be raised or lowered using the same technique as the load termination. The reflection coefficient of the source should be zero.

Sometimes logic devices have varying output impedance while operating. If the output impedance varies, but is very high (1–10 kΩ), then the variance will not have a significant impact. If the output impedance is in the same range as the transmission line (10–100 Ω), then it can pose a significant problem. This type of device should be avoided or a buffer used nearby with a more reliable impedance. If the output impedance is very low (1–10 Ω), this variance might be acceptable.

For a low output impedance of the driver, a resistor can be placed in series to raise the source impedance to the same level as that of the transmission line. The value of this resistor is the difference between the transmission line and the source impedances. For a large output impedance, a resistor can be placed in parallel to the transmission line. This resistor has usually the same value as the transmission line.

When using a source termination, the voltage injected into the transmission line will be half the voltage of the source. This voltage will travel to the load and reflect back. Since the reflection coefficient of the source is zero, the voltage will not reflect again. The load must have a reflection coefficient of +1 to bring the voltage on the load to the same voltage level produced by the source. A reflection coefficient of +1 means that the load impedance must be very high relative to the characteristic impedance of the transmission line.

Power Consumption

A terminating resistor is necessary to prevent reflections in our circuits, but there are some drawbacks of using them. For example, if the load impedance is very high, a terminating resistor to ground is needed. If the voltage from the source is a stable 0 V, then no current will be flowing through the transmission line. If the voltage from the source is a stable 5 V, then a current will flow from the source through the transmission line, and through the terminating resistor to ground. The value of the terminating resistor must be the same as the transmission line, so there is no flexibility in that value. If the resistor must be 50 Ω, then 100 mA of current must be provided by the source. This is very high current for a logic device and will probably make it fail. This will only happen when the source is driving a high voltage.

One important note about terminating resistors is that they need to be connected to an ac ground. This means any plane which is 0 V at ac. A power plane qualifies because it should have no frequency components, and therefore can be used to terminate resistors.

If the previous example had a terminating resistor connected to the 5 V power plane, there would be no current flowing through the terminating resistor when the source was set to 5 V. But there would be 100 mA of current when the source was set to 0 V. So the situation has not improved.

A compromise could be made by having two terminating resistors. One resistor would connect to the power plane, and one would connect to ground. This is called a *split termination*. A few criteria determine the values of these resistors. The combination of these two resistors must still equal the characteristic impedance of the transmission line. At a high frequency, these resistors appear to be in parallel. The maximum high-level output current ($I_{\mathrm{OH\,max}}$) and the maximum low-level output current ($I_{\mathrm{OL\,max}}$) must not be exceeded. These criteria can best be expressed using equations

$$\frac{R_1 R_2}{R_1 + R_2} = Z_0 \tag{2.74}$$

$$\frac{(V_{\mathrm{CC}} - V_{\mathrm{OH}})}{R_1} - \frac{(V_{\mathrm{OH}} - V_{\mathrm{EE}})}{R_2} > I_{\mathrm{OH\,max}} \tag{2.75}$$

$$\frac{(V_{\mathrm{CC}} - V_{\mathrm{OL}})}{R_1} - \frac{(V_{\mathrm{OL}} - V_{\mathrm{EE}})}{R_2} < I_{\mathrm{OL\,max}}. \tag{2.76}$$

For given V_{CC} and V_{EE}, there may be no solution for resistors to meet these criteria. The transmission line impedance could be designed higher to compensate, or the difference of V_{CC} and V_{EE} could be lowered by using a different technology. Fig. 2.11 shows a sample of different technologies and their voltage swings.

With split terminators, a path from power to ground is formed through two small resistances. This means that the current through these terminators can be significant which consumes large amounts of power. One resistor will always have the entire voltage swing across it, so the amount of consumed power is

$$P_T = \frac{(V_{\mathrm{CC}} - V_{\mathrm{EE}})^2}{2 Z_0}. \tag{2.77}$$

The split termination can be transformed into a Thevenin equivalent circuit. A dc voltage will appear between the two resistors which is between V_{CC} and V_{EE}. This voltage is called the termination voltage. A single terminating resistor can replace the two split resistors which will have the same value as the transmission line impedance. The voltage source will be the same as

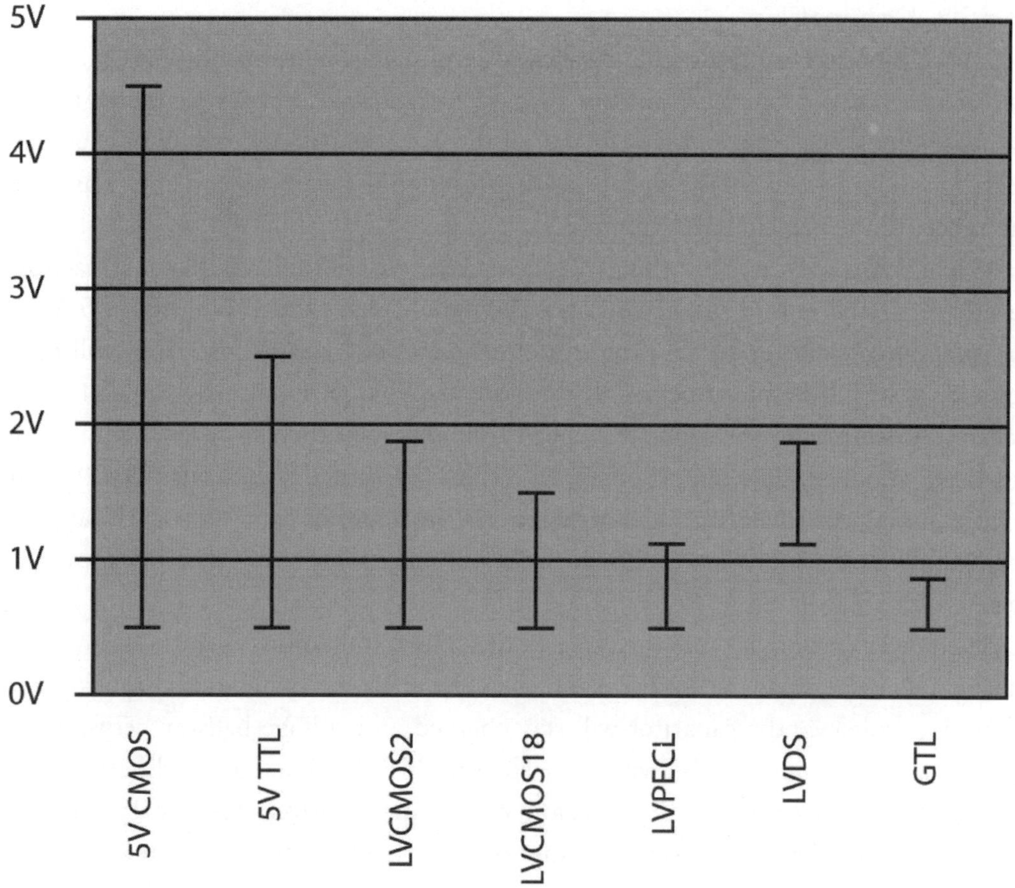

FIGURE 2.11: Comparison of voltage swings for various technologies

the dc voltage present between the split terminations, specifically

$$V_{\mathrm{TT}} = \frac{R_1 V_{\mathrm{EE}} + R_2 V_{\mathrm{CC}}}{R_1 + R_2}. \tag{2.78}$$

The terminating voltage is usually halfway between the high and low voltage levels. Therefore the power consumption will be

$$P_T = \frac{\left(V_{\mathrm{CC}} - V_{\mathrm{EE}}/2\right)^2}{Z_0} = \frac{(V_{\mathrm{CC}} - V_{\mathrm{EE}})^2}{4Z_0}. \tag{2.79}$$

The power consumption will be half of the split termination, but the overall power consumption may still be more than the power system can supply. Therefore, the dc current through the terminating resistor needs to be minimized.

Capacitive Termination

Sometimes a capacitor is used with a resistor to terminate a transmission line. The capacitor is designed to block dc current, but not to have a significant impact on the signals in the transmission line. The time constant of the terminating resistor and capacitor must be large compared to the frequency of the signals passing through the transmission line. This capacitor is placed between the resistor and ground.

While the capacitor is blocking the dc current, it still behaves like a capacitor storing a charge. If the output of the source is 5 V, the capacitor will build up a 5 V charge. When the source switches to 0 V, the capacitor is still charged to 5 V. This will act like a power supply at 5 V before it begins to dissipate the charge. Eventually the voltage on the capacitor will decrease to the source voltage. While this is happening, a large amount of current is being supplied. The opposite happens when the source transitions from 0 V to 5 V. Therefore, during the steady state the capacitor prevents current from flowing through the capacitor, but while the source is switching, the driver must source/sink a significant amount of current.

The power consumption is the same as the split termination when the source switches and decreases as the capacitor discharges. If the voltage source switches very quickly between high and low voltages, the capacitor will stay charged at a voltage halfway between the two voltages. The power consumption will be constant in this situation, but the difference between the voltage of the capacitor and the high/low voltage of the driver will be half of the voltage swing. This means that the power consumption will be the same as the single terminating resistor connected to a terminating voltage.

To achieve the best power consumption, the driver should maintain the voltage on the capacitor halfway between the high and low voltage. The driver should spend half of the time at each voltage level, so the data stream must ensure an equal number of 1s and 0s. This is called a dc-balanced data stream.

Differential Termination

Often logic devices transmit signals in differential mode. In this case, the voltages on the transmission lines will always be opposite of each other. These lines can be terminated using the previous methods. The general form of terminating differential transmission lines looks like Fig. 2.12.

If both lines are terminated using the Thevenin equivalent model, with one voltage source and terminating resistor, the configuration would look like Fig. 2.13(a). The transmission lines would be terminating individually. However, the terminating voltages would be the same for each signal, which means the voltage at the ends of each terminating resistor would be the same.

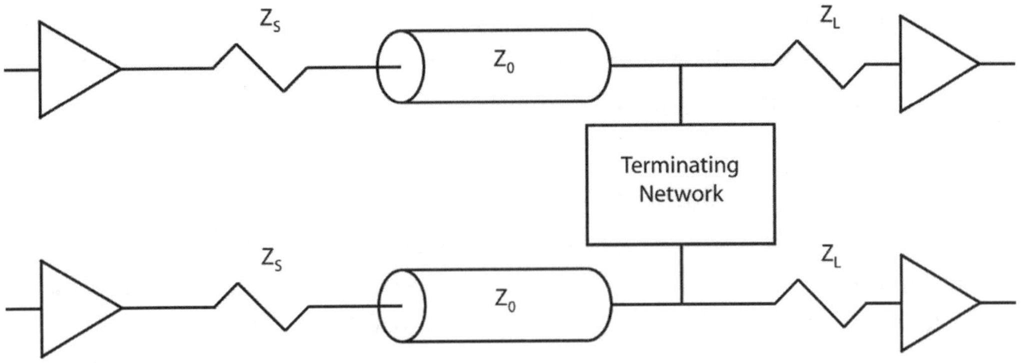

FIGURE 2.12: General differential termination configuration

This means that the ends of the resistors are effectively connected as long as the voltage at that point is the same as in Fig. 2.13(b).

One special feature of differential signals that can be used to improve the termination is the opposing voltages on each line. Since the voltages are always exactly opposite, the voltage across both terminating resistors is the entire voltage swing. If the resistors have the same value, the voltage between the two resistors will always be exactly halfway between the two voltages. This is also the terminating voltage. Therefore, if the voltage between the two resistors is the terminating voltage without the actual voltage supply, the voltage supply can be removed from the circuit entirely. The two terminating resistors can be combined into one resistor with a value of $2Z_0$ as in Fig. 2.13(c).

Any delay between the received differential signals, caused by differences in transmission line length or impedance discontinuities, can cause the center terminating voltage to vary from the ideal center. To help maintain the center voltage, a capacitor can be placed between the

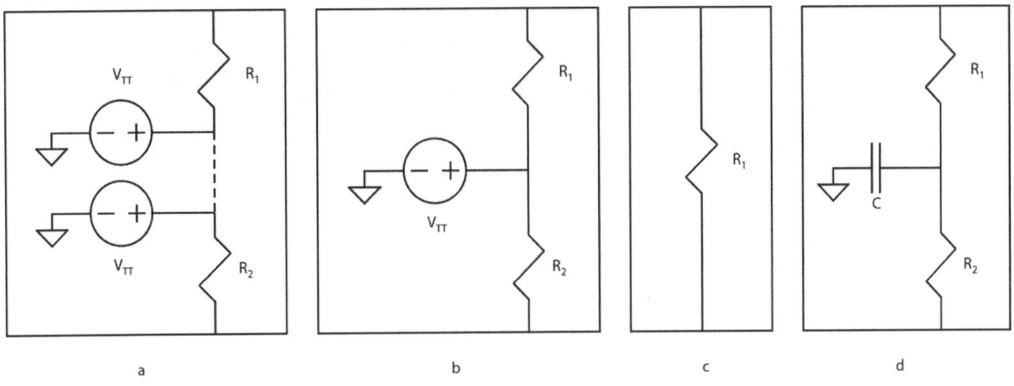

FIGURE 2.13: Differential termination options

resistors connected to ground. Once this capacitor charges, it will draw very little current because it will never discharge unless there is a delay between the differential voltage waves. Fig. 2.13(d) shows the termination with a capacitor.

Capacitive and Inductive compensation

Sometimes discontinuities are not only resistive, but capacitive and inductive. Stubs in a transmission line appear as capacitors, and vias can appear as either capacitors or inductors. Since a capacitive discontinuity will reflect a negative voltage, and an inductive discontinuity will reflect a positive voltage, the discontinuities can be compensated by creating the opposing discontinuity in the transmission line.

The width of the transmission line can be changed to provide extra capacitance or inductance. A thinner section of transmission line will provide extra inductance, while a thicker section will provide extra capacitance. Fig. 2.14 shows an example of compensating for a capacitive discontinuity. When increasing the capacitance, the widest transmission line possible should be used. When increasing the inductance, the thinnest transmission line possible should be used. The following ratio corresponds to these limits,

$$k = \frac{Z_A}{Z_0} \tag{2.80}$$

where Z_A is the impedance of the transmission line at the adjusted section in Ω.

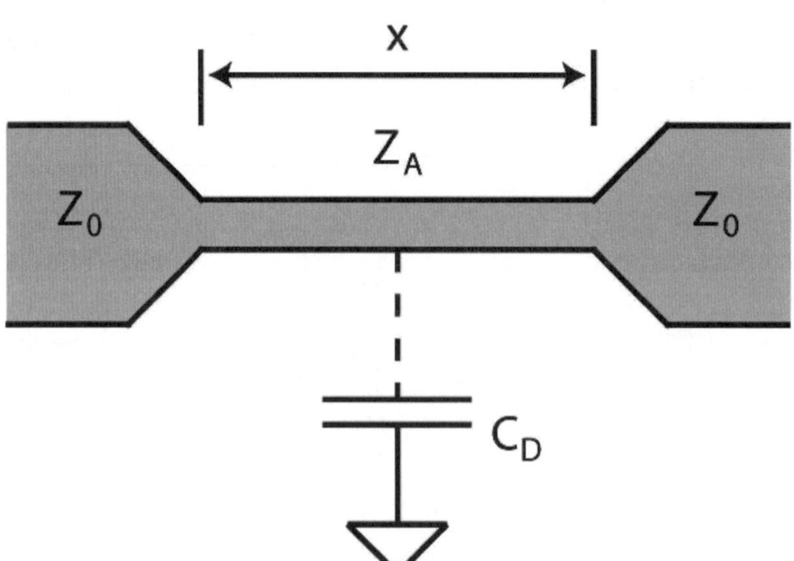

FIGURE 2.14: Compensating for a capacitive discontinuity

The characteristic impedance around the discontinuity must match the surrounding transmission line. Therefore, the equation for characteristic impedance is modified to

$$Z_0 = \sqrt{\frac{C_A + C_D}{L_A + L_D}} \tag{2.81}$$

where

Z_0 is the characteristic impedance of the surrounding trace in Ω;
C_A is the capacitance of the transmission line in the adjusted section in F;
C_D is the measured capacitance of the discontinuity in F;
L_A is the inductance of the transmission line in the adjusted section in H;
L_D is the measured inductance of the discontinuity in H.

Either C_D or L_D will be zero depending on the type of discontinuity. The capacitance and inductance of the adjusted trace is

$$C_A = \frac{x}{v}\frac{1}{Z_0 k} \tag{2.82}$$

$$L_A = \frac{x}{v} Z_0 k \tag{2.83}$$

where x is the length of the adjusted section of line in meters, and v is the velocity of the wave through the adjusted section in m s^{-1}.

The length of the adjusted line for a capacitive discontinuity depends on how small the width is. A thinner width will make a shorter adjusted section:

$$x = Z_0 C_D v \left(\frac{k}{k^2 - 1}\right). \tag{2.84}$$

The length of the adjusted line for an inductive discontinuity depends on how wide the width is. A wider line will make a shorter adjusted section:

$$x = \frac{L_D}{Z_0} v \left(\frac{k}{1 - k^2}\right). \tag{2.85}$$

This technique is a stop-gap type of fix for the discontinuity. It will only work if the effective delay of the adjusted segment is less than the rise time of the transmitted signal. The effective delay can be computed with equation

$$t_d = \frac{x}{v}. \tag{2.86}$$

As discussed in the previous chapter, vias have both parasitic capacitance and inductance. The impedance of a via is determined by the same balance as a transmission line,

$$Z_v = \sqrt{\frac{L_v}{C_v}}. \tag{2.87}$$

If the impedance of the via matches the impedance of the signal trace, then no reflection will occur. If the impedance of the via is larger, the inductance is too large. If the impedance of the via is smaller, the capacitance is too large. This excess capacitance or inductance can be used to solve for the adjusted trace width and length.

CHAPTER 3

Realistic Transmission Lines

The purpose of this chapter is to give a more practical model of a real transmission line and describe how to design traces to compensate for their drawbacks. This chapter assumes that the reader is familiar with analog components, simple circuit analysis, basic printed circuit board (PCB) design, digital circuits, differential signaling, and ideal transmission lines.

3.1 LEARNING OBJECTIVES

After reading this chapter, you will be able to perform the following tasks:

- Determine which model to use for a given length of PCB trace.
- Decide if a low-loss dielectric material should be used instead of FR-4.
- Use pre-emphasis and equalization techniques to counteract lossy transmission lines.
- Decide on a maximum trace length based on its attenuation at high frequency.

3.2 TELEGRAPHER'S EQUATIONS

The first long distance communication had problems with "high-speed" transmissions. Telegraph wires stretched across miles. The behavior of signals on these wires behaved strangely, so a model was created to understand this behavior. The result is the telegrapher's equations.

In a uniform transmission line, electric and magnetic fields are transverse to the direction of wave propagation; therefore, transmission line fields are called transverse electromagnetic (TEM) waves. The equations for these waves have two variables: time and location. The time-domain equations are

$$\frac{\partial v(z,t)}{\partial z} = -Ri(z,t) - L\frac{\partial i(z,t)}{\partial t} \tag{3.1}$$

$$\frac{\partial i(z,t)}{\partial z} = -Gv(z,t) - C\frac{\partial v(z,t)}{\partial t}. \tag{3.2}$$

These equations assume that any transmission line can be modeled as an infinite series of small independent elements. Ideal transmission lines are an infinite series of elements as well with one series inductor and one parallel capacitor. A more realistic model has a resistance in series with the inductor and an admittance in parallel with the capacitor. The resistance and admittance are functions of frequency which make the equation for characteristic impedance much more complex:

$$Z_C(\omega) = \sqrt{\frac{R(\omega) + \mathrm{j}\omega L(\omega)}{G(\omega) + \mathrm{j}\omega C(\omega)}}. \tag{3.3}$$

The resistance and admittance also introduce loss to the transmission line. This is a per-unit-length loss which is dependent on frequency. This is called a "lossy" model since the signal will attenuate as it passes through the transmission line. The ideal transmission line is called a lossless model because the resistance and admittance components are removed. Sometimes the characteristic impedance is written for only one frequency or a small range of frequencies. In this case, term Z_0 is used as in the following equation:

$$Z_0 = Z_C(\omega_0). \tag{3.4}$$

In some reference texts, the two terms, Z_C and Z_0, are used interchangeably, but there is a significant difference. Term Z_C refers to all frequencies, but this equation breaks down at very high frequencies because the effects within the transmission line can no longer be modeled by the telegrapher's equations.

Since the signal is attenuated by the transmission line, a second equation is needed to describe the loss. The attenuation factor $H(\omega, l)$ is called the *propagation function* which varies with frequency and length of wire. This describes how the signal is modified as it travels through a wire. It has a real and imaginary term which describes how the signal is delayed, and how the signal is attenuated. This is an exponential function, so the natural log of this function γ called the *propagation coefficient*:

$$H(\omega, l) = \mathrm{e}^{-l\cdot\gamma(\omega)} \tag{3.5}$$

$$\gamma(\omega) = \sqrt{[R + \mathrm{j}\omega L][G + \mathrm{j}\omega C]} = \alpha + \mathrm{j}\beta. \tag{3.6}$$

The real part of this equation α describes the attenuation per unit length, while the imaginary part β describes the phase shift per unit length.

Ideal transmission line equations are derived from the previous equations assuming that the resistance and admittance are zero. The equations become

$$Z_C = \sqrt{\frac{\mathrm{j}\omega L}{\mathrm{j}\omega C}} = \sqrt{\frac{L}{C}} \tag{3.7}$$

$$\gamma = \sqrt{(\mathrm{j}\omega L)(\mathrm{j}\omega C)} = \mathrm{j}\omega\sqrt{LC}. \tag{3.8}$$

The real part of the propagation coefficient, which describes the loss, is zero. The imaginary part is the phase delay, so the delay per unit length is $\sqrt{LC}$. This can be inverted to give the velocity of the wave

$$v_p = \frac{1}{\sqrt{LC}}, \tag{3.9}$$

which is the same equation given in the previous chapter.

For realistic transmission lines, the resistive component is not zero. Any wire will have at least some dc resistance. This is given in resistance per unit length. For PCB traces, this resistance depends on the thickness of the trace. This thickness is rated in plating weight, which is usually in ounces. This is the number of ounces of material deposited on a one foot square flat surface. One ounce plating is a thickness of 34.8 μm. For standard copper traces, the dc resistance of any trace can be found using the equation

$$R_{\mathrm{dc}} = \frac{0.00048}{w t_{\mathrm{oz}}} \tag{3.10}$$

where R_{dc} is the resistance across the length of the trace in $\Omega\ \mathrm{m}^{-1}$, w is the width of the line in meters, and t_{oz} is the plating weight of the line in oz. ft^{-2}.

The plating weight can be converted into thickness by using the equation

$$t_{\mathrm{th}} = 3.48 \times 10^{-5} t_{\mathrm{oz}} \qquad \text{(in meters)} \tag{3.11}$$

$$t_{\mathrm{th}} = 1.37 t_{\mathrm{oz}} \qquad \text{(in mils)}. \tag{3.12}$$

Example 3.1. I want to find out what plating was used on a circuit board I have on hand. Most of the traces on the PCB are very skinny, which makes them difficult to measure. So I use my RLC meter to find the dc resistance across the largest trace on the PCB. The trace width is 1 mm as accurately as I can measure it using calipers. I measure the resistance to be 0.99 $\Omega\ \mathrm{m}^{-1}$:

$$0.99 = \frac{0.00048}{0.001t} \tag{3.13}$$

$$t = \frac{0.00048}{(0.99)(0.001)} = 0.485. \tag{3.14}$$

Since PCB manufacturers usually only do plating in 1/2 oz., 1 oz., or 2 oz., I can predict that this board used 1/2 oz. copper.

The conductance of the traces on a PCB is usually very close to zero. This represents how easily electrons can pass between the trace and ground. Since there is a good insulator between them, no current passes. The only way a current could pass is with extremely high voltages, which hopefully none of your digital circuits will experience.

3.3 RC AND LC REGIONS

A major question every person asks is "at what frequency of my digital signals do I have to start worrying about my traces acting like transmission lines?" The problem with this question is the part about frequency. A digital signal ideally has an infinite rise and fall time. An infinite rise and fall time has an infinite frequency range. Therefore, for ideal digital signals, traces are always transmission lines. A digital clock with a frequency of 10 kHz is a slow clock, but if its transitions are infinitely fast, then the traces must be treated as transmission lines.

Fortunately, no digital signals are infinitely fast with the exception of those taught in the classroom. A digital signal has a finite rise and fall time which is usually measured from the 10% to the 90% level. While the rise and fall times may be the same, most material in these chapters refers only to rise time assuming it is the shorter of the two. At very slow rise times, no noticeable reflections are measured on the PCB traces, but reflections are occurring. The amplitude of those reflections may be so small that they cannot be detected. The reflections will dampen out before the signal can fully transition from a low to a high voltage. With an extremely long trace, a very slow rise time is needed for this to happen. Therefore, signal rise time and signal trace length are the two key factors in determining if a trace needs to be treated as a transmission line. An approximation is

$$l < \frac{\lambda}{10} \tag{3.15}$$

where l is the length of the trace, and λ is the wavelength of the signal.

When the length of the PCB traces is less than about one-tenth of the wavelength, reflections will be difficult or impossible to detect. This assumes the wavelength of the signal is switching full amplitude as fast as possible. The effective frequency of a signal based on its rise time is

$$f \approx \frac{0.35}{t_r}. \tag{3.16}$$

The wavelength depends on the velocity of the signal through the wire. As long as the dielectric constant of the board is known, the velocity is simple to compute:

$$\lambda = \frac{v_p}{f} = \frac{c}{f\sqrt{\varepsilon_r}}. \tag{3.17}$$

Combining the above equations will give the maximum trace length based on rise time and dielectric constant:

$$l < \frac{c\,t_r}{3.5\sqrt{\varepsilon_r}}. \tag{3.18}$$

This is a very rough estimate and the error can be significant, but it works well before the PCB design has begun. More accurate predictions require the trace geometries.

Lumped-Element Region

If the traces do not need to be treated as a transmission line, the traces must still be treated as a lumped-element circuit because a digital step can create a ringing with the right conditions. A PCB trace still has inductance and capacitance. A load may be placed on the circuit, which could cause a resonance at a specific frequency.

A better approximation for when the PCB traces can be treated as a lumped-element circuit is based on the resistance, capacitance, and inductance of the trace. A few equations determine how well a trace can be modeled as a lumped-element circuit. The following two equations specify the maximum operating frequency given a specific PCB length,

$$\omega_{\mathrm{LE}} < \left(\frac{\Delta}{l}\right)^2 \frac{1}{RC} \qquad \text{when } l > \frac{\Delta}{R}\sqrt{\frac{L}{C}} \tag{3.19}$$

$$\omega_{\mathrm{LE}} < \left(\frac{\Delta}{l}\right) \frac{1}{\sqrt{LC}} \qquad \text{when } l < \frac{\Delta}{R}\sqrt{\frac{L}{C}} \tag{3.20}$$

where

ω_{LE}	is the maximum operating frequency in rad s^{-1};
R	is the series dc resistance of the trace in $\Omega\ m^{-1}$;
C	is the parallel capacitance of the trace in $F\ m^{-1}$;
L	is the series inductance of the trace in $H\ m^{-1}$;
l	is the length of the PCB trace in meters;
Δ	is the constant usually equal to about 0.25.

Example 3.2. Can I treat my printed circuit board traces as simple lumped elements? My logic devices have a rise time of 1 ns. I will start with the same PCB dimensions as in Example 2.1. The characteristic impedance is 53.61 Ω, the capacitance is 2.64 pF in.^{-1}, and the inductance is 7.587 nH in.^{-1}. First I will convert these to metric units:

$$C_0 = \left(2.64\ \frac{\text{pF}}{\text{in.}}\right)\left(39.37\ \frac{\text{in.}}{\text{m}}\right) = 104\ \frac{\text{pF}}{\text{m}} \tag{3.21}$$

$$L_0 = \left(7.587\ \frac{\text{nH}}{\text{in.}}\right)\left(39.37\ \frac{\text{in.}}{\text{m}}\right) = 298.7\ \frac{\text{nH}}{\text{m}}. \tag{3.22}$$

Next I can find the dc resistance of my traces, but I will also need to convert my 1.37-mil-thick traces in ounces. The width of my traces is 15 mil, so I will also convert it to the metric unit:

$$w = (15 \text{ mil}) \left(2.54 \times 10^{-5} \frac{\text{m}}{\text{mil}}\right) = 0.000381 \text{ m} \tag{3.23}$$

$$t_{oz} = (1.37 \text{ mil}) \left(\frac{1}{1.37} \frac{\text{oz.}}{\text{mil}}\right) = 1 \text{ oz.} \tag{3.24}$$

$$R_{dc} = \frac{0.00048}{(0.000381)(1)} \tag{3.25}$$

$$R_{dc} = 1.26\Omega \, \text{m}^{-1}. \tag{3.26}$$

The length of the longest trace on my circuit board is 15 cm, or 0.15 m. First I have to determine which equation to use:

$$0.15 < \frac{0.25}{1.26}\sqrt{\frac{298.7 \times 10^{-9}}{104 \times 10^{-12}}} \tag{3.27}$$

$$0.15 < (0.198)\sqrt{2872} \tag{3.28}$$

$$0.15 < 10.63. \tag{3.29}$$

So I must use the second equation to find the maximum operating frequency:

$$\omega_{LE} < \left(\frac{0.25}{0.15}\right) \frac{1}{\sqrt{(298.7 \times 10^{-9})(104 \times 10^{-12})}} \tag{3.30}$$

$$\omega_{LE} < 2.99 \times 10^{8} \tag{3.31}$$

$$f_{LE} < \frac{\omega_{LE}}{2\pi} \tag{3.32}$$

$$f_{LE} < 47.6 \, \text{MHz}. \tag{3.33}$$

This relates to a signal rise time of

$$t_r > \frac{0.35}{47.6 \times 10^{6}} \tag{3.34}$$

$$t_r > 7.35 \, \text{ns}. \tag{3.35}$$

The predicted maximum trace length using the formula from the previous section is

$$l < \frac{(7.35 \times 10^{-9})(3 \times 10^{8})}{3.5\sqrt{4.5}} \tag{3.36}$$

$$l < 29.7 \, \text{cm}. \tag{3.37}$$

This predicted that trace length is almost double the 15 cm trace used in this example. Therefore the error associated with the previous section is significant.

If the PCB traces qualify as lumped-element circuits, then two conditions must be met to ensure that the trace does not cause any change in the signal quality. First, the source impedance of the driver must be much smaller than the capacitance of the trace. The ideal source impedance is zero, so with many drivers the trace impedance will be much larger:

$$|Z_S| \ll \left|\frac{1}{l \times \mathrm{j}\omega C}\right|. \tag{3.38}$$

Second, the load impedance must be much greater than the series impedance of the trace:

$$|Z_L| \gg \left|l \times (R + \mathrm{j}\omega L)\right|. \tag{3.39}$$

If either of these conditions does not hold true, then the signal can resonate. Even very short traces can cause the signal to ring given certain source and load impedances.

RC Region

If a transmission line cannot be modeled as a lumped-element circuit, then a number of models exist to describe how the characteristic impedance changes over frequency. The characteristic impedance has a different model within certain frequency ranges. The equation for the real characteristic impedance can be modified to remove the admittance since it is very close to zero:

$$Z_C(\omega) = \sqrt{\frac{R(\omega) + \mathrm{j}L(\omega)}{\mathrm{j}\omega C}} = \sqrt{\frac{L(\omega)}{C}}\sqrt{1 - \mathrm{j}\frac{R(\omega)}{\omega L(\omega)}}. \tag{3.40}$$

At low frequencies, the inductance is much smaller than the dc resistance, $R \gg \omega L$, and therefore the inductance can be ignored. Since the characteristic impedance only depends on the resistance and capacitance in this range, it is called the RC region. As the operating frequency increases, the inductance will eventually exceed the resistance. This frequency defines the transition into the LC region. The border between these two regions is defined as

$$\omega_{\mathrm{LC}} = \frac{R_{\mathrm{dc}}}{L} \tag{3.41}$$

where ω_{LC} is the frequency which defines the border between the RC and LC regions, R_{dc} is the dc series resistance in $\Omega\ \mathrm{m}^{-1}$, and L is the series inductance in $\mathrm{H\ m}^{-1}$.

For PCB designers, the RC region is almost never encountered because the length of traces required to be in this range is longer than the largest imaginable PCB. Given the previous example,

$$\omega_{\mathrm{LC}} = \frac{1.26}{298.7 \times 10^{-9}} = 4.21\,\mathrm{MHz}. \tag{3.42}$$

The longest PCB trace, which will not qualify for the lumped-element region at this frequency is over 10 m. These lengths are only encountered in cabling between systems, so designing for them is still needed. The characteristic impedance in this region is

$$Z_{\mathrm{RC}} \approx \frac{1-\mathrm{j}}{\sqrt{2}}\sqrt{\frac{R(\omega)}{\omega C}}. \tag{3.43}$$

Note that the real and imaginary parts have the same magnitude. This equation is only a rough approximation because the inductance increases with frequency. Attenuation in this region varies with the square root of frequency. In other words, the speed of operation varies inversely with the square of the length of the wire. A tradeoff must be made between length of the wires and operating frequency. Terminating this transmission line will be difficult because of its frequency dependence.

LC Region

Above frequency ω_{LC} the characteristic impedance behaves differently since the inductance factor has increased to approach the value of the dc resistance. This region is easier to design a termination for because the attenuation does not vary significantly with frequency. The characteristic impedance for this region is

$$Z_C = \sqrt{\frac{R + \mathrm{j}\omega L}{\mathrm{j}\omega C}} \tag{3.44}$$

$$Z_C = \sqrt{\frac{L}{C}}\left(1 - \mathrm{j}\frac{1}{2}\frac{R(\omega)}{\omega L(\omega)}\right). \tag{3.45}$$

The real and imaginary terms do not always have the same magnitude. Since term $R(\omega)$ is proportional to the square root of frequency, when the frequency increases far above ω_{LC}, the dc resistance becomes negligible, which makes this equation predominantly real. This reduces the equation to the ideal form of characteristic impedance:

$$Z_0 = \sqrt{\frac{L}{C}}. \tag{3.46}$$

This equation has less than 5% error when the frequency is 10 times above ω_{LC}.

The propagation coefficient changes in the LC region as well. In the RC region, the propagation coefficient has the same magnitude for its real and imaginary parts. The real part represents the attenuation, while the imaginary part represents the phase shift. In the LC region, the attenuation is constant, while the phase shift increases with frequency. This is why the LC region is sometimes referred to as the constant-loss region:

$$\gamma(\omega) = \mathrm{j}\omega\sqrt{LC}\sqrt{1 + \frac{R_{\mathrm{dc}}}{\mathrm{j}\omega L}}. \tag{3.47}$$

At frequencies far above ω_{LC}, this equation can be approximated as

$$\gamma(\omega) = \mathrm{j}\omega\sqrt{LC}\left(1 + \frac{1}{2}\frac{R_{\mathrm{dc}}}{\mathrm{j}\omega L}\right) \tag{3.48}$$

$$\gamma(\omega) = \mathrm{j}\omega\sqrt{LC} + \frac{1}{2}\frac{R_{\mathrm{dc}}}{\sqrt{LC}} \tag{3.49}$$

$$\gamma(\omega) = \mathrm{j}\omega\sqrt{LC} + \frac{1}{2}\frac{R_{\mathrm{dc}}}{Z_0}. \tag{3.50}$$

The imaginary part approaches $\omega\sqrt{LC}$, which is the ideal form of the propagation coefficient. The real part remains constant at

$$\alpha = \frac{R_{\mathrm{dc}}}{2Z_0}. \tag{3.51}$$

For a transmission line with a characteristic impedance of 50 Ω, it will have an attenuation lower than the dc resistance by a factor of 100. Terminating a transmission line in the LC region uses the same techniques as an ideal transmission line. The most effective method is the end termination because it is least sensitive to the dc resistance of the transmission line.

3.4 SKIN EFFECT

As stated many times so far, every simple wire has a parasitic inductance, capacitance, and resistance. The inductance becomes a problem when more current is passing through the wire. Higher frequencies means that the current is moving back and forth along the wire at higher speeds. At very high frequencies, the wire stops acting like a uniform inductor. The magnetic field which forms because of the inductance starts to effect how the electrons are moving through the wire.

Much like how a changing current produces a magnetic field, a changing magnetic field can produce a current. The magnetic field surrounds the wire, but also penetrates the wire. Fig. 3.1(a) shows how a magnetic field is produced around a wire. These magnetic field lines circle around the wire according to the right-hand rule. This magnetic field will reverse direction when the current is reversed.

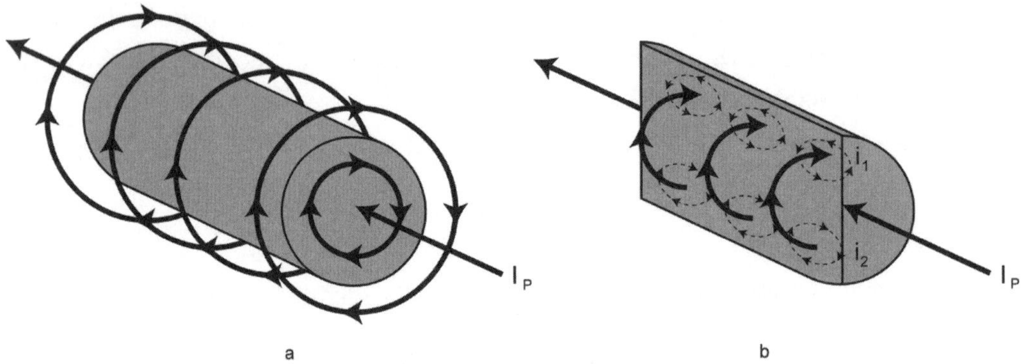

FIGURE 3.1: Magnetic field effects through a wire

Fig. 3.1(b) shows how the magnetic field can produce smaller currents within the wire. These smaller currents are called eddy currents. The eddy currents circle around the magnetic field lines in the figure. Near the surface of the wire, the eddy currents i_1 and i_2 flow with the direction of the primary current I_P. In the middle of the wire, the eddy currents i_1 and i_2 flow against the direction of the primary current I_P, which will tend to cancel it out. In reality, the magnetic field will be uniform, and the eddy currents will occur in all places at once. As the amplitude of I_P increases, the magnetic fields increase, which increase the eddy currents. The eddy currents in the center of the wire are flowing opposite of the original current. The end result is that the current in the center of the wire approaches zero while the current around the outside of the wire approaches I_P. This means that the current is only flowing through a small section of the wire. This is called the *skin effect* since the current is only flowing around the "skin" of the wire.

The problem with the skin effect is the resistance of the wire. The cross-section of a wire has a fixed resistance given a specific material. Specifically,

$$R_{dc} = \frac{k_a}{\sigma A} \tag{3.52}$$

where

R_{dc} is the low-frequency resistance of the wire in $\Omega\ m^{-1}$;
σ is the conductivity of the wire in $S\ m^{-1}$;
A is the cross-sectional area of the wire in m^2;
k_a is a constant dependent on the return path of the current (for PCB traces with a ground plane, this is about 1).

As this formula indicates, the overall resistance of a wire decreases as the cross-sectional area through which current flows increases. This is similar to running a current through two resistors of the same value instead of one: the value of the resistance is halved. If current is moving through a smaller area because of the skin effect, the resistance will increase. Therefore, the resistance of a wire increases with the amount of changing current flowing through it, and therefore, with frequency.

This effect is only noticeable above a specific frequency for a given cross-section of wire. At this frequency, the current only flows through the wire at a certain depth. The depth at which the skin effect occurs is

$$\delta = \frac{1}{\sqrt{\pi f \mu \sigma}} \tag{3.53}$$

where

- δ is the skin depth which the current density decays to $1/e$ (about 0.37) in meters;
- f is the frequency of operation in Hz;
- μ is the absolute magnetic permeability of the wire in H m^{-1};
- σ is the conductance of the wire in Ω m^{-1}.

If the wire has a smaller radius than the skin depth, the current will flow through the entire wire.

When the skin depth is very small, the resistance of the wire depends significantly on the outer geometry of the wire. The area through which the current passes is the perimeter of the wire times the skin depth. This area can be substituted into the low-frequency equation to find the high-frequency resistance,

$$R_{\text{ac}} = \frac{k_p k_r}{p \delta \sigma} \tag{3.54}$$

where

- R_{ac} is the high-frequency resistance in Ω m^{-1};
- p is the perimeter of the wire in meters;
- δ is the skin depth in meters;
- σ is the conductance of the wire in Ω m^{-1};
- k_p is the correction factor based on the proximity effect discussed in the next section;
- k_r is the correction factor based on the roughness effect discussed in the next section.

The geometries of two different types of wires, round and rectangular, are shown in Fig. 3.2. The perimeter of a circle is $2\pi r$, so the area through which the current would flow is $2\pi \delta r$. The perimeter of a rectangle is $2(w + t)$, so the area through which the current would flow is $2\delta(w + t)$.

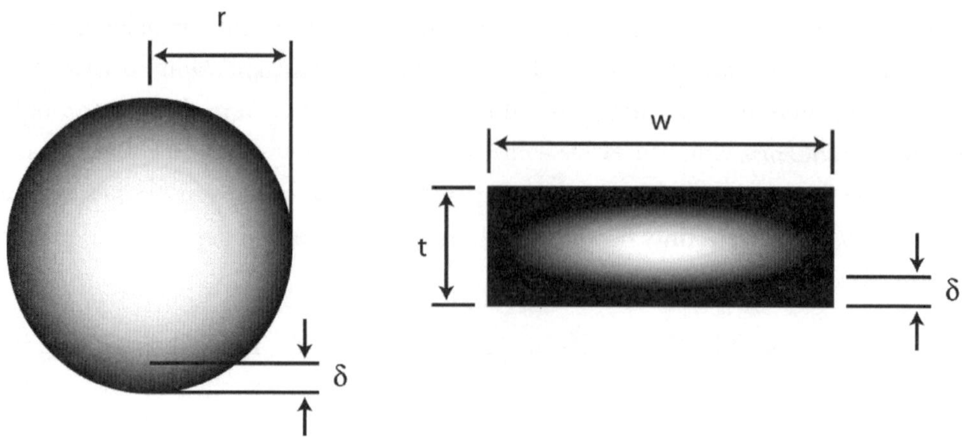

FIGURE 3.2: Skin depth of round and rectangular wires

Substituting the equation for skin depth into the high-frequency resistance equation gives

$$R_{ac} = \frac{k_p k_r \sqrt{\pi f \mu}}{p\sqrt{\sigma}}. \qquad (3.55)$$

This equation demonstrates how the resistance varies proportionally to the square root of frequency. This is only an approximation because the current is not uniform throughout the skin depth.

The low-frequency and high-frequency resistance equations coincide at a specific frequency. The intersection can be defined by the equation

$$f_\delta = \frac{1}{\pi \mu \sigma}\left(\frac{k_a p}{k_p A}\right)^2 \qquad (3.56)$$

where

f_δ is the frequency which marks the onset of the skin effect in Hz;
p is the perimeter of the wire in meters;
μ is the absolute magnetic permeability of the wire in H m^{-1};
σ is the conductance of the wire in $\Omega\ m^{-1}$;
k_p is the correction factor based on the proximity effect discussed in the next section;
k_a is a constant dependent on the return path of the current (for PCB traces with a ground plane, this is about 1).

For a trace on a printed circuit board, the perimeter of a rectangle can be substituted giving the equation

$$f_\delta = \frac{4}{\pi \mu \sigma} \left(\frac{w+t}{wt} \right)^2 \left(\frac{k_a}{k_p} \right)^2 . \tag{3.57}$$

If the width of the trace is large compared to the thickness, then this equation can be significantly simplified to

$$f_\delta = \frac{4}{\pi \mu \sigma t^2} \left(\frac{k_a}{k_p} \right)^2 . \tag{3.58}$$

This frequency occurs above the LC region. For PBCs, the onset often occurs between 10 and 100 MHz. The LC region is therefore usually very small since the onset of the LC region is often around 5 MHz.

The increase in resistance will affect the characteristic impedance and propagation coefficient. The resistance increases with the square root of frequency, but the inductance is increasing directly proportional to frequency. Since the resistance at the skin effect onset region is already small, the resistive term in the characteristic impedance can still be disregarded at frequencies well into the skin effect region. This means that the characteristic impedance will be the same as in the LC region. Terminations in the skin effect region are the same as in the LC region:

$$Z_C = \sqrt{\frac{L}{C}} . \tag{3.59}$$

The propagation coefficient is not constant in the skin effect region. In the RC region, the real and imaginary parts increase with the square root of frequency. In the LC region, the real part starts to approach a constant while the imaginary part increases linearly with frequency. In the skin effect region, the imaginary part continues to be linear with frequency, but the real part increases with the square root of frequency. Since the LC region is so small, the real part does not normally have enough time to stabilize at a constant level before the skin effect region starts. The equation for the propagation coefficient can be reduced to

$$\gamma(\omega) = \sqrt{(R_{ac} + j\omega L)(j\omega C)} . \tag{3.60}$$

If $R_{ac} \ll \omega L$, the square root can be changed to

$$\gamma(\omega) = j\omega\sqrt{LC} + \frac{1}{2}\frac{R_{ac}}{Z_0} . \tag{3.61}$$

The imaginary part of this equation is the phase delay which is the same as the LC region delay. The bulk propagation delay is $\sqrt{LC}$ in seconds per meter. This overall delay is dependent on the length of the wire. Doubling the length of the wire doubles the delay. The real part of this equation can be transformed to show its frequency dependence. For the ac resistance, a frequency ω_0 is chosen well above the skin effect onset frequency. The value of R_0 represents the real part of the skin effect impedance at that frequency. The real part of the propagation coefficient, also the attenuation in the skin effect region, is defined by the equation

$$\alpha_r = 4.34\frac{R_0}{Z_0}\sqrt{\frac{\omega}{\omega_0}} \tag{3.62}$$

where

α_r is the skin effect loss coefficient in dB m^{-1};
ω_0 is an arbitrary frequency well above the onset of the skin effect region;
R_0 is the computed value of the ac resistance at ω_0 in $\Omega\ m^{-1}$;
Z_0 is the characteristic impedance at ω_0 in $\Omega\ m^{-1}$.

This equation shows how the attenuation varies with the square root of frequency. The coefficient implies a low-pass filter propagation function in dB m^{-1} of the form

$$|H(\omega, l)| = e^{-l \cdot 4,34\frac{R_0}{Z_0}\sqrt{\frac{\omega}{\omega_0}}}. \tag{3.63}$$

The transfer gain varies in proportion to the length of the wire and the square root of frequency. Doubling the distance doubles the loss in dB. Doubling the frequency multiplies the loss by $\sqrt{2}$. Loss of more than 3 dB can cause significant errors in a digital transmission.

The major drawback of the skin effect region is how it modifies the step response of a signal being sent through the wire. Since the transfer function looks like a low-pass filter, the step response will look like a curve which rises quickly, but does not reach its maximum value for a very long time. This can cause problems when quickly switching between the high and low states. A system may perform well as long as it is quickly switching data, such as when the data are dc balanced. When the data settles at a high or low for a long time, when it begins switching again it may encounter an error.

Surface Roughness

When the operating frequency is well beyond the skin effect onset frequency, the current is flowing through a very small band around the perimeter of the wire. So far, only perfect geometric structures can be used. In reality, the wires are not so perfect. Small imperfections can be found on the surface of the wire. This can occur from many sources in the PCB manufacturing process. The copper layers may be purposefully etched to facilitate adhesion to the core and prepreg layers (called toothing profiles). The layers may be mechanically pressed together which can leave indentation in the metal. These imperfections occur on the microscopic level. They are also difficult to predict and therefore difficult to model, but the worst case can be identified

FIGURE 3.3: Worst-case surface roughness

which gives an upper bound to the surface roughness effect. Since the current is flowing only very near the surface, the current will bend around these imperfections.

The worst case of surface roughness is bands of steep mountains on the surface of the wire. In Fig. 3.3, the low-frequency current would normally flow beneath the ridges and through the central part of the wire. The current would be moving in a straight line through this section of wire which is $4\times$ lengths. At high frequency, the current is moving along the surface of the wire and following the contours of the mountains. The distance the signal must travel is doubled to $8\times$ and, therefore, the total resistance is increased.

Surface roughness is measured by the room-mean-squared (RMS) height of the surface bumps. If the skin depth decreases to less than the RMS height, then the current begins to follow the surface contours. Surface roughness can increase series resistance 10% to 50%. The surface roughness can be estimated for a given material and process in constant k_r.

A number of polishing options are available to minimize the RMS height of the surface roughness. The inside layers are the most difficult to control roughness. The outer layers, since they are exposed, can be more easily modified. From worst to best are the reverse-treat foil process, sulfuric peroxide treatments, oxide treatments, and double-treat process. While none of these creates a perfectly smooth surface, they can minimize the surface roughness.

Proximity Effect

The skin effect causes high-frequency current to only flow around the outer edge of the transmission line. The changing magnetic fields on the outside of the wire tend to distribute this current nonuniformly around the perimeter. The current adjusts to minimize the inductance between the transmission line and the current return path. The currents are pulled toward each other inside the wire. This is called the *proximity effect*.

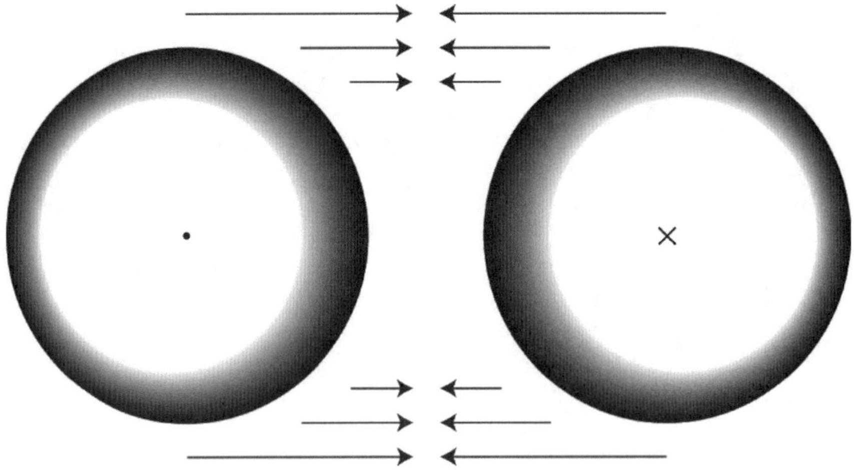

FIGURE 3.4: Proximity effect on a transmission line

The high-frequency current in one wire creates a changing magnetic field. This magnetic field interacts with the second wire by creating eddy currents. These eddy currents are stronger on the side closer to the first wire; therefore, the current density near the first wire is higher than on the opposite side. Fig. 3.4 shows how the current is distributed between two wires.

The proximity effect is different from the skin effect, but both have similar causes. Magnetic fields cannot penetrate a conductor, but they cause current to flow within the conductor. With the skin effect, the magnetic field caused by its own currents push the current to the edge of the conductor. With the proximity effect, magnetic fields from an external source (namely the return current flowing in a nearby wire) push the currents to the edge of the conductor. The proximity effect only matters when the current is already flowing near the surface of the conductor. At low frequencies, there is no skin effect, so the proximity effect does not matter. Also, the magnetic fields at low frequency are not strong enough to measurably affect the current flow. The frequency at which the proximity effect starts to matter is the same frequency at which the skin effect starts to matter.

The proximity effect increases the ac resistance above what the skin effect alone would cause. Constant k_p is used to signify the adjustment which needs to be made to the skin effect computation. This constant is dependent on a number of factors. First, if the current and the return current paths are not close together, the proximity effect is negligible ($k_p = 1$). As the current paths are moved closer together, the constant increases. For round wires, the constant is dependent on the ratio of the separation of the wires to the wire diameter, s/d. The constant approaches 2 as this ratio increases.

The proximity effect also takes place in a ground plane which returns the current on a PCB. For low-frequency currents, the return current will follow the path of least resistance. On a ground plane, the current will spread out as much as possible across the ground plane. For high-frequency currents, the return current will follow the path of least inductance. The current will flow directly beneath the trace to minimize the inductance. As frequency increases, the current is pulled from over the entire ground plane to a narrow band. This is because current in a conductor at high frequencies distributes itself to minimize the internal magnetic forces. This is similar to why slots in the ground plane increase the inductance of the trace. The return current cannot flow directly underneath the trace, but must flow around the ground slot.

The specific values for the proximity effect can be calculated using a field solver. For microstrip traces, the value of k_p is usually between 1.9 and 1.5. For stripline traces, the value of k_p is usually between 1.7 and 1.5. As the height of the trace over the ground plane decreases, the constant decreases. As the width of the trace decreases, the constant decreases. Since signal quality is best with minimum ac resistance, lower constants are better. Therefore, small, stripline traces very close to the ground planes minimize the proximity effect.

3.5 DIELECTRIC LOSSES

As the frequency of the signals passing through a transmission line increases well beyond the skin effect region, another effect begins to take place. Ceramic materials absorb some electromagnetic power. This power is turned into heat. For example, a typical capacitor is charged up to a specific voltage and then removed from the circuit completely. In theory, this voltage will stay on the capacitor indefinitely until a load is placed on the capacitor leads to discharge it. In practice, if this experiment is performed with a typical off-the-shelf capacitor, the capacitor will eventually lose all of its charge even without a load applied to it. The capacitor industry calls this effect the *dissipation factor* and is based on the relative permittivity of the insulating material.

The permittivity for any insulating material is measured as a ratio of two capacitances. The geometries of both capacitors are exactly the same. The first capacitor uses the insulating material to separate the two capacitor plates. The second capacitor has a perfect vacuum between the two plates. The capacitance is measured for both and the ratio of the two capacitances equals the relative permittivity of the insulating material. The insulating material increases the effective capacitance. The permittivity of vacuum is 1, and the permittivity of any other material is greater than 1.

The same effect in the PCB industry is called the *dielectric loss tangent*. The relative permittivity of a material is a complex number. The real part of the permittivity is called the *dielectric constant*. The dielectric loss tangent, or sometimes simply the loss tangent, is the ratio of the real and imaginary parts of the permittivity:

$$\tan\theta = \frac{-\mathrm{Im}(\varepsilon)}{\mathrm{Re}(\varepsilon)}. \tag{3.64}$$

The dielectric constant is not quite a constant: it varies with frequency. For all calculations which require the dielectric constant, this is an important fact to keep in mind. One of the most important equations which this affects is the propagation velocity of a signal through a transmission line:

$$v_p = \frac{c}{\sqrt{\varepsilon_r}}. \tag{3.65}$$

This equation implies a specific speed at which a signal passes through a transmission line. If the dielectric constant varies with frequency, then the velocity of the signal varies with frequency. This can cause a significant problem for the signal quality. An ideal unit step incident on a transmission line has all frequencies. The frequencies spread out across the transmission line since they all travel at different speeds. This is called *dispersion*. The farther the pulse travels, the more dispersed and distorted it becomes.

Since the velocity of signals depends on their frequency, the dielectric loss in a transmission line scales in proportion to both frequency and length. The dielectric loss is very small at low frequencies. These losses become noticeable when they rise as high as the resistive losses of the skin effect. While the losses are low, they increase in direct proportion to frequency. Since the skin effect increases with the square root of frequency, the dielectric losses will eventually exceed the skin effect losses.

For PCBs using a dielectric such as FR-4, the onset of the dielectric loss region begins in the mid-MHz range. For PCBs which will operate above 500 MHz, a different dielectric material should be used to minimize the dielectric losses and increase the onset frequency into the multi-GHz range. To compute the specific onset frequency of the dielectric loss region, an arbitrary frequency is chosen to compute the ac resistance, the characteristic impedance, the velocity of propagation, and the loss tangent. These factors determine the onset frequency

$$\omega_\theta = \frac{1}{\omega_0}\left(\frac{v_0}{Z_0}\frac{R_0}{\theta_0}\right)^2 \tag{3.66}$$

where

ω_θ	is the onset frequency of the dielectric loss region;
ω_0	is an arbitrary frequency chosen to compute the remaining variables;
v_0	is the velocity of propagation at ω_0 in m s^{-1};
Z_0	is the characteristic impedance at ω_0 in Ω;
R_0	is the series ac resistance at ω_0 in Ω m^{-1};
$\tan\theta$	is the loss tangent of the dielectric material at ω_0.

The characteristic impedance in the dielectric loss region behaves similarly to the skin effect and LC regions. The capacitance is relative to the frequency since it is dependent on the dielectric constant. The dielectric loss increases the capacitance with frequency. At the onset frequency of the dielectric loss region, the characteristic impedance increases slightly with frequency, but the termination method follows the same rules as the LC region.

The imaginary term for the propagation coefficient is the phase delay which was already stated to vary with frequency. This is because the capacitance varies with frequency. The delay also varies with the transmission line length. The delay varies much more with length than frequency.

The real term for the propagation coefficient is the attenuation which follows the equation

$$\alpha_d = 4.34\frac{\theta_0 \omega}{v_0}\left(\frac{\omega}{\omega_0}\right)^{-\theta_0/\pi}. \tag{3.67}$$

The overall signal loss is then represented by the transfer function in dB m^{-1},

$$|H(\omega, l)| = \mathrm{e}^{-l \cdot 4.34\frac{\theta_0 \omega}{v_0}\left(\frac{\omega}{\omega_0}\right)^{-\theta_0/\pi}}. \tag{3.68}$$

From this equation, the signal loss varies in proportion to the length of the line and to the square root of frequency. Doubling the length doubles the loss. Doubling the frequency increases the loss by $\sqrt{2}$. Digital signals will encounter errors at about 3 dB. Fig. 3.5 shows the frequency response of a variety of stripline trace lengths. The traces are designed with 50 Ω characteristic impedance, 6 mil trace width, and ½ oz. of copper weighting.

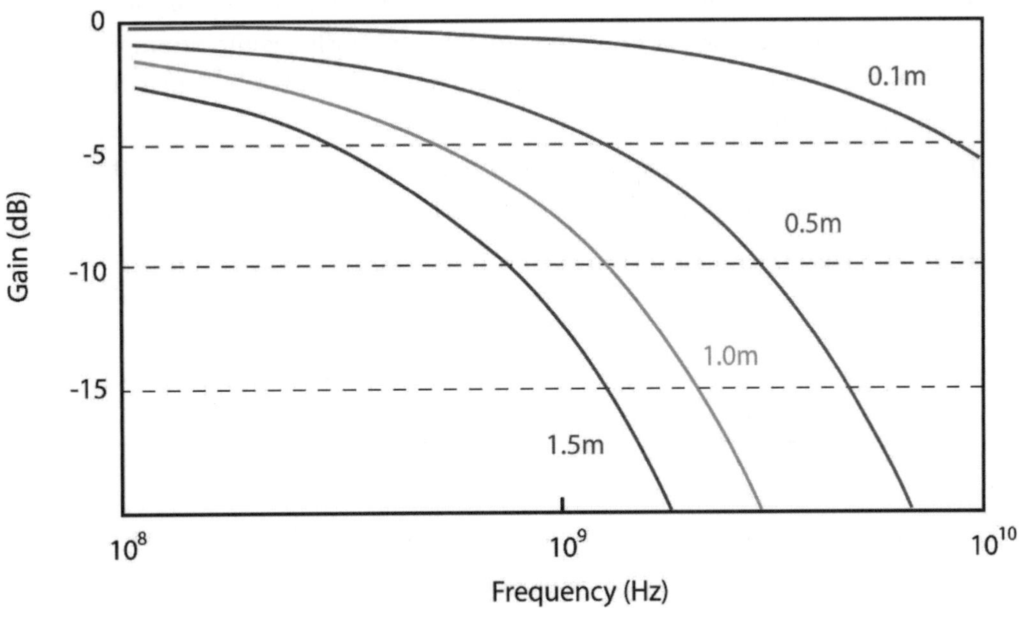

FIGURE 3.5: Transmission line attenuation at high frequencies

The dielectric loss region represents the behavior of the transmission line until extremely high frequencies (~140 GHz). These frequencies are outside the range of normal digital signals as of today. Sometimes these frequencies occur in RF applications. The transmission line theory begins to break down, and a whole new set of rules starts to apply to signals. This region is called the waveguide dispersion region and will not be covered in this book.

3.6 COMPENSATING TECHNIQUES

The skin effect and dielectric losses cause significant degradation of signal transmitted across PCBs. These losses act like a low-pass filter on the transmission line; therefore, each line has an associated bandwidth. After the onset frequency of the skin effect region, the gain of the higher frequency signals begins to decrease quickly. A sharp pulse generated at the source will take a long time to reach its maximum voltage at the load. Sometimes the pulse width is so short that the voltage received does not cross the receiver's threshold to register a bit transition. In this case, the receiver does not detect the bit transitions.

Fig. 3.6 shows the response of a signal which is suffering degradation due to skin effect and dielectric losses. The received signal is slowed down so that it barely passes the receiver's threshold for detecting a transition. The received pulse is sometimes called a "runt pulse." For normal binary communication, the amplitude of the runt pulse should never be below 70% of the maximum amplitude.

The data stream represented in Fig. 3.6 is a long series of 0s, followed by a single 1, followed by another long series of 0s. This represents the worst case for the losses in the transmission lines. The best case is when the signal is constantly toggling from a 0 to a 1. The signal will never reach the maximum or minimum amplitude, but will bounce back and forth across the receiver threshold.

In a lossy line, the time where the receiver detects the transition from a 0 to a 1 occurs slightly after the intended time. This is called jitter. There are many different types of jitter, and

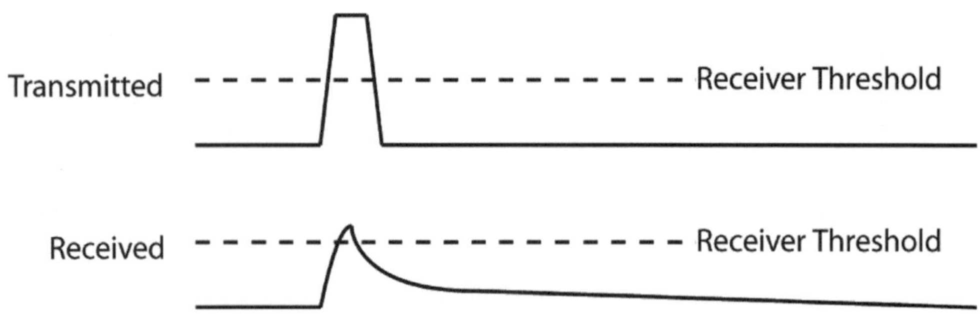

FIGURE 3.6: Effects of lossy transmission lines on transmitted pulse

they are defined by their cause. The type of jitter caused by the effects of a lossy line is called intersymbol interference (ISI), which is a subtype of data-dependent jitter.

The ISI jitter is minimized by having a dc-balanced data stream. An ideal data stream will have a constant 0101 pattern, but then will not be able to transmit any meaningful data. Typically, a run of no more than four or five consecutive 1s or 0s is allowed in a high-speed data stream. Some methods to ensure this include bit stuffing or encoding of the data stream. If more than five consecutive 1s or 0s is sent, then the next bit may not be received correctly. When a data stream does not need to send any data, it will usually constantly toggle the bits to ensure that the voltage does not settle at the maximum or minimum. If the transmission line ever does settle to the maximum or minimum, such as during the startup of the system, a certain amount of time must pass while sending the 0101 pattern to ensure no errors due to ISI.

Transmitter Pre-emphasis

Since a lossy transmission line acts like a low-pass filter, the low-pass effect can be canceled out by increasing the gain of the frequencies which get attenuated. This is called equalization. This is very similar to an audio equalizer to increase or decrease the volume of certain frequencies. Since the high frequencies are being attenuated, if the gain of the transmitted signal increases for the high frequencies, the response measured at the receiver end will be flat.

The first type of equalization employed by logic devices is called transmitter pre-emphasis. Fig. 3.7 shows a simple binary waveform $x[n]$ and its first difference waveform $x[n] - x[n-1]$. The first difference computation is similar to the derivative in calculus. The difference waveform shows how the original waveform changes over time. At every transition in $x[n]$, the difference waveform has a transition either higher or lower. Note that the difference waveform is not a binary signal because there are more than two logic states. The pre-emphasis

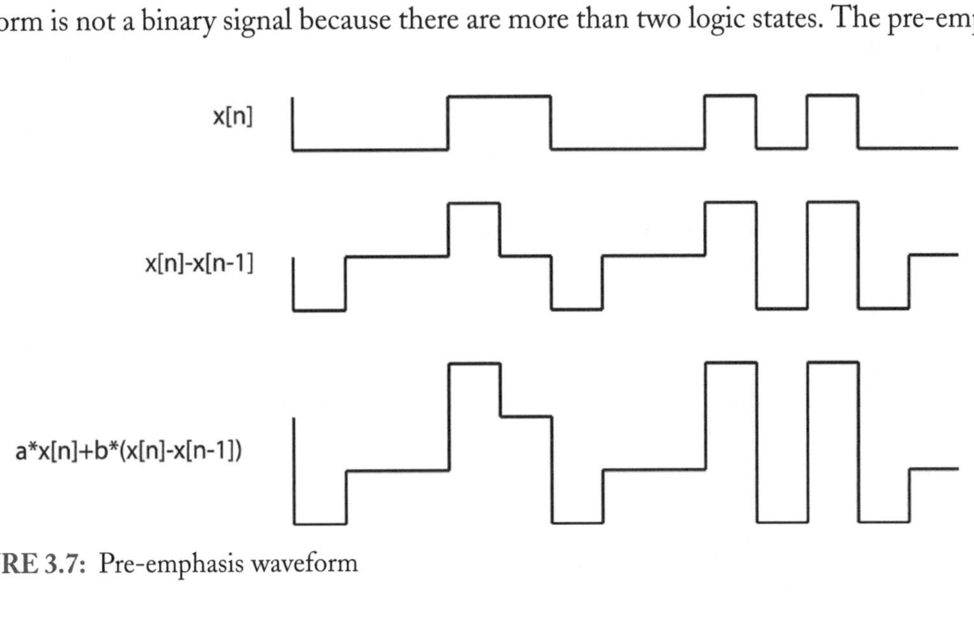

FIGURE 3.7: Pre-emphasis waveform

circuit will combine the $x[n]$ and $x[n] - x[n-1]$ in a specific ratio resulting in the composite waveform.

The above example uses only the first difference, but the second difference can also be used to further increase the pre-emphasis. The second difference is added to the composite waveform after being multiplied by it own coefficient. The coefficient for any of the waveforms can be positive or negative.

The resulting waveform transmitted boosts the high-frequency components without increasing the low frequencies. The difference waveform is a type of high-pass filter for binary data streams. By using a high-pass filter, the low-pass filter of the transmission line is compensated. The amount of boost given to the high-pass filter depends on the amount of low-pass filtering which depends on the length of the transmission line. The knee frequency of the high-pass filter created by the pre-emphasis should be at the highest frequency being transmitted across the trace. The resulting frequency response at the receiver with pre-emphasis is shown in Fig. 3.8.

The goal of pre-emphasis is to create a flat frequency response through the maximum signal frequency. The maximum bandwidth is the frequency where the signal is attenuated by -3 dB. The bandwidth of the lossy line is where the curves begin in Fig. 3.8. The bandwidth is extended to just over 1 GHz.

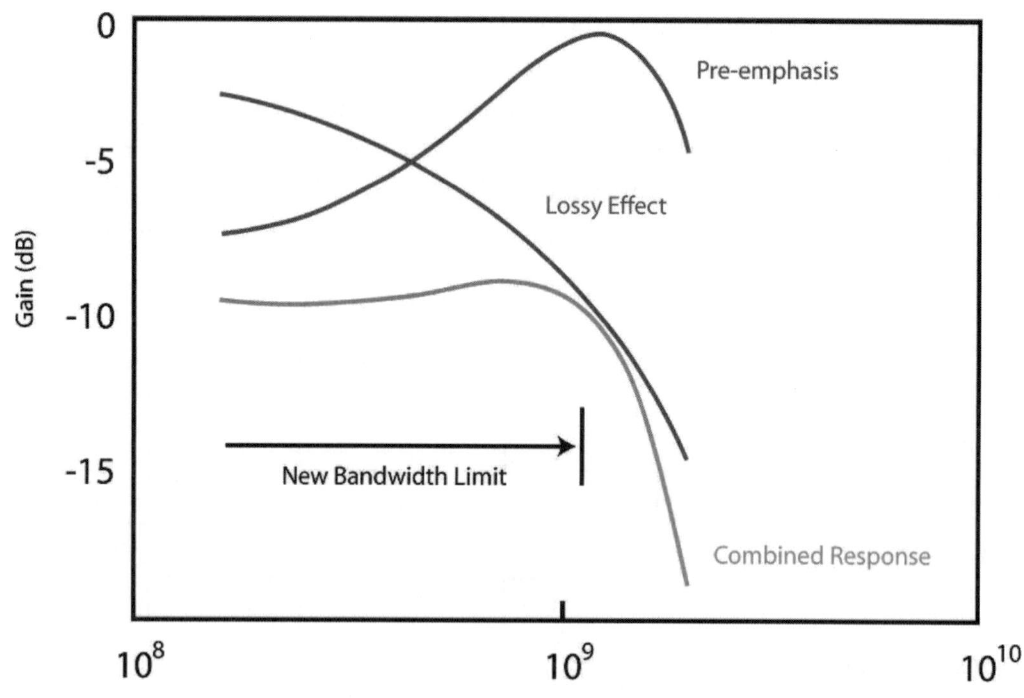

FIGURE 3.8: Pre-emphasis frequency response

The main disadvantage of pre-emphasis is the increased crosstalk in the traces because of the increase in the initial voltage level of the transmitted pulses. The more the pre-emphasis on the transmitted waveform, the more crosstalk there will be. Fortunately, the high frequencies which are being boosted are also being attenuated because of the losses. The losses in the transmission line will partially attenuate the crosstalk.

Receiver Equalization

The receiver can also compensate for losses in the transmission line. The same technique is used as in the pre-emphasis circuit. A high-pass filter is used to attenuate the low-frequency signals. The resulting signal will then be amplified to return the signal to its original amplitude. The final frequency response should be similar to the pre-emphasis circuit.

One advantage of receiver equalization over transmitter pre-emphasis is the ability to adapt to the conditions of the attenuation. The receiver can be selected to automatically tune itself to achieve the best fit. The transmitter pre-emphasis would require feedback from the receiver to know how to adjust.

One disadvantage of the receiver equalization is the decrease of signal-to-noise ratio. Since the low frequencies in the signal are being attenuated, some signal is lost. When the composite signal is amplified, the noise in the signal is amplified as well.

The best solution is to combine both transmitted pre-emphasis and receiver equalization. Often these solutions will offset the attenuation from long transmission lines. These techniques have proven to be successful beyond 10 GHz. Reducing the trace length will increase the frequency at which these techniques will work.

Let us sum up the different approaches to improving signal quality:

1. Reduce skin effect loss by widening traces or using thinner traces. The current will spread over a larger area which will keep the ac resistance low.
2. Reduce the dielectric loss by shortening trace lengths. Shorter lines will decrease the overall attenuation.
3. Reduce the dielectric loss by using a low-loss dielectric material in the PCB fabrication. Materials such as GETEK, Nelco 4000-13, or Rogers 4003 have lower loss tangents than FR-4.
4. Use driver pre-emphasis by boosting the initial voltage amplitude of each edge to increase the high-frequency gain.
5. Use receiver equalization to reduce the low-frequency voltage amplitude to match the high-frequency amplitude, and amplify the resulting balanced signal to normal voltage levels.

3.7 ROUTING SIGNALS THROUGH VIAS

High-frequency signals are best routed through the internal layers of a printed circuit board. Since signals usually come from logic devices or connectors, they must pass through vias to reach the internal layers. In order to minimize the reflections caused by vias, the impedance of the via must match the characteristic impedance of the traces. Altering the radius of the via can change the impedance of the via. Smaller holes reduce the capacitance while increasing the inductance. Back-drilling vias reduces the capacitance.

The return current follows the path of least inductance at high frequency. For PCBs with multiple ground planes, the return current flows on the ground plane closest to the signal trace. If the signal is routed through a via to a different signal layer, the return current must find a path to the new closest ground plane through a different via. This causes the return current to flow away from the trace which increases the loop inductance. For any high-speed signal which must traverse between planes through vias, placing another via nearby connected to all ground planes will reduce the loop inductance. The loop inductance can be calculated from the equation

$$L_v = 5.08d\left(2\ln\left(\frac{x}{r}\right)\right) \tag{3.69}$$

where

L_v is the loop inductance of the via in nH;
d is the distance through the via the signal must travel in inches;
x is the separation of the signal and ground vias in inches;
r is the radius of the via in inches.

Multiple vias further reduce the loop inductance. If two ground vias are placed on either side of the signal, the return current is split between the two ground vias. Four ground vias can be placed around the signal via to further reduce the loop inductance as in Fig. 3.9.

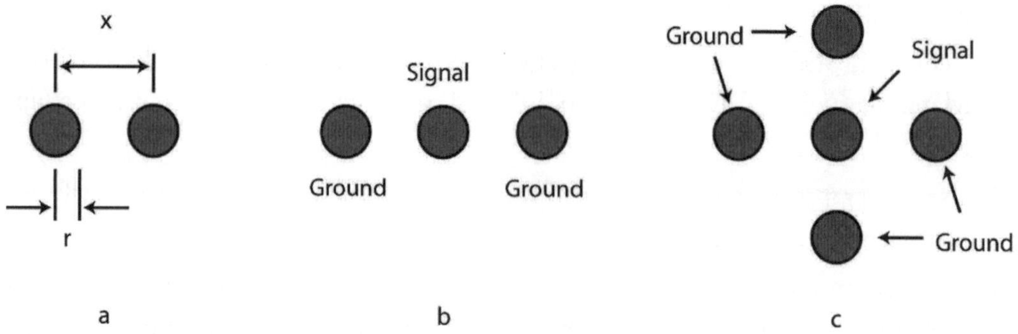

FIGURE 3.9: Via configurations for return current paths

For two ground vias,

$$L_v = 5.08d\left(\frac{3}{2}\ln\left(\frac{x}{r}\right) - 0.347\right). \tag{3.70}$$

For four ground vias,

$$L_v = 5.08d\left(\frac{5}{4}\ln\left(\frac{x}{r}\right) - 0.347\right). \tag{3.71}$$

CHAPTER 4

Signal Quality Degradation

The purpose of this chapter is to explain the causes and effects of crosstalk and how to minimize it. This chapter assumes that the reader is familiar with analog components, simple circuit analysis, basic printed circuit board (PCB) design, digital circuits, differential signaling, and transmission lines.

4.1 LEARNING OBJECTIVES

After reading this chapter, you will be able to perform the following tasks:

- Determine the amount of crosstalk between two traces.
- Identify the type of crosstalk from the measured voltage waveform.
- Minimize crosstalk by adjusting spacing and setting spacing rules.
- Route differential lines so that crosstalk is not injected unequally.

4.2 CROSSTALK IN LUMPED-ELEMENT MODELS

Crosstalk is the undesired capacitive, inductive, or conductive coupling from one transmission line to another. High-frequency signals through a wire generate large magnetic fields, and those magnetic fields can create a current in other nearby wires. Usually the inductive crosstalk is the largest factor in digital systems.

The amount of crosstalk between two wires can be found from their mutual inductance and the signal rise time. The analysis of transmission lines in the lumped-element region is different from the analysis in the LC region. Assuming that the two wires are lumped-element circuits with resistive terminations, the voltage and current flowing through the wires will be proportional to each other. The amount of crosstalk from one wire to the other will follow the equation

$$XT = \frac{L_M}{2R_T t_r} \tag{4.1}$$

where

XT is the amount of crosstalk induced in the opposing wire;
L_M is the mutual inductance between the two wires in H;
R_T is the resistance of the termination in Ω;
t_r is the signal rise time in s.

Since ground planes are usually used in high-frequency digital circuits, they reduce the inductive coupling between traces; however, this ground plane can be a source of crosstalk as well. Low-frequency return current on the ground plane spreads out across the plane because it follows the path of least resistance. High-frequency return current follows the path of least inductance which is directly beneath the signal trace. This minimizes the total loop area between the outgoing and return current paths. The current density beneath the signal trace balances between these two forces:

$$i(x) = \frac{i_0}{\pi h} \frac{1}{1 + \left(\frac{x}{h}\right)^2} \tag{4.2}$$

where

$i(x)$ is the current density on the ground plane in A in.$^{-1}$;
i_0 is the total current in A;
h is the height of the trace over the ground plane in inches;
x is the distance on the ground plane away from the trace in inches.

The highest current density on the ground plane is directly beneath the trace, and the lowest is the maximum distance away from the trace. The current density ramps down away from the trace as shown in Fig. 4.1. This return current can pass beneath other traces which can

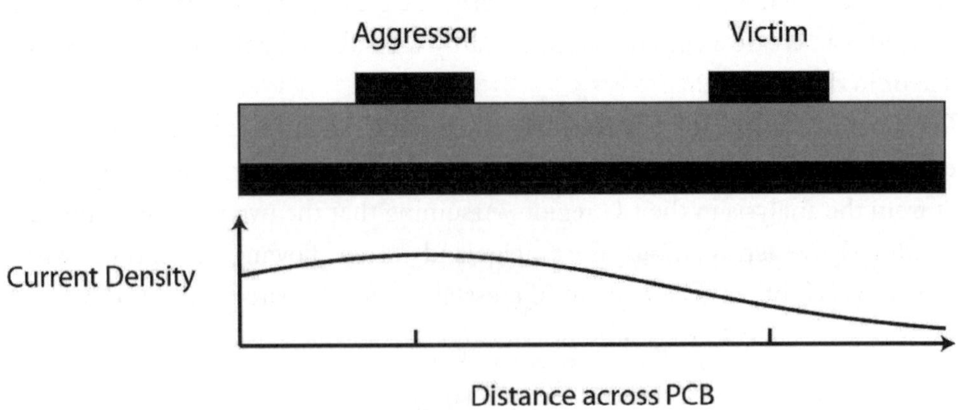

FIGURE 4.1: Crosstalk between two traces from the return current

cause a reverse current. The induced current in the second trace is proportional to the current density below the trace and the height of the trace above the ground plane. As the height of the traces above the ground plane increases, the crosstalk from the ground plane decreases, but the victim trace loses the magnetic shielding provided by the ground plane. Therefore, as the height of the ground plane increases significantly, the crosstalk will actually increase.

A second example of crosstalk caused by ground planes is when long slots are present. If a trace passes over a ground slot, the high-frequency return current cannot flow directly beneath it. It will pass around the slot creating a large loop which increases the inductance of the signal path. If multiple traces pass over the same slot, the return currents all flow around the ground slot and overlap near the edge of the slot. This overlap causes a mutual inductance between the traces. The mutual inductance is

$$L_M \approx 5x \ln\left(\frac{x}{w}\right) \tag{4.3}$$

where

L_M is the mutual inductance between traces in nH;
x is the slot length in inches;
w is the trace width in inches.

If the traces are on opposite ends of the slot, then they will have less mutual inductance. Also, if the slot length is short, then little coupling will occur. The voltage induced from one trace to the other is given by

$$v_X = \frac{v_r L_M}{t_r Z_0} \tag{4.4}$$

where

v_X is the voltage amplitude induced by the ground slot in V;
v_r is the voltage amplitude of the source pulse in V;
L_M is the mutual inductance between the traces in H;
t_r is the rise time of the voltage pulse in s;
Z_0 is the characteristic impedance of the traces in Ω.

In general, crosstalk between traces can be minimized by placing them far apart; however, this is often not possible on tightly packed PCBs. A 10% increase in separation between the traces, or a 10% decrease in height over the ground plane, will decrease crosstalk by 20%. Doubling the separation decreases crosstalk by a factor of 4. These equations are only approximations. For a more precise estimation of crosstalk a field solver is needed; however, those estimates do not consider ground slots. Ensuring that the traces do not pass over ground slots minimizes the inductance. Since most ground slots occur because of via clearances on the ground plane, careful planning of vias can minimize the crosstalk.

4.3 NEAR-END AND FAR-END CROSSTALK

For transmission lines in the LC region, the crosstalk involves both inductive and capacitive coupling. A transmission line on a PCB acts as a distributed series of inductors and parallel capacitors. The mutual coupling between two transmission lines is also modeled as a distributed series of segments with series inductance and parallel capacitance. As the incident signal travels down the transmission line, each segment of the victim line will have some crosstalk.

The easiest way to describe the effects of crosstalk is by having two parallel transmission lines. The transmission line which has the source signal propagated is called the aggressor line. The transmission line which carries the crosstalk is called the victim line. The aggressor line is terminated normally to prevent reflections. The victim line is terminated at both ends with no other loads. The crosstalk will propagate in both the forward and reverse directions. The voltage response measured on the victim line near the original source is called near-end crosstalk. The voltage response at the other end is called the far-end crosstalk. Each type has very different characteristics.

The near-end crosstalk (sometimes referred to as NEXT) can be represented as a series of crosstalk events associated with each segment. A forward propagating signal will create a blip as it passes each segment which returns toward the source in the victim line. Each segment will create a similar blip. Therefore, at the near end on the victim line a series of blips are measured. The blips will travel on the transmission line at a speed corresponding to the velocity of those lines. The last blip will be created as the initial signal wave reaches the far end of the aggressor line. If the time of travel across the aggressor line is t seconds, then the last blip will take $2t$ seconds to be measured at the near end. Fig. 4.2 shows a bounce diagram of how these blips will be received. The transmission lines represented are modeled with four segments.

In a real transmission line, the segments will be infinitely small. Therefore, the blips will overlap and appear as a steady voltage step for a duration of $2t$ at which time it will return to zero. The amplitude of this pulse is very difficult to compute mathematically and a field solver is needed to give even an approximate answer; however, it will be less than the amplitude of the voltage wave in the aggressor. The polarity of this pulse will be the same as the original voltage wave.

Near-end crosstalk varies with the length of the parallel section of overlap of the two transmission lines. As the length of the line increases, the delay of the line increases. The duration of the crosstalk increases with the delay. The duration will always be $2t$ except for very short parallel sections of transmission lines. When the delay associated with the overlap decreases to half the rise time of the original voltage pulse, the amplitude of the crosstalk will

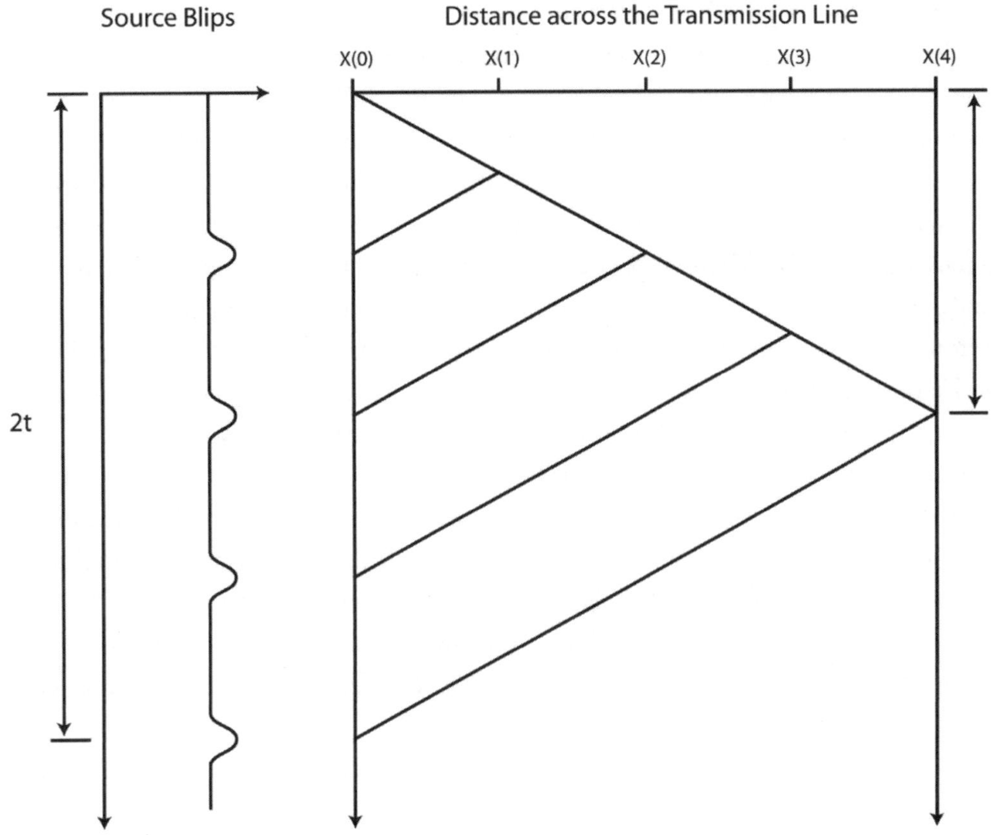

FIGURE 4.2: Near-end segmented crosstalk

begin to decrease from its maximum. The amplitude approaches zero as the parallel section of the overlap approaches zero. Therefore, anytime the line delay of the parallel section is larger than half the rise time, crosstalk will always be at its maximum value. On most PCBs, this associated length is very small which means that the crosstalk will always reach its maximum value.

Far-end crosstalk (sometimes referred to as FEXT) looks significantly different from near-end crosstalk. In each segment, a forward traveling blip is produced. This blip travels at the same speed as the original voltage wave. As the original voltage wave passes through each segment, the blips from the previous segments are added to the current blip. This increases the amplitude of the total blip. Once the voltage wave reaches the far end, the large blip is measured as a single pulse with duration equal to the rise time of the original voltage wave. In other words, the shape of the far-end crosstalk is the derivative of the original voltage wave.

The amplitude of the far-end crosstalk is proportional to the length of the parallel sections of the transmission lines. Each individual blip has amplitude proportional to the amount of mutual inductance and capacitance. Since the blips are added together along the length of the transmission line, the amplitude depends on how many blips there are. The final amplitude of the far-end crosstalk requires a field solver to accurately predict.

The polarity of the far-end crosstalk depends on the differences in the mutual inductance and capacitance. Mutual inductance causes a pulse with the opposite polarity, while mutual capacitance causes a pulse with the same polarity. For configurations such as stripline, the mutual inductance and capacitance are equal, and therefore the two polarities cancel out any forward moving crosstalk. Microstrip traces do not have balance between the inductance and capacitance. Microstrip has electric field lines which travel through air instead of the insulator, and therefore, it has less capacitive crosstalk than inductive crosstalk. This causes a small blip with the opposite polarity. If the traces cross over a slot in the ground plane, the mutual inductance is much larger producing a large blip with the opposite polarity. The waveforms for both the near-end and far-end crosstalk are shown in Fig. 4.3.

In practical digital designs, all transmission lines will have a source and a load. Any signals which have some crosstalk will have a driver and receiver. If the receiver on a victim line detects a voltage change because of crosstalk, this could cause an unwanted bit transition.

The near-end and far-end crosstalk may encounter impedance discontinuities along the transmission line, or the transmission lines may not be properly terminated. Either case will

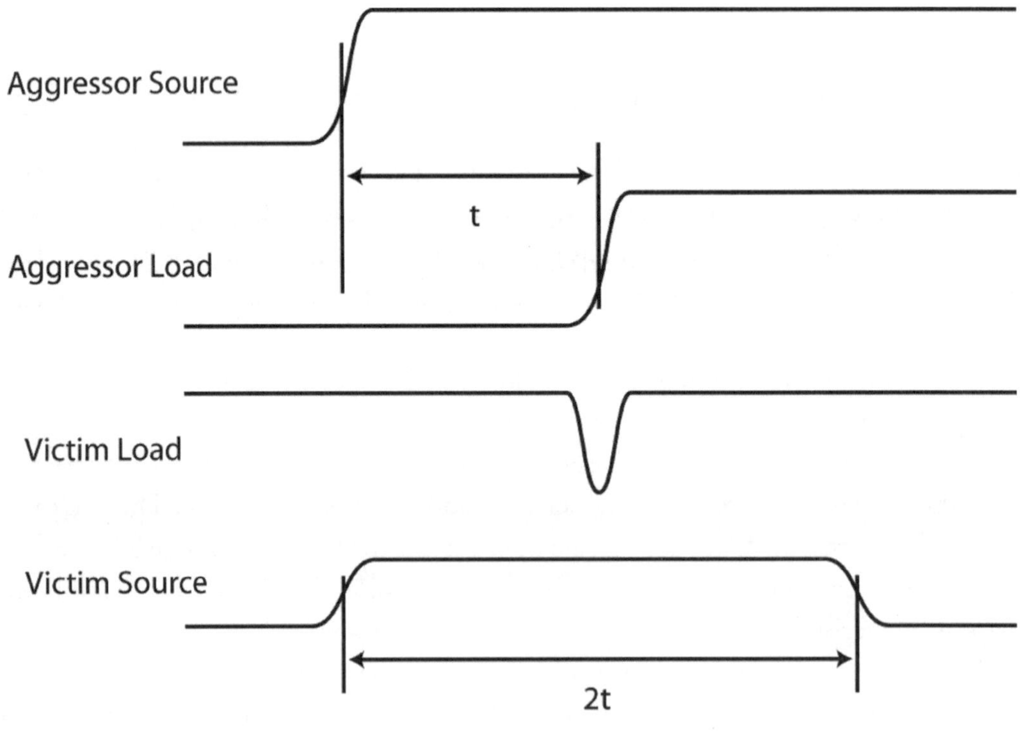

FIGURE 4.3: Near-end and far-end crosstalk waveforms

cause reflections of the crosstalk signal. If the only termination was provided at the load of the transmission lines, the far-end crosstalk may not reflect, but the near-end crosstalk probably will reflect. If no source termination was used, the impedance of the driver will probably be very low causing a reflection coefficient close to −1. This will cause the entire near-end crosstalk to change polarity and travel down the transmission line toward the load. Using source termination in addition to load termination will reduce reflections. The reflections caused by discontinuities in the middle of the transmission line will still occur. These discontinuities can be handled as discussed in the previous chapters.

4.4 CROSSTALK IN VIAS

Vias do not have shielding for their magnetic lines since they are perpendicular to the metal planes. The magnetic field generated permeates the dielectric. The magnetic permeability of most core and prepreg material is very small, which means that the dielectric does not interfere with the magnetic field. The magnetic field can then couple into other vias. This creates a large mutual inductance between vias on the PCB.

The magnetic field can be shielded by placing ground vias around the signal via. Since high-frequency signals need a path for the return current, ground vias should already be placed nearby the signal via. Additional ground vias surrounding the signal via will reduce the mutual inductance caused by the magnetic field from the signal trace.

Sometimes a ground via is shared between multiple signal vias. If a ground via is placed between two signal vias, the return current may flow on this ground via. The return current from an aggressor via generates a magnetic field. If a victim via is on the other side of the ground via, the magnetic field from the return current will couple with it and create crosstalk. The amount of mutual inductance between the three vias is

$$L_M = 5.08 \cdot d \cdot \ln\left(\frac{x_{ag} x_{vg}}{x_{av} r}\right) \qquad (4.5)$$

where

L_m is the mutual inductance in nH;
d is the distance through the via the signal travels in inches;
r is the radius of the via in inches;
x_{ag} is the distance between the aggressor and ground vias in inches;
x_{vg} is the distance between the victim and ground vias in inches;
x_{av} is the distance between the aggressor and victim vias in inches.

The total amount of crosstalk induced in the victim is based on the rise time of the signal and the difference in current. For a step voltage, the peak voltage in the victim is

$$v_{\text{victim}} = \frac{v_{\text{aggr}}}{Z_C}\frac{L_M}{t_r}. \tag{4.6}$$

4.5 CROSSTALK IN DIFFERENTIAL SIGNALS

Differential signals do an excellent job reducing common-mode noise, but imbalances in noise distribution can cause significant problems. If a nearby aggressor trace passes close to a pair of differential lines, crosstalk in the near trace will be higher than the far trace. The receiver will not be able to compensate for the imbalance. If the crosstalk is equal in both traces, then the receiver will be able to subtract out the noise.

Increasing the distance of the aggressor trace from the differential pair is the best way to reduce this crosstalk. Routing the differential lines close together will also reduce the amount of crosstalk, but to a much lesser degree. Often in PCB tools, the traces can be identified as differential lines. One attribute which can be defined in the toolset is the minimum separation from other traces. Any violation of this separation will generate a warning. More complex rules can be set for different kinds of traces. If certain traces are going to be noisy, they can be defined within the tools as such. The separation rules can be increased for those particular traces. Therefore, some traces are allowed to be closer to the differential lines than others. The specific mechanism to define the separation rules differs between software packages.

Biography

Justin Stanford Davis received his Ph.D. in Electrical Engineering from the Georgia Institute of Technology in August 2003, as well as his M.S. and B.E.E. degrees in 1999 and 1997. During the summers of 1998 and 1999, he worked at Hewlett-Packard (now Agilent Technologies). In fall of 2003, he joined the faculty in the Department of Electrical Engineering at Mississippi State University as an Assistant Professor. His research interests include digital testing for high-speed systems, SoCs, and SoPs, as well as signal integrity, systems engineering, and fault-tolerant design. He is currently working on the development of low-cost test support processors using programmable devices.

Introduction to Logic Synthesis using Verilog HDL

Robert B. Reese
Mississippi State University

Mitchell A. Thornton
Southern Methodist University

MORGAN & CLAYPOOL PUBLISHERS

ABSTRACT

Introduction to Logic Synthesis Using Verilog HDL explains how to write accurate Verilog descriptions of digital systems that can be synthesized into digital system net lists with desirable characteristics. The book contains numerous Verilog examples that begin with simple combinational networks and progress to synchronous sequential logic systems. Common pitfalls in the development of synthesizable Verilog HDL are also discussed along with methods for avoiding them. The target audience is any one with a basic understanding of digital logic principles who wishes to learn how to model digital systems in the Verilog HDL in a manner that also allows for automatic synthesis. A wide range of readers, from hobbyists and undergraduate students to seasoned professionals, will find this a compelling and approachable work. This book provides concise coverage of the material and includes many examples, enabling readers to quickly generate high-quality synthesizable Verilog models.

KEYWORDS

Verilog, Digital System Design, Digital Logic Synthesis, HDL (Hardware Description Language), Combinational Logic, Sequential Logic.

Contents

Preface

Modern digital logic design flows heavily utilize Hardware Description Languages (HDLs) for specification, simulation, automatic synthesis, and validation of target digital systems. The use of HDLs is an important skill for all digital designers to have and this book serves as a concise introduction to one of the most popular HDLs in common use today, Verilog. We introduce the Verilog language through examples as we review basic building blocks of combinational and sequential digital system design. The focus will be to restrict the discussion to synthesizable Verilog and to mention common pitfalls and how to avoid them.

The book is divided into two main chapters. The first chapter is devoted to a quick review of digital logic building blocks and their corresponding Verilog descriptions. Verilog descriptions are shown that correctly describe various logic elements as are those that show common errors and the unintentional digital logic that is produced by a synthesis tool.

A discussion of the basic internal operation of a discrete event simulator is included to help the reader gain an understanding of how the Verilog HDL is simulated. By understanding how a simulator operates, insight is gained into how to write and debug Verilog HDL.

The second chapter focuses on synchronous sequential circuits and their use as controllers for a datapath. Algorithmic State Machine (ASM) charts are described and emphasized throughout the chapter since they are easily understandable by designers and can be directly translated into a Verilog HDL module. An example design of a memory-zeroing circuit is used to illustrate the topics of controller modeling and synthesis.

CHAPTER 1

Digital Logic Review with Verilog Quickstart

This chapter assumes that the reader is already familiar with binary number representation and digital logic principles, and provides a refresher course on these topics while introducing the reader to the Verilog hardware description language (HDL), logic synthesis, and event-driven simulation.

1.1 LEARNING OBJECTIVES

After reading this chapter, you will be able to perform the following tasks:

- Implement combinational gate networks and commonly used combinational building blocks in Verilog.
- Express sequential storage elements and commonly used sequential building blocks in Verilog.
- Discuss the role of logic synthesis in digital design.
- Discuss the principles behind event-driven simulation.

1.2 LOGIC SYNTHESIS INTRODUCTION AND MOTIVATION

Recall that a digital system performs logical and/or arithmetic computations on binary-valued (0, 1) inputs using a mixture of combinational gates and sequential storage elements and produces binary-valued outputs. Fig. 1.1 shows two representations for the combinational and sequential elements of a digital system: (a) a schematic with gate symbols (graphical), and (b) a Boolean equation (text). While a schematic offers a visual representation of the system's operation that can be intuitively easier to understand than a symbolic model, it becomes unwieldy and confusing once the schematic's symbol count exceeds a few 10s of symbols. Design entry and data sharing for schematics is also problematic. Schematic entry tools supplied by different vendors have different graphical user interfaces (GUIs), which must be relearned if a user switches design entry tools. Furthermore, there is no standard data format for schematics, and thus a schematic

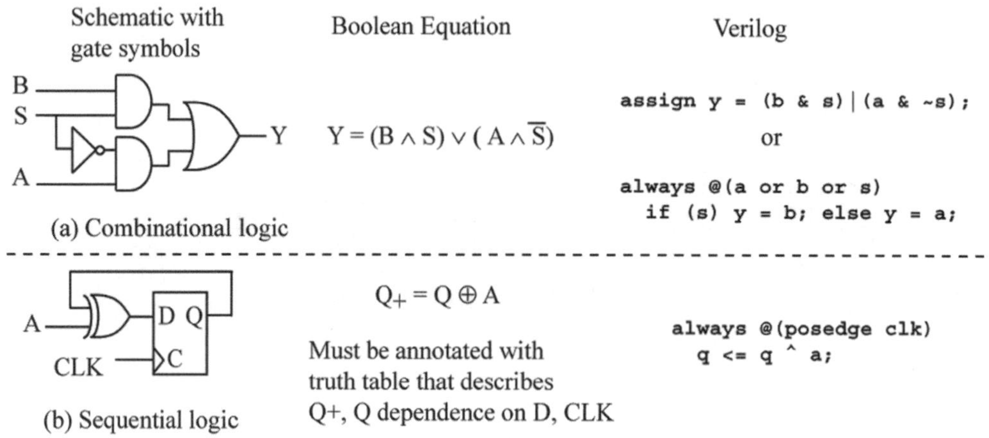

FIGURE 1.1: Combinational and sequential logic representations

file created by one schematic entry tool is not easily portable to a schematic entry tool produced by a different vendor. This hinders design sharing between users who select different tools for design entry.

A text format solves the data format portability problem, as all text editor programs can process files that use the American Standard Code for Information Interchange (ASCII) for character encoding. However, the syntax used to express a digital system's operation must be agreed upon by all parties for the design to be portable between different design tools and users. There are several shortcomings in using Boolean equations for expressing complex digital systems:

- Boolean equations are a low-level description of a system's operation; it can be difficult for an external reader to grasp a system's functionality when it is specified as a sequence of Boolean equations.
- There is no standard method for representing sequential behavior using Boolean equations, as many different notations exist.
- There is no standard method for representing Boolean operations on groups of signals (busses), which is needed for reducing the number of Boolean statements that describe a system.
- Arithmetic operations such as addition, subtraction, multiplication, etc. have to be expressed as their component Boolean equations, increasing the number of Boolean statements for describing a system.

In the late 1970s and throughout the 1980s, as computer tools became essential for designing complex digital systems, many *hardware description languages* (HDLs) were created by different companies and universities for describing the structure and operation of digital systems. Two of these HDLs, Verilog and VHDL (a nested acronym that stands for VHISC HDL, where VHSIC is very high speed integrated circuit), eventually emerged as the standard HDLs for digital system representation. These two HDLs are adept at specifying digital system operation, with Verilog being the more popular language within the United States. This book uses Verilog as its HDL of choice. Verilog was initially a proprietary language, but was transitioned to the public domain by Cadence Design Systems and was made an IEEE standard in 1995, with a revised standard released in 2001 [1]. VHDL is an IEEE standard as well.

In Fig. 1.1(a), the Verilog representation of the combinational gate network is shown in two forms. One form uses a one-to-one mapping of the Boolean operations to Verilog logical operators (& is AND, | is OR, ~ is NOT), which are the same logical operators used in the C and C++ programming languages, and demonstrates that Boolean equations are easily represented in Verilog. The second form that contains the `always` block is an alternate view of the logic network's operation, as it uses an `if` statement that specifies the output `y` as a choice between the inputs of `a`, `b`. It is easier for an external reader to grasp the behavior of the logic network when it is expressed in this form, as it is a familiar representation for any person who has programmed in a high level language (HLL). Digital systems expressed in Verilog typically use these behavioral-type statements instead of Boolean equations, because of the increased clarity for external readers. The sequential network of Fig. 1.1(b) is also represented in Verilog by using an `always` block, using the syntax rules defined in the Verilog standard for expressing sequential behavior. Verilog syntax details are discussed on an *as needed* basis in the following review sections on combinational and sequential logic. Only the Verilog subset required by the design examples in this book is covered as the language contains many features that are outside the scope of these design examples.

There are many ways to implement digital logic; the most common methods in use today are field programmable gate arrays (FPGAs) or standard cells within an application specific integrated circuit (ASIC). In brief, an FPGA contains programmable gates and routing that can be configured to implement a digital system of the user's choice. A *logic synthesis tool* is used to convert a user's digital system specified in an HDL to an FPGA implementation. The logic synthesis tool performs optimizations to meet user-specified constraints concerning circuit speed and gate count. Fig. 1.2 shows that the role of a logic synthesis tool is similar to that of a compiler, whose function is to map a program specified in a high level language to an implementation on a particular central processing unit (CPU).

With respect to Fig. 1.2 and the syntactical similarities between Verilog and common high level languages, it is inaccurate to think of an HDL as a programming language. There are fundamental differences between HDLs and high level programming languages that are discussed in more detail in Sections 1.3 and 1.6.

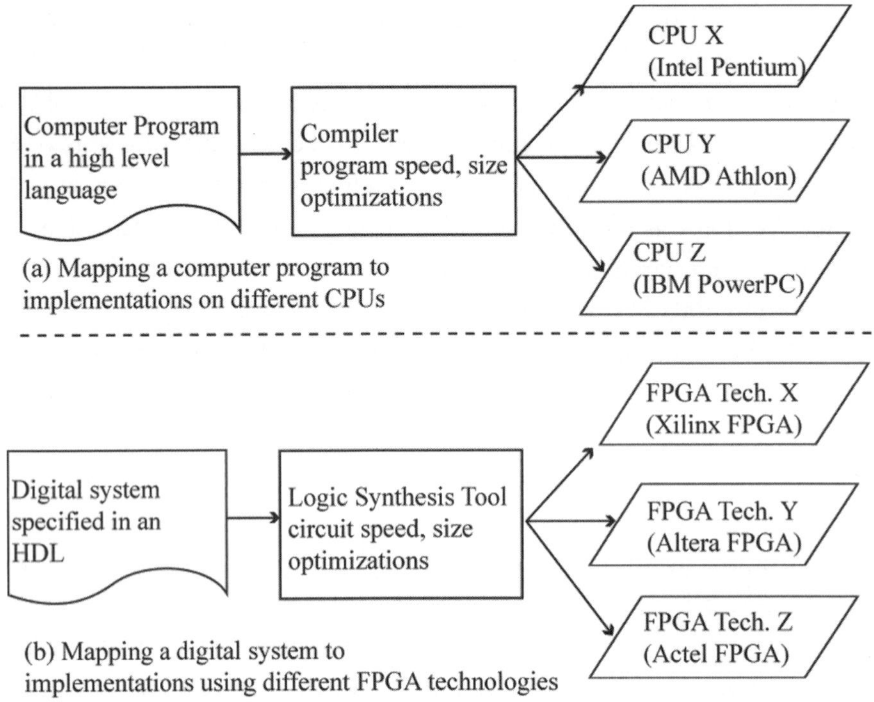

FIGURE 1.2: HDLs and logic synthesis

1.3 COMBINATIONAL LOGIC IN VERILOG

Fig. 1.3 shows the common Boolean functions in their truth table, gate symbol, Boolean equation, and Verilog representations. The Verilog Boolean operators are the same as the bitwise logic operators used in the C and C++ programming languages. Verilog also possesses some gate-level primitives that implement the standard Boolean functions, but this book's examples use the Boolean operators exclusively.

1.3.1 *Assign* Statements

Fig. 1.4 shows the gate logic and symbol for a 1-bit 2-to-1 multiplexer (mux) function; recall that a 2-to-1 multiplexer passes one of the two inputs to the output based upon a select input. In Fig. 1.4, the output *Y* is equal to the input *B* if the select input *S* is "1," else the output *Y* is equal to *A*.

Three different Verilog implementations of the 2-to-1 mux are used as an introduction to Verilog combinational logic. Verilog is case sensitive with all keywords in lowercase; keywords are italicized for emphasis in Fig. 1.5. The basic design unit in Verilog is the *module*, which contains the module's interface signals and statements that describe its behavior. The `module` statement contains a list of the module interface signals, also known as *ports*, which can be given in any order. The three variations of Fig. 1.5 only differ in how the module's behavior is

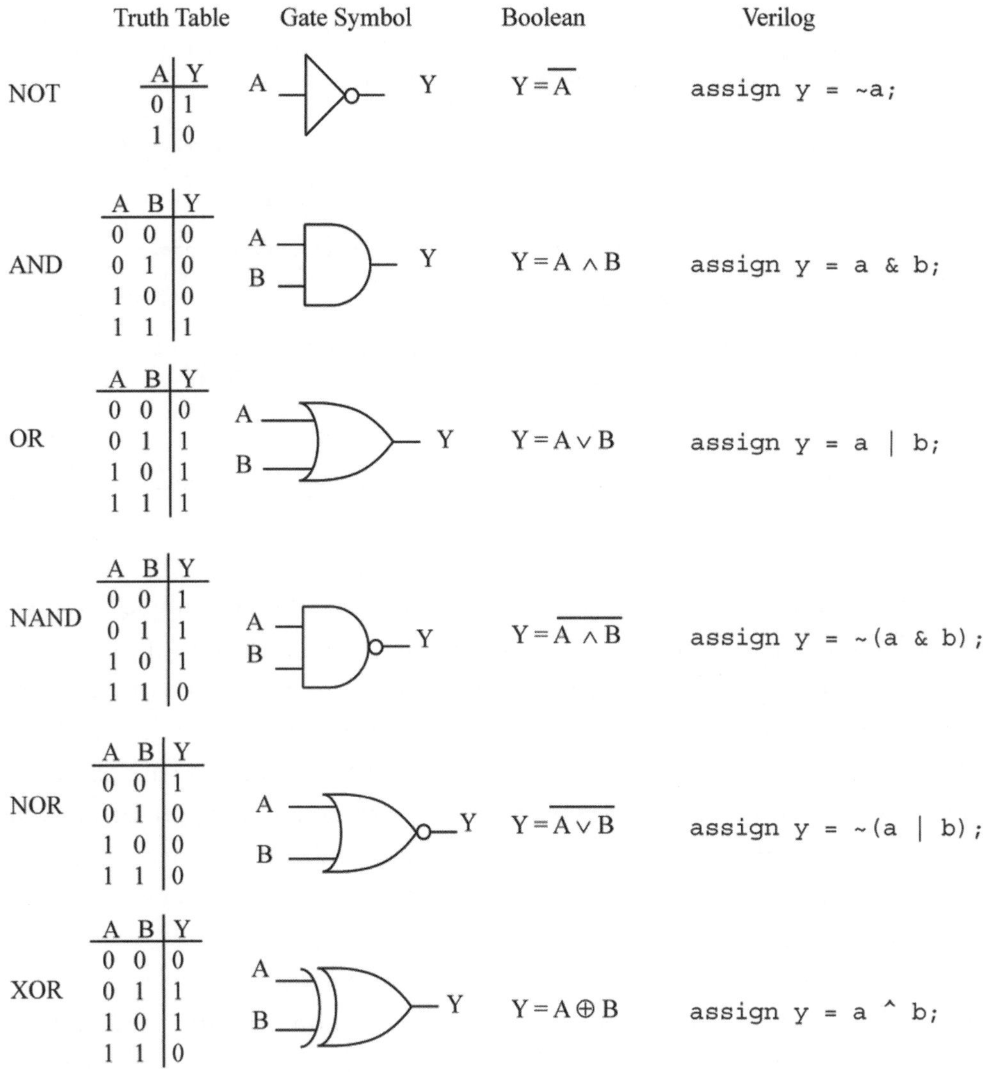

FIGURE 1.3: Boolean logic functions

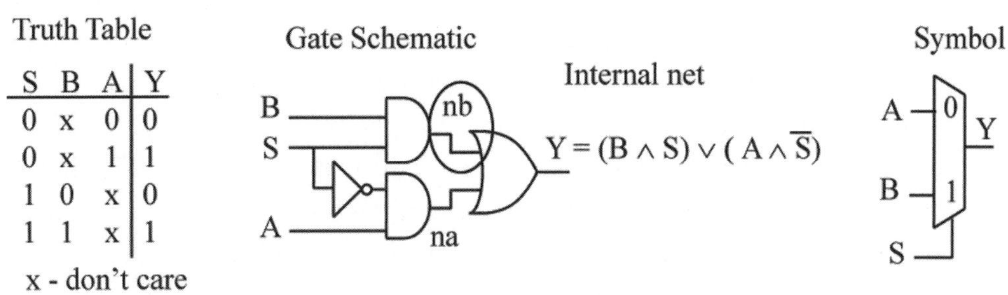

FIGURE 1.4: One-bit 2-to-1 multiplexer

(a) Single `assign` statement using boolean operators

```
module mux2to1(s,a,b,y);
input  s,a,b;
output y;

//this is a comment
assign y=(b & s)|(a & ~s);


endmodule
```

(b) Multiple `assign` statements using Boolean operations and intermediate values

```
module mux2to1(s,a,b,y);
input  s,a,b;
output y;

wire na, nb;

//this is a comment
assign nb = b & s;
assign na = a & ~s;
assign y = na | nb;

endmodule
```

(c) Single `assign` statement using a conditional operator

```
module mux2to1(s,a,b,y);
input  s,a,b;
output y;

//conditional statement
//                s=1 s=0
assign y = s ? b : a ;


endmodule
```

FIGURE 1.5: One-bit 2-to-1 multiplexer implementations using `assign` statements

written, so the interface of the three modules remains the same. These examples have the input signals first, followed by the output signal, but the order is arbitrary. The `input` and `output` statements that follow the module declaration define the direction of each interface signal. A port is declared bidirectional by using the `inout` keyword. These examples list all of the input ports using a single `input` statement; individual `input` statements could have also been used. Our examples use lowercase for user-defined names such as module and port names; uppercase is commonly used as well. It is suggested that mixed case not be used for user-defined names, as the tools that operate on the design data after they have been transformed from Verilog to some intermediate format may have difficulty with mixed-case names.

An `assign` statement makes a continuous assignment of the right-side expression to the net or port that appears on the left side, using the "=" operator. The best way to think of a continuous assignment is that the right-side expression defines a block of combinational logic, whose output is continuously connected, and thus continuously drives, the net or the port on the left side. Fig. 1.5(a) uses a single `assign` statement to specify a 2-to-1 multiplexer using Boolean operators. Fig. 1.5(b) divides the Boolean operators across multiple `assign` statements, using internal *nets* (gate ports joined by a wire) `na` and `nb` in the last `assign` statement to implement the final *or* gate. The `wire` statement is used to declare nets `na` and `nb`; this is not strictly necessary as only multibit nets (busses) have to be explicitly declared. Fig. 1.5(c) uses the *conditional* operator "?" in a single assignment to specify the multiplexer logic. The first operand in the conditional operator is the select operand (i.e., `s`); if this operand evaluates as nonzero then the second operand (i.e., `a`) is evaluated and returned, else the third operand is evaluated and returned. A module is terminated using the `endmodule` keyword. All of these implementations result in the same logic; none of them have any inherent advantages over the others.

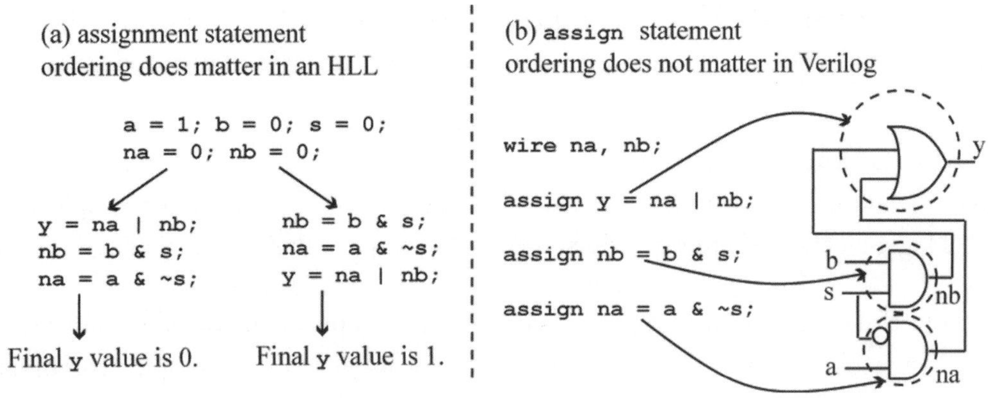

FIGURE 1.6: Assignment statement ordering

It was previously stated that it is inaccurate to think of an HDL as just another form of high level programming language where wires are variables and a sequence of HDL statements are semantically equivalent to a sequence of HLL statements. Fig. 1.6 shows assignment statements in a HLL versus Verilog. In Fig. 1.6(a), the ordering of the assignment statements affects the final y value as the statements are evaluated sequentially. In Fig. 1.6(b), it is seen that Verilog `assign` statements describe hardware (gates) that operate concurrently, so the statement ordering does not affect the final hardware that is generated. Another way of stating this is that the arrangement of the gate symbols in a schematic does not affect the circuit function as long as the connections between the gate symbols are the same.

Fig. 1.7 shows another difference between an HLL and the Verilog HDL. In an HLL, multiple assignments to the same variable result in the variable's value being the result of the last assignment. In Verilog, multiple `assign` statements to the same wire causes the gate outputs of the logic implementation to be connected together. The only time that this is allowed is for tri-state logic implementation, in which only one of the tri-state gates driving the wire is

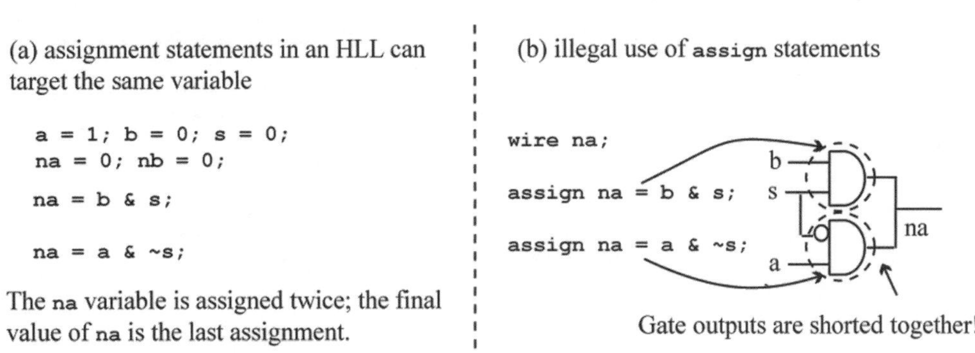

FIGURE 1.7: Continuous assignments to the same wire

asserted at a given time. The Verilog `assign` statements in Fig. 1.7(b) are illegal as this does not implement tri-state drivers; correct implementation of tri-state logic is discussed in Section 1.4.

1.3.2 *Always* Procedural Blocks

While combinational logic can be described using `assign` statements, higher complexity combinational logic blocks are better described using `always` procedural blocks for several reasons:

- Powerful statements like `if`, `if-else`, `case` and looping constructs can only be used in `always` blocks; these statements are useful for implementing complex combinational blocks with greater clarity and in a more concise manner than is possible with `assign` statements.
- Multiple output nets can be assigned within a single `always` block.
- Sequential logic can only be specified within `always` blocks.

Fig. 1.8 serves as an introduction to `always` blocks by showing the implementation of the 2-to-1 multiplexer of Fig. 1.5 using three variations that each use an `always` block. For combinational logic, an `always` block header contains an *event list* that is designated by "@(*net1* or *net2* or... *netN*)" where changes (events) on nets 1, 2, ..., *N* cause the logic represented by the `always` block to evaluate their inputs. For combinational logic, any net that appears on the right side of an "=" operator in the `always` block should be included in the event list. The body

(a) Combinational `always` block using an `if` statement.

```
module mux2to1(s,a,b,y);
input  s,a,b;
output y;

reg y;

//use an if statement
always @(a or b or s)
begin
 if (s) y = b;
else y = a;
end

endmodule
```

(b) Combinational `always` block using Boolean operations.

```
module mux2to1(s,a,b,y);
input  s,a,b;
output y;

reg y;

//use boolean ops
always @(a or b or s)
begin
 y = (b & s)|(a & ~s);
end

endmodule
```

(c) Combinational `always` block using Boolean operations and intermediate values.

```
module mux2to1(s,a,b,y);
input  s,a,b;
output y;

reg y, na, nb;

//use intermediates
//and implicit event
//list
always @*
begin
 nb = b & s;
 na = a & ~s;
 y = na | nb;
end

endmodule
```

FIGURE 1.8: 1-bit 2-to-1 multiplexer using `always` procedural blocks

of an `always` block can be one or more statements; `begin` and `end` keywords are used to group multiple statements. The statement "`reg y;`" is included in each of the three modules of Fig. 1.8 as any net that is assigned within an `always` block must be declared as a `reg` type; this does not imply that this net is driven by a register or sequential logic.

Fig. 1.8(a) implements the multiplexer using an `if-else` statement, with the `if`-body evaluated for a nonzero (true) conditional expression and the `else` clause evaluated otherwise. Keywords `begin` and `end` can be used to place multiple statements in an `if`-body or `else` clause. Fig. 1.8(b) implements the multiplexer using Boolean operators, while Fig. 1.8(c) distributes the Boolean operators using intermediate nets and multiple assignments. The "=" operator when used in an `always` block is called a *blocking* assignment, this terminology is discussed in more detail in Section 1.6, which covers event-driven simulation principles. The `always` block of Fig. 1.8(c) uses an *implicit* event list designated by "`@*`," which means that all nets on the right side of the assignments are included in the event list.

The semantics of the net assignments in an `always` blocks differs *significantly* from `assign` statements in that statements in an `always` block use the same sequential execution model as the statements in an HLL.

- The logic synthesized for an `always` block duplicates the assignments' behavior assuming that the assignments are evaluated *sequentially*. This means that the *order in which assignments are written* in an `always` blocks affects the logic that is synthesized.
- Because of the sequential nature of an `always` block, the same net can be assigned multiple times in an `always` block; the last assignment takes precedence.

Fig. 1.9 shows a case in which multiple blocking assignments are made to a net in the same `always` block, with assignment ordering affecting the synthesized logic. In Fig. 1.9(a), the `clr` input takes precedence over the `ld` input if both are "1," while in Fig. 1.9(b) the opposite holds true. Observe that in the two `always` blocks, if both `ld` and `clr` are "0," then the initial assignment of `q=q_old` sets the value of `q`. This is a common coding style used in combinational logic `always` blocks in that a default assignment is made to an output at the block's beginning, which is then overridden by later assignments in the block based upon the assertion of other inputs.

The use of a default assignment to the output net of a combinational `always` block *guarantees* that the output net is the target of an assignment for any input combination. If there is some logic path through the `always` block that does not assign a value to the output net, then a latch is inferred on that output as shown in Fig. 1.10. A latch is a sequential logic element, and should not be synthesized in an `always` block that is meant to implement combinational logic. Inferred latches are a common coding mistake for users who are new to Verilog and logic

(a) `clr` takes precedence over `ld` if both are '1'

```
always @(ld or clr or d or q_old)
begin
  q = q_old;
  if (ld) q = d;
  if (clr) q = 0;
end
```

ld	clr	q
0	0	q_old
0	1	0
1	0	d
1	1	0

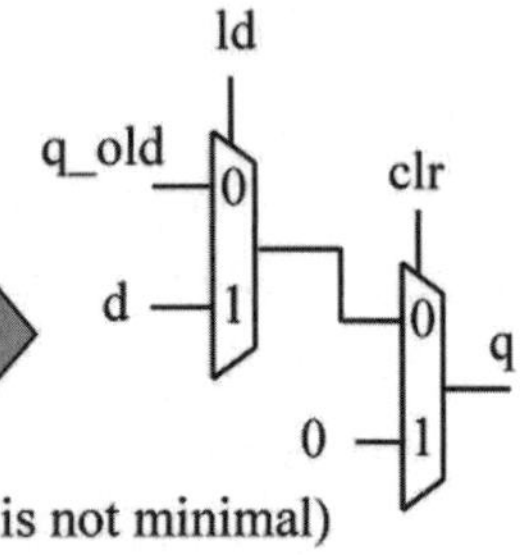

(logic is not minimal)

(b) `ld` takes precedence over `clr` if both are '1'

```
always @(ld or clr or d or q_old)
begin
  q = q_old;
  if (clr) q = 0;
  if (ld)  q = d;
end
```

ld	clr	q
0	0	q_old
0	1	0
1	0	d
1	1	d

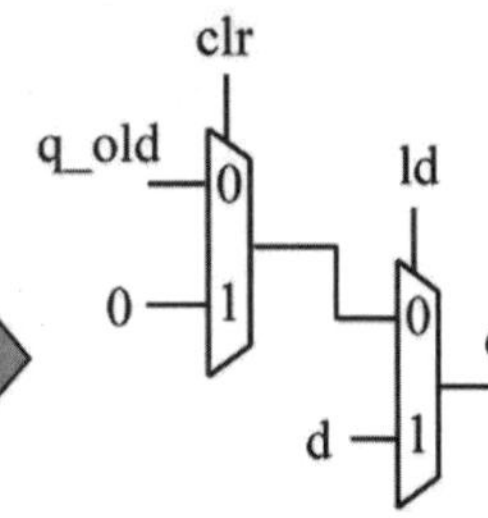

FIGURE 1.9: The sequential nature of `always` blocks

synthesis in general. Fig. 1.10(a) shows that an inferred latch is placed on the net output q because there is no assignment to q if `ld` is "0." Fig. 1.10(b) corrects this mistake by assigning a default value to the q output net. Fig. 1.10 gives a peek at how to specify sequential logic in Verilog; Section 1.5 covers this in detail.

Another viewpoint of an `always` block's functionality is that it is a complex form of an `assign` statement. A module can have multiple `always` blocks, with each `always` block representing discrete logic blocks, just as a module can have multiple `assign` statements. The ordering of `always` blocks in a module does not affect the logic that is synthesized, for the same reason that `assign` statement ordering does not affect synthesized logic. Except for tri-state logic implementations, the same output net cannot be driven from multiple `always` blocks, just as the same net cannot be driven from multiple `assign` statements, because the gate drivers synthesized for the output net in each `always` block will be connected together.

(a) Incorrect, produces an inferred latch as no assignment is made to `q` if `ld` is '0'

```
always @(ld or d)
begin
  if (ld) q = d;
end
```

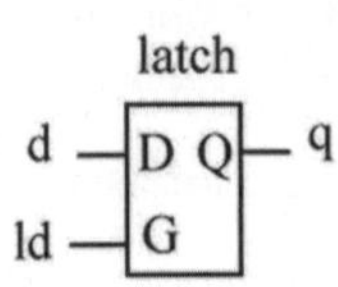

(b) Correct, produces combinational logic

```
always @(ld or d or q_old)
begin
  q = q_old;
  if (ld) q = d;
end
```

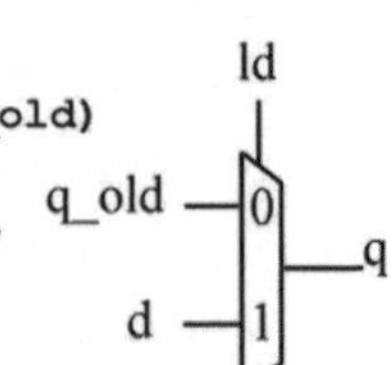

FIGURE 1.10: A common mistake with combinational `always` blocks: inferred latches

1.4 COMBINATIONAL BUILDING BLOCKS IN VERILOG

This section discusses common combinational building blocks and their implementation in Verilog, with new Verilog concepts introduced as they are required.

1.4.1 Multibit/Multiinput Muxes, Verilog Hierarchical Design, and Bus Notation

The 1-bit 2-to-1 mux of Fig. 1.4 can be extended to an N-bit 2-to-1 mux by paralleling N of the 1-bit muxes as shown in Fig. 1.11 for $N = 4$. The inputs a, b and the output y now become 4-bit wide busses, labeled as a[3:0], b[3:0], and y[3:0], respectively. The individual bits of these busses are labeled as a[0], a[1], etc., and connect to the appropriate data ports on each of the 1-bit 2-to-1 muxes. The s control input of the four 1-bit muxes are tied together so that the s input selects the four data bits of the a and b inputs in parallel. Specifying this in Verilog requires introducing the Verilog syntax for busses and hierarchical modules. In Verilog, a port or wire that is greater than 1 bit in width is officially referred to as a *vector*, but is informally referred to as a *bus* in this book. The bus indices can be arbitrary values in Verilog, but our examples assume that a bus carries an N-bit quantity whose most significant bit is $N - 1$ and whose least significant

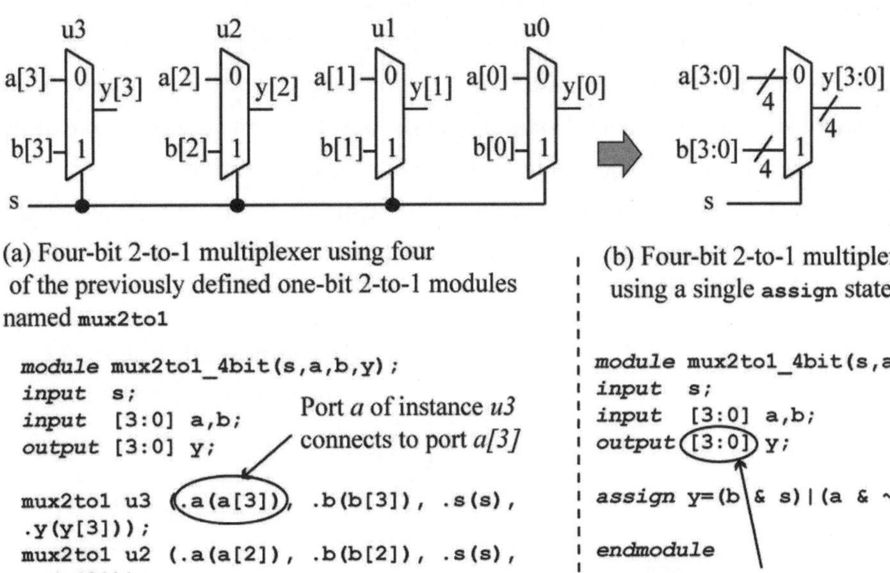

(a) Four-bit 2-to-1 multiplexer using four of the previously defined one-bit 2-to-1 modules named **mux2to1**

```
module mux2to1_4bit(s,a,b,y);
input  s;
input  [3:0] a,b;
output [3:0] y;

mux2to1 u3 (.a(a[3]), .b(b[3]), .s(s),
.y(y[3]));
mux2to1 u2 (.a(a[2]), .b(b[2]), .s(s),
.y(y[2]));
mux2to1 u1 (.a(a[1]), .b(b[1]), .s(s),
.y(y[1]));
mux2to1 u0 (.a(a[0]), .b(b[0]), .s(s),
.y(y[0]));

endmodule
```

Port *a* of instance *u3* connects to port *a[3]*

User-defined instance name, must be unique within this module.

Component's module name

(b) Four-bit 2-to-1 multiplexer using a single **assign** statement

```
module mux2to1_4bit(s,a,b,y);
input  s;
input  [3:0] a,b;
output [3:0] y;

assign y=(b & s)|(a & ~s);

endmodule
```

Bus definition for a 4-bit bus; most significant bit is *y[3]* and least significant bit is *y[0]*.

FIGURE 1.11: 4-bit 2-to-1 multiplexer implementations

bit is 0, to form a range $N-1$ to 0. In Verilog, this range is declared as "$[N-1:0]$," so a 4-bit bus has the range "[3:0]." Individual wires in the bus are labeled as *netname*[0], *netname*[1], etc. The 4-bit 2-to-1 mux in Fig. 1.11 thus has Verilog port names for the busses of `a[3:0]`, `b[3:0]`, and `y[3:0]`. Observe that in the `module` interface of Fig. 1.11(a), the `a`, `b` inputs use a separate `input` statement from the `s` input because their widths are different.

The module in Fig. 1.11(a) is a *hierarchical* module, because it uses four *instances* of the previously described `mux2to1` module to implement the 4-bit 2-to-1 mux. Each instance statement contains the module name (i.e., `mux2to1`), a user-defined instance name (i.e., `u3`), and a terminal list that describes how the instance's ports connect to nets or ports within the hierarchical module. This example uses *named association* within the terminal list, using the syntax "*.instance_port*(*netname*)" to connect an *instance_port* to the *netname* in the hierarchical module. A shorthand notation can be used in which the instance port names are not specified, with the hierarchical nets in the terminal list assumed to connect to instance ports in the order in which the instance ports are declared within the instance module. Named association is clearer and less prone to careless error, so all examples in this book used named association for terminal lists in instances.

Fig. 1.11(b) shows the 4-bit 2-to-1 mux implemented with the same `assign` statement used for the 1-bit version in Fig. 1.5(a), by simply changing the module interface to accommodate the 4-bit data ports. This illustrates Verilog's capability of specifying N-bit wide operations as easily as 1-bit wide operations. This is a powerful mechanism that improves a designer's productivity, allowing designers to focus on higher complexity designs.

Fig. 1.12 shows a 4-bit 4-to-1 mux implemented using 4-bit 2-to-1 muxes. While this is one of the traditional implementation methods for higher level muxes and the Verilog

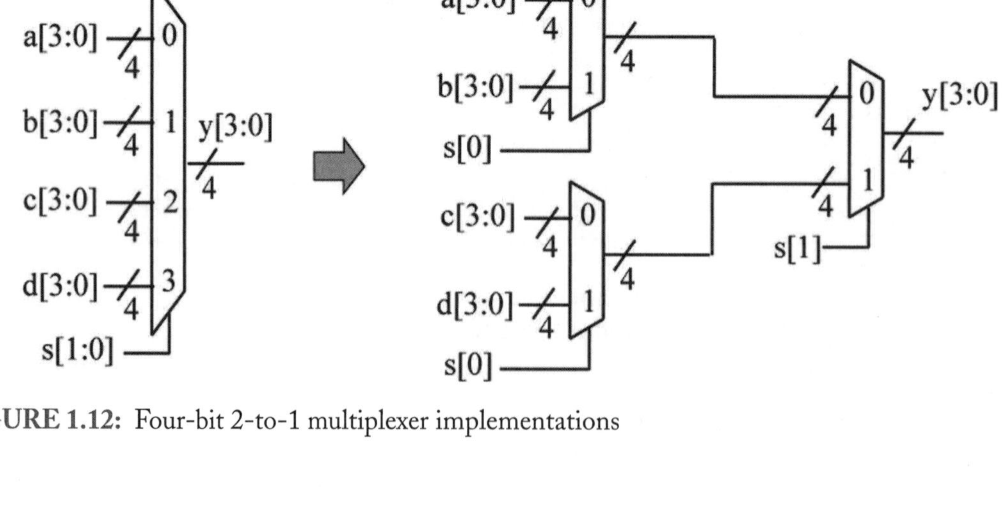

FIGURE 1.12: Four-bit 2-to-1 multiplexer implementations

(a) Four-bit 4-to-1 mux using `if-else` chain

```
module mux4to1_4bit(s,a,b,c,d,y);
input  [1:0] s;
input  [3:0] a,b,c,d;
output [3:0] y;

reg [3:0] y;

always @*
 begin
  if (s == 2'b00) y = a;
  else if (s == 2'b01) y = b;
  else if (s == 2'b10) y = c;
  else y = d;
 end

endmodule
```

2-bit binary constant

equality operator

(b) Four-bit 4-to-1 mux using a `case` statement

```
module mux4to1_4bit(s,a,b,c,d,y);
input  [1:0] s;
input  [3:0] a,b,c,d;
output [3:0] y;

reg [3:0] y;

always @*
 begin
  case (s)
   2'b00 :   y = a;
   2'b01 :   y = b;
   2'b10 :   y = c;
  default:   y = d;
  endcase
 end
endmodule
```

FIGURE 1.13: Verilog implementations of a 4-bit 2-to-1 multiplexer

specification could be structured in this manner, it may or may not be an efficient implementation for a particular target technology, such as an FPGA. It is the function of the logic synthesis tool to map a Verilog specification to a target technology using the most efficient gate structure for that technology, so the Verilog descriptions of combinational building blocks that offer clarity and leave out implementation details are usually preferred.

Fig. 1.13 gives two Verilog specifications for this 4-bit 4-to-1 bit mux that only specify the multiplexer's functionality, not the implementation. Both Verilog modules use `always` blocks, but Fig. 1.13(a) uses an *if-else* chain for the logic while Fig. 1.13(b) uses a *case* statement. In Fig. 1.13(a), the equality operator `==` is used to compare the `s` select input against the 2-bit binary constants `2'b00`, `2'b01`, `2b'10` in the *if-else* chain, corresponding to the assignments of `y=a`, `y=b`, and `y=c`. If none of these comparisons are true, then the last *else* clause makes the assignment `y=d`. The `case` statement used in Fig. 1.13(b) is a shorthand notation for an *if-else* chain; there is no advantage to either form in terms of the logic that is synthesized. The `default` clause in the `case` statement is the clause that is chosen if `s` does match any of the previous values given for `s`. Observe that in both `always` blocks, the output `y` is guaranteed to be assigned some value due to the use of the last *else* clause in Fig. 1.13(a) and the *default* clause in Fig. 1.13(b). This is important, as it prevents a latch from being inferred on the `y` output.

This is the first time that a Verilog module has used a constant that was greater than a single bit in width. The general format of an unsigned integer constant is *[size]'baseDDDD..D*, where size is the number of bits in the constant, *base* is a single letter designating the base, and *DDDD..D* are digits in the specified base. Supported bases are d/D (decimal), b/B (binary), o/O (octal), and h/H (hex). The default base is decimal; all other bases must use a base specification. The size specification is optional; it is sized by the logic synthesis tool to match the sizes of other operands in the associated expression. A constant is left padded with zeros if necessary.

1.4.2 Addition, Subtraction

Fig. 1.14(a) shows the truth table, Boolean equations, and logic symbol for a binary full adder. The A, B, and Ci (carry-in) inputs are summed using binary arithmetic to produce the S (sum) and Co (carry-out) outputs. Fig. 1.14(b) shows a 4-bit adder using a *ripple-carry* adder configuration in which four 1-bit binary full adders are used, with the carry-out of full adder i as the carry-in to full adder $i+1$. While this is the most gate-efficient implementation for an N-bit adder, it is also the slowest as the longest delay path passes through the carry chain of all N full adders. There are many other adder structures, such as *carry lookahead* and *carry select*, that are better suited for larger values of N given a particular maximum delay constraint,

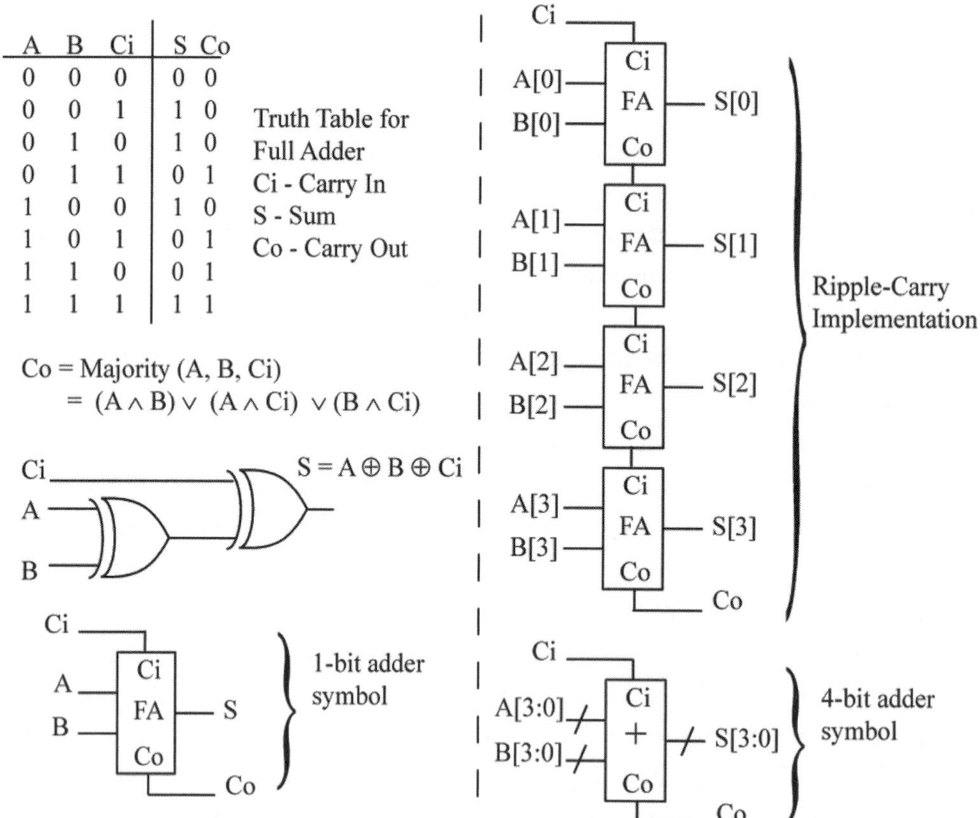

(a) Full Adder Truth table, Boolean equations, Symbol

(b) 4-bit adder symbol and Ripple Carry Implementation

FIGURE 1.14: Binary full adder and a ripple-carry 4-bit adder

(a) Four-bit adder with no carry-in or carry-out

```
//4-bit adder
// no carry-in, carry-out
module add4bit ( a, b, s);

input [3:0] a,b;
output [3:0] s;

assign s = a + b;

endmodule
```

(b) Four-bit adder with carry-in, carry-out

```
//4-bit adder with carry-in, carry-out
module add4bit (ci, a, b, s, co);

input ci;
input [3:0] a,b;
output [3:0] s;
output co;

wire [4:0] y;

//do 5-bit sum so that we
// have access to carry out
assign y = {1'b0,a} + {1'b0,b} + {4'b0,ci};
assign s = y[3:0]; //four-bit output
assign co = y[4]; //carry-out

endmodule
```

$0a_3a_2a_1a_0$
$+\,0b_3b_2b_1b_0$
$+\,0\,0\,0\,0\,c_i$
$y_4y_3y_2y_1y_0$

{ } is the concatenation operator

FIGURE 1.15: Verilog 4-bit adder implementations

but the N value at which one adder implementation becomes better than another is dependent on the implementation technology. For example, one FPGA vendor may implement fast carry chains so a ripple-carry structure may have the same delay as a carry lookahead structure for $N < 12$, while for a different FPGA vendor this may be true for $N < 8$. As such, for small to medium sized adders it is generally better to specify a behavioral implementation of the addition using the Verilog "+" operator instead of a structural implementation, and let the logic synthesis tool choose the best adder structure. For large adders, FPGA vendors will typically supply a parameterized library element that offers performance and gate count tradeoffs for their particular architecture. Increased performance comes at increased gate count, which is a well-known tradeoff in digital system design.

The Verilog module of Fig. 1.15(a) implements a 4-bit addition using the continuous assignment `s = a + b`. The use of the "+" operator leaves the logic synthesis tool free to choose the adder structure that will best meet any user-specified constraint, such as delay or area. The Verilog module of Fig. 1.15(b) also implements a 4-bit addition, except that it also provides carry-in and carry-out ports, similar to the adder of Fig. 1.14(b). The addition performed in the first assignment statement "`y =` " is a 5-bit sum with the fifth bit used as the carry-out signal. The concatenation operator {} is used to transform the 4-bit `a`, `b` operands to 5-bit operands by appending a most significant bit of 0. The carry-in bit is concatenated with four leading "0"s by the `{4'b0,ci}` operation and is also part of the sum in the first assignment. The sum output `s` is the lower 4 bits of the 5-bit `y` internal sum.

A *full subtractor* and an N-bit subtractor can be derived in a similar manner to that of the full adder and N-bit ripple-carry adder. As with addition, subtraction is supported within

Even though these two implementations are functionally equivalent, some synthesis tools may produce a slightly more efficient implementation for (b) than (a).

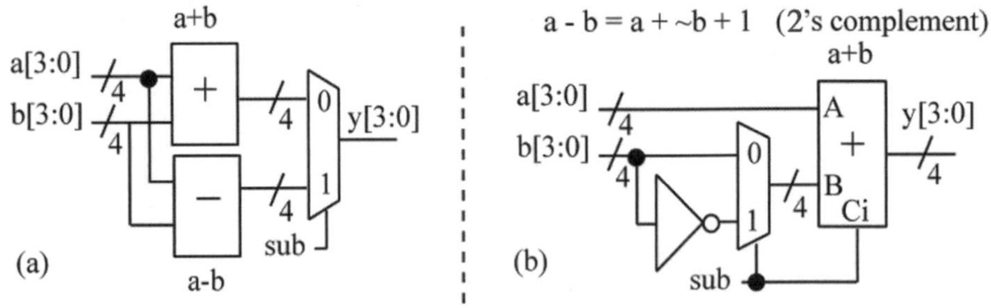


```
//adder/subtractor
module addsub4bit(sub,a,b,y);

input sub;
input [3:0] a,b;
output [3:0] y;
//                    sub=1   sub=0
assign y = sub ?(a-b):(a+b);

endmodule
```

```
//adder/subtractor
module addsub4bit ( sub, a, b, y);

input sub;
input [3:0] a,b;
output [3:0] y;

wire [3:0] btmp;
//                        sub=1 sub=0
assign btmp = sub ? ~b   : b;
assign y = a + btmp + {3'b0,sub};

endmodule
```

FIGURE 1.16: Adder/subtractor implementations

Verilog by the subtraction operator "−." Fig. 1.16(a) shows an adder/subtractor implementation that uses a multiplexer to select between addition "+" and subtraction "−" operations based on the sub input. While this may be the most intuitive way to construct an adder/subtractor, it is not necessarily the most efficient as it implies separate subtraction and addition blocks. The traditional implementation of an adder/subtractor is shown in Fig. 1.16(b), which uses the fact that a-b is equal to a+∼b+1, that is, add the 2's complement of b to a to perform the subtraction. It is obvious that Fig. 1.16(b) can be implemented in fewer gates than that of Fig. 1.16(a) if both structures are mapped in a straightforward manner to gate primitives. However, because the two Verilog modules are functionally equivalent, when the Verilog operators are mapped to Boolean equations and logic optimization is done, one might expect that the same implementation is created for both. However, it is dependent on the particular logic synthesis tool that is used as to whether one of the two Verilog modules in Fig. 1.16 is implemented in a more efficient manner than the other. This example is given to illustrate that logic synthesis is not a magic panacea that removes all responsibility from the designer for creating efficient implementations. The manner in which the Verilog RTL (*Register Transfer Level*) is written can affect the efficiency of the resulting implementations, and a good designer is always aware of this.

								Binary	Decimal
multiplicand				r_2	r_1	r_0		1 1 1	7
multiplier			X	s_2	s_1	s_0		x 1 0 1	x 5
partial product				s_0*r_2	s_0*r_1	s_0*r_0		1 1 1	35
			s_1*r_2	s_1*r_1	s_1*r_0			0 0 0	
+		s_2*r_2	s_2*r_1	s_2*r_0				+ 1 1 1	
	p_5	p_4	p_3	p_2	p_1	p_0	product	1 0 0 0 1 1 = 35	

FIGURE 1.17: The multiplication operation for unsigned numbers

1.4.3 Multiplication, Division

Fig. 1.17 shows an unsigned 3 × 3 multiplication. A k-bit multiplicand and an m-bit multiplier produce a $(k + m)$-bit product. Typically, the operands in a multiplication are of the same length, so two n-bit operands form a $2n$-bit product. Depending on the implementation, some of the product bits may have to be discarded; choosing which bits to discard depends on the fixed-point representation chosen for the two operands.

As seen in Fig. 1.17, a binary multiplication operation is an AND operation. Each bit of the multiplier is AND'ed with the multiplicand to form *partial products*, which are summed to form the product. An intuitive implementation of a multiplier that directly maps from the operations of Fig. 1.17 is shown in Fig. 1.18(a). This is a combinational *array multiplier* and while it is intuitive and functionally correct, there are other array multiplier structures that offer higher performance as the operand size increases. Fig. 1.18(b) shows the 3 × 3

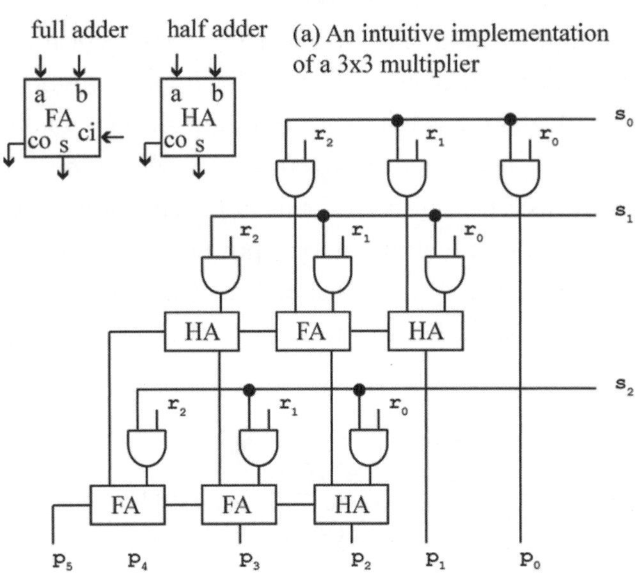

(b) A Verilog implementation of a 3x3 multiplication using the '*' operator

```
//3x3 multiplier
//using the '*' operator

module mult3x3 (a, b, y);

input [2:0] a,b;
output [5:0] y;

//do 3x3 multiply
assign y = a * b;

endmodule
```

FIGURE 1.18: (a) An intuitive 3 × 3 array multiplier and (b) a 3 × 3 multiplier in Verilog

multiplication implemented in a Verilog module using the "∗" operator. For the same reasons as the "+" operation, it is usually advantageous to use the "∗" operator for small to medium width multipliers and let the logic synthesis tool choose the best multiplier structure. For multipliers with large width operands, most logic synthesis tools have a library of multiplier implementations that offer performance versus gate count tradeoffs. It must also be noted that the multiplier implementation also depends on whether the operands are unsigned or signed (2's complement); the multiplier synthesized by the Verilog "∗" operator is an unsigned multiplier.

The division operation is only mentioned briefly here as it is similar to multiplication in terms of logic synthesis. Verilog has a division operator ("/") that will synthesize to an unsigned combinational division implementation. However, as with multiplication, most logic synthesis tools offer a parameterized library module for division that provides choices for unsigned versus signed operands, the sizes of the dividend, remainder and quotient, and performance versus gate count tradeoffs.

1.4.4 Shifting

Right and left shift operations by a single bit position on 8-bit values are shown in Figs. 1.19(a) and 1.19(b), respectively. The s_i value is the bit shifted into the vacant bit position, which is the most significant bit for a right shift, and the least significant bit for a left shift. A 1-bit shift implementation is simply wiring, as shown by the Verilog concatenation statements that accomplish the shift. The right shift operation copies the upper 7 bits of a to the lower 7 bits of y, with the most significant bit of y filled by s_i. The left shift operation copies the lower 7 bits of a to the upper 7 bits of y, with the least significant bit filled by s_i.

Fig. 1.20(a) shows a multiplexer implementation of an 8-bit shift block that can shift left by 1 bit, right by 1 bit, or pass its input through unchanged. The Verilog implementation of this shift block is shown in Fig. 1.20(b). Verilog also has left (<<) and right shift (>>) operators that can be used for shifting; both of these operators use a zero for the shift input bit. The Verilog module in Fig. 1.20(b) allows the user to control the value of the shift input bit (s_i).

The shifter implemented in Fig. 1.20 only shifts by one position; a shifter that can shift multiple positions is called a *barrel shifter*. Fig. 1.21(a) shows the traditional multiplexer implementation of a 32-bit barrel shifter that shifts left by 0 to 31 positions based on the 5-bit *s*

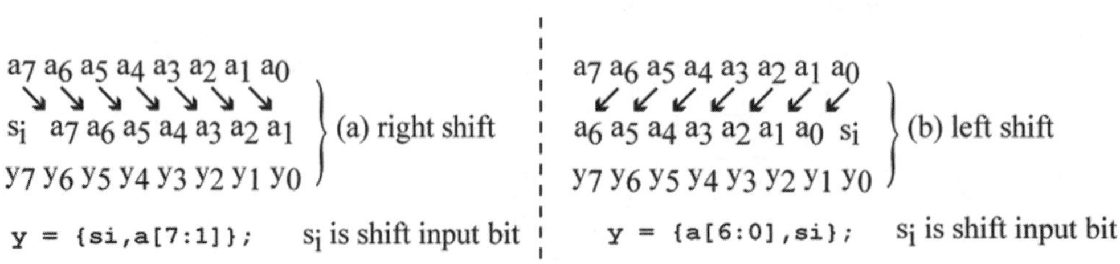

FIGURE 1.19: Right/left shift operations

(a) left/right shift with a multiplexer

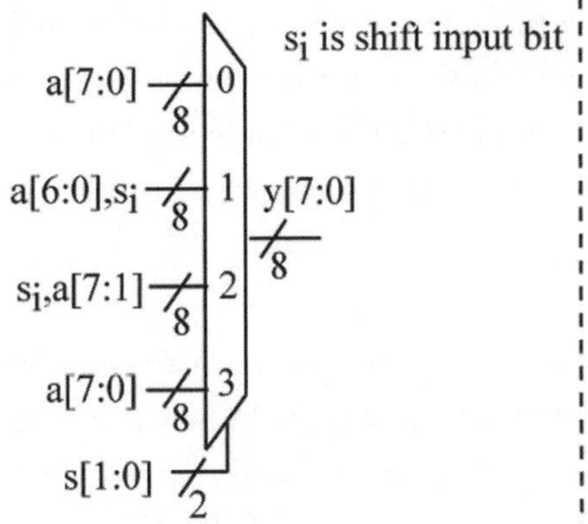

s[0]= sleft (shift left)
s[1]= sright (shift right)

s	*y*
00	*y* = *a*
01	*y* is *a* shifted to left
10	*y* is *a* shifted to right
11	*y* = *a*

(b) Verilog left/right shift

```
module lrshift_8bit (si,sleft,sright,a,y);

input si,sleft,sright;
input [7:0] a;
output [7:0] y;

reg [7:0] y;
wire [1:0] s;

assign s[0] = sleft;
assign s[1] = sright;

always @(s or a) begin
 case (s)
  2'b01   : y = {a[6:0],si}; //left shift
  2'b10   : y = {si,a[7:1]}; //right shift
  default: y = a;
 endcase
end

endmodule
```

FIGURE 1.20: Verilog left/right shift by one example

(a) Multiplexer implementation of a 32-bit left-shift barrel shifter

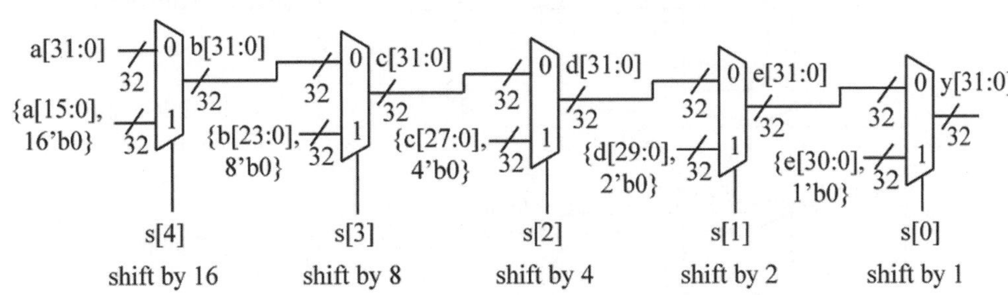

(b) Verilog implementation

```
module bshift_32bit (s, a, y);
input [4:0] s;
input [31:0] a;
output [31:0] y;

assign y = a << s;

endmodule
```

FIGURE 1.21: Barrel shifter (32-bit, left shift)

input. Each stage of the barrel shifter corresponds to a fixed shift by a power of 2 (16, 8, 4, 2, 1) controlled by a bit in *s*. The composition of the five stages produces a shift amount between 0 and 31. The Verilog implementation of this shifter is implemented by a single continuous assignment statement of `y = a << s`. This again illustrates the power of an HDL to hide implementation details from a designer and to specify complex logic blocks in a very compact manner.

1.4.5 Tri-State Logic

Fig. 1.22(a) shows the operation of a *tri-state* buffer with a high true enable. A tri-state buffer can be thought of as a noninverting buffer with a switch on its output. When the buffer is enabled, the output is driven by the noninverting buffer. When the buffer is disabled, the output is disconnected from the noninverting buffer, and the output is left floating or in the *high-impedance* logic state. The high-impedance logic state is traditionally indicated by the "*Z*" value. Two common uses of tri-state logic are shown in Figs. 1.22(b) and Figs. 1.22(c). The half-duplex link shown in Fig. 1.22(b) between blocks 1 and 2 allows communication in either

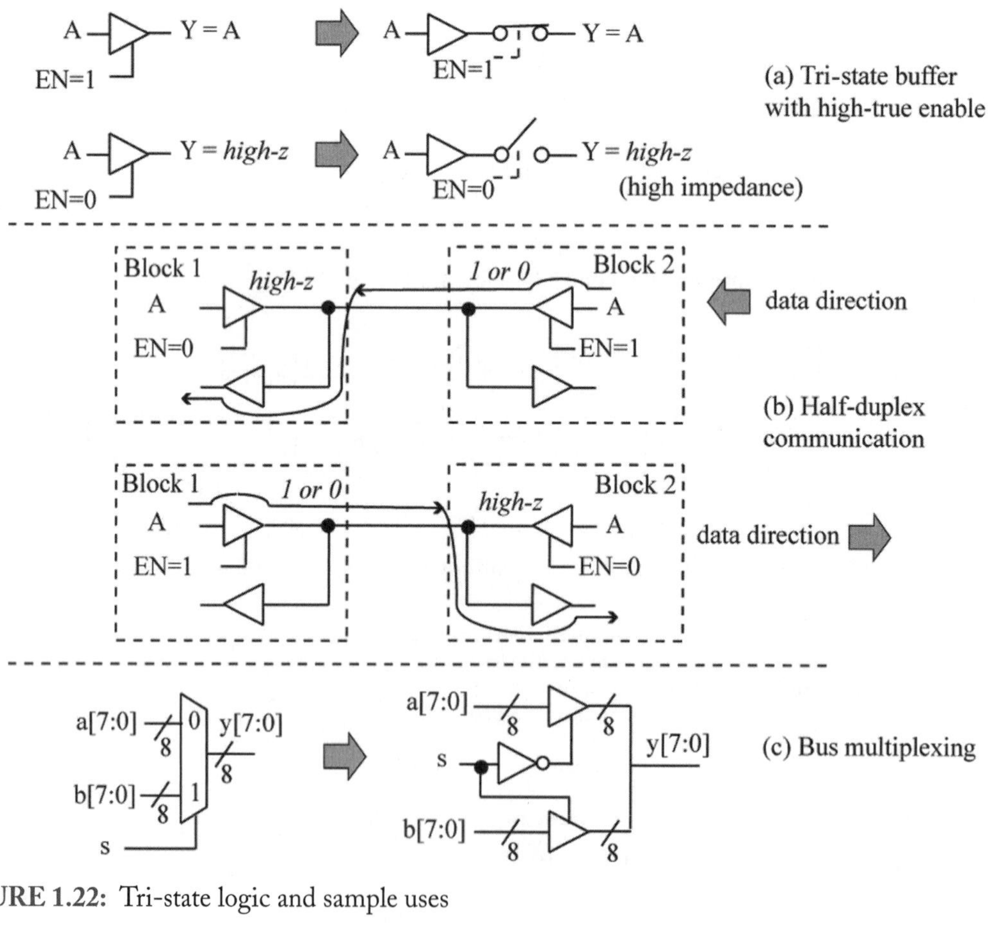

FIGURE 1.22: Tri-state logic and sample uses

(a) Bus multiplexing with tri-state drivers

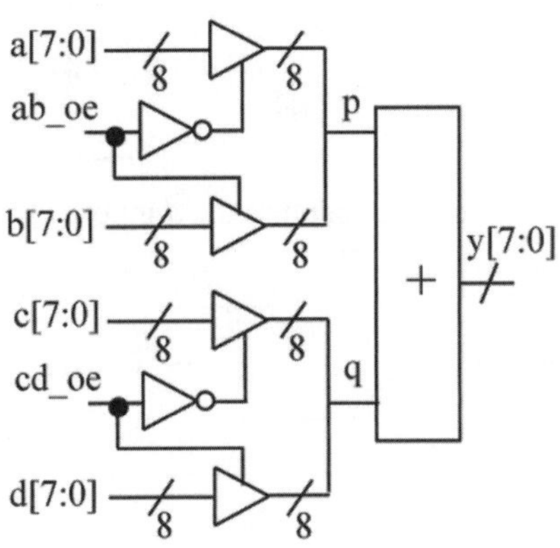

(b) Verilog implementation

```
//tri-state buffer test
module tsb (a, b, c, d, ab_oe, cd_oe, y);

input ab_oe,cd_oe;
input [7:0] a,b,c,d;
output [7:0] y;
wire [7:0] p,q;

//                 ab_oe=1 ab_oe=0
assign p = ~ab_oe ? a : 8'bzzzzzzzz;
assign p = ab_oe ? b : 8'bzzzzzzzz;
//                 cd_oe=1 cd_oe=0
assign q = ~cd_oe ? c : 8'bzzzzzzzz;
assign q = cd_oe ? d : 8'bzzzzzzzz;

assign y = p + q;

endmodule
```

'z' is an allowed logic value and implies a tri-state driver for synthesis

Assignments to the same wire allowed because of tri-state drivers

FIGURE 1.23: Tri-state logic in Verilog

direction over a single wire, but only one of the blocks can be driving the wire at a particular time. Tri-state buffers can also be used to implement bus multiplexing as shown in Fig. 1.22(c). Tri-state logic is most commonly used for signals that drive off-chip to a shared bus, but some FPGA vendors implement on-chip tri-state logic as well.

An example of tri-state bus multiplexing and its implementation in Verilog is given in Fig. 1.23. Each adder input in Fig. 1.23(a) is selected from two different 8-bit values, where the multiplexing is done by tri-state buffers. The Verilog implementation in Fig. 1.23(b) uses internal wires `p` and `q` as the adder inputs. Wire `p` has two continuous assignments, with each assignment using a conditional statement that selects between an input port value (`a`, `b`) and an 8-bit tri-state value given as `8'bzzzzzzzz`. The assignment of "z" to a bit value in Verilog implies a tri-state driver when the Verilog is synthesized. Inferred tri-state drivers are the only time that multiple continuous assignments can be made to the same wire in Verilog. Some synthesis tools may replace tri-state bus multiplexing with equivalent logic multiplexing if the target implementation technology does not support tri-state drivers.

1.5 SEQUENTIAL LOGIC IN VERILOG

This section reviews sequential systems, 1-bit storage elements, and common sequential building blocks, along with their Verilog implementations.

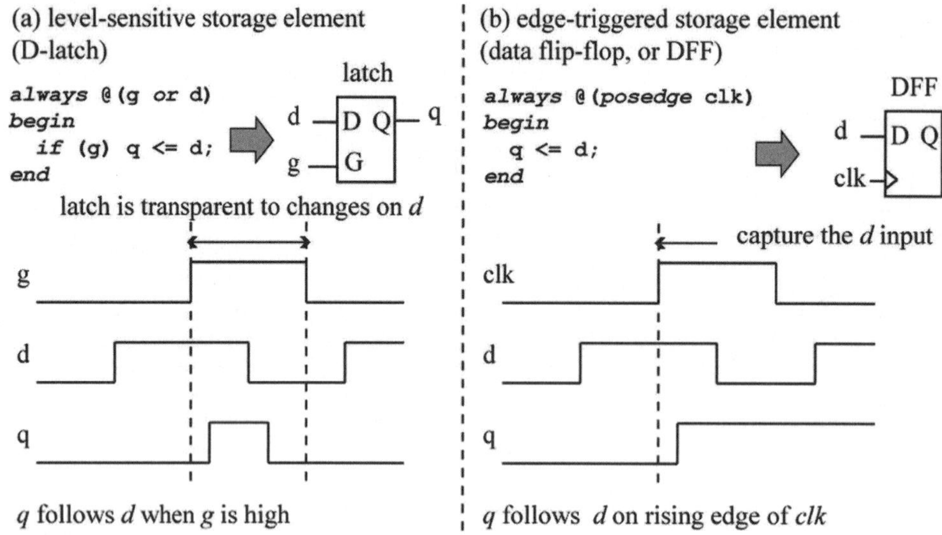

FIGURE 1.24: Level-sensitive and edge-triggered devices in Verilog

1.5.1 One-bit Storage Elements

A *sequential* system differs from a combinational system in that the sequential system has internal state (or memory) and thus its outputs are dependent upon both its external inputs and internal state. One-bit storage elements are used for state storage in a sequential system. Fig. 1.24(a) shows the logic symbol, Verilog implementation, and timing diagram for a *level-sensitive* 1-bit storage element known as a *D-latch*. The device is termed *level-sensitive* because when the gate (G) signal is high, then the Q output follows the D input, i.e., the latch is in *transparent* mode. When G drops low, then the latch's internal state becomes equal to the last D input value when G was high. The Verilog implementation of a D-latch is an `always` block that makes a *nonblocking* assignment ("<=") of `d` to `q` when the `g` input is nonzero. Observe that if the `g` input is zero, then the `always` block does not make any assignment to `q`, causing the synthesis tool to infer a latch on the `q` output as the `q` output must retain its last known `d` value when `g` was nonzero. Nonblocking assignments ("<=") as opposed to blocking assignments ("=") should be used in `always` blocks that are used to synthesize sequential logic; this is discussed in more detail in Section 1.6.

Fig. 1.24(b) gives the logic symbol, Verilog implementation, and timing diagram for a *rising edge-triggered* 1-bit storage element known as a data flip-flop (DFF). The DFF is said to be edge-triggered as the DFF's internal state is only affected on the active edge of the clock (*clk*) input, which causes the internal state to become equal to the D input value. The `q` output always reflects the internal state of the DFF. The DFF shown in Fig. 1.24(b) is rising-edge triggered; the logic symbol for a falling-edge triggered DFF has a bubble on the clock input.

The *D* input of the DFF is said to be *synchronous* with respect to the clock input as the *D* input only affects the *Q* output on the active clock edge. The Verilog DFF implementation indicates the sensitivity of the DFF to the rising clock edge by using the keyword *posedge* (positive edge) in the event list of the `always` block. The `d` input does not appear on the event list of the `always` block as it is the `clk` input that triggers a possible change to the `q` output, not the `d` input. The body of the `always` block simply makes a nonblocking assignment of `d` to `q`.

In general, edge-triggered storage elements are preferred to level-sensitive storage elements because of simpler timing requirements, and as such, this book uses DFFs exclusively in its designs. You may be familiar with other types of edge-triggered storage elements such as the JK flip-flop or T (toggle) flip-flop; these devices are DFFs with extra logic placed around them. The 1-bit edge-triggered storage elements provided by FPGA vendors are DFFs because of their simplicity and speed.

1.5.2 DFF Chains

The `always` blocks of Fig. 1.25(a) can be somewhat confusing to readers who are new to logic synthesis in that all of these Verilog code fragments synthesize to the same chain of DFFs as shown. This occurs because each nonblocking assignment synthesizes to a single DFF whose

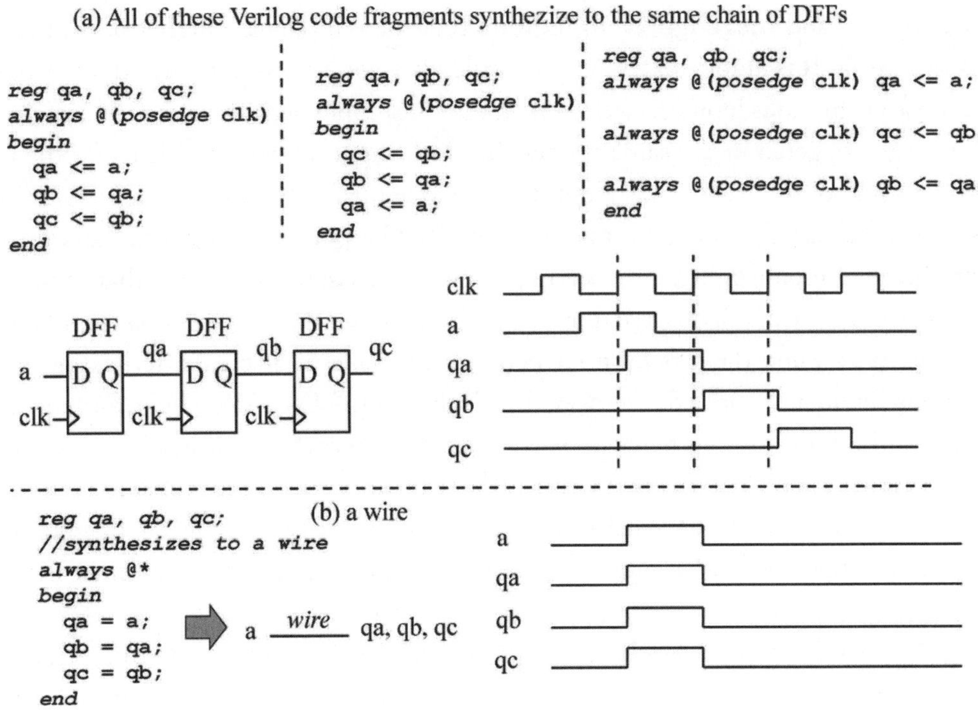

FIGURE 1.25: Synthesizing a chain of DFFs versus a wire

input happens to be the output of another nonblocking assignment. The ordering of these nonblocking assignments within an `always` block does not matter since they are assignments to different outputs, and these nonblocking assignments can be either in a single `always` block or in separate `always` blocks as shown. The key factor is that each of the nonblocking assignments is protected by "`posedge clk`" in the event list of the various `always` blocks, causing a DFF to be synthesized for that assignment. The timing diagram in Fig. 1.25(a) shows that the edge-triggered nature of the DFF causes the `qa`, `qb`, and `qc` outputs to change to the DFF input value at the rising clock edge. Contrast this to the blocking assignments in Fig. 1.25(b) that are not protected by "`posedge clk`" in the event list. This causes combinational logic to be synthesized, which simplifies to a wire from the input `a` to the three outputs `qa`, `qb`, and `qc`. Changes on the input `a` are immediately propagated by the wire to the `qa`, `qb`, and `qc` outputs as shown in the timing diagram.

1.5.3 Asynchronous Versus Synchronous Inputs

Fig. 1.26 illustrates the differences between asynchronous and synchronous inputs to a DFF. An asynchronous input affects the DFF's internal state independent of the active clock edge, while a synchronous input requires an active clock edge. The DFF in Fig. 1.26(a) has high-true reset (R) and low-true set (S) asynchronous inputs; the polarity of these inputs is arbitrary and opposite polarities were chosen for example purposes. The event list of the Verilog `always` block uses "`posedge r`" and "`negedge s`" to indicate that these inputs are high-true and low-true inputs, respectively. It is unfortunate that the Verilog keywords `posedge`/`negedge` are required for use with asynchronous inputs as well as with the clock input, as it implies that asynchronous inputs are edge-triggered in the same way as the clock input. However, this is not the case as asynchronous inputs are *level-sensitive* inputs, and force the DFF output either low or high as long as they are asserted, overriding the clock input. The `posedge`/`negedge` keywords used with asynchronous inputs indicate the leading edge of the assertion level for that input. In the `always` block of Fig. 1.26(a), observe that the *if-else* chain has the asynchronous input behavior specified first (`r`, `s`), and the synchronous behavior in the last *else* clause, indicating that the asynchronous inputs take precedence over the clock input. This *if-else* chain format must be followed with the asynchronous behavior specified first and synchronous behavior specified last for a logic synthesis tool to correctly infer a DFF with asynchronous inputs. The ordering of asynchronous inputs in the *if-else* chain determines their priority. The timing diagram of Fig. 1.26(a) indicates that the DFF's internal state is affected upon assertion of an asynchronous input.

Fig. 1.26(b) shows a DFF with synchronous clear (`clr`) and preset (`pre`) inputs. The naming of these inputs is arbitrary; in this example different names were chosen from the asynchronous set and reset inputs of Fig. 1.26(a) to avoid confusion. Observe that the only

(a) DFF with asynchronous Set/Reset

s is low-true, *r* is high-true

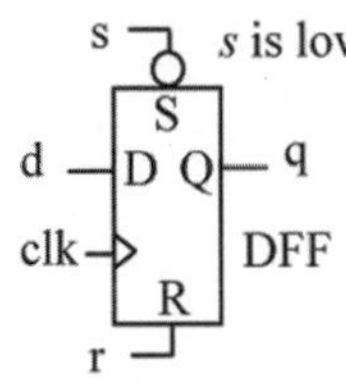

r	*s*	*clk*	*q*
1	x	x	0
0	0	x	1
0	1	?	d
0	1	1 or 0	q_{old}

```
always @(posedge clk or
         posedge r or
         negedge s)
begin
  if (r) q <= 1'b0;
  else if (!s) q <= 1'b1;
  else q <= d;
end
```

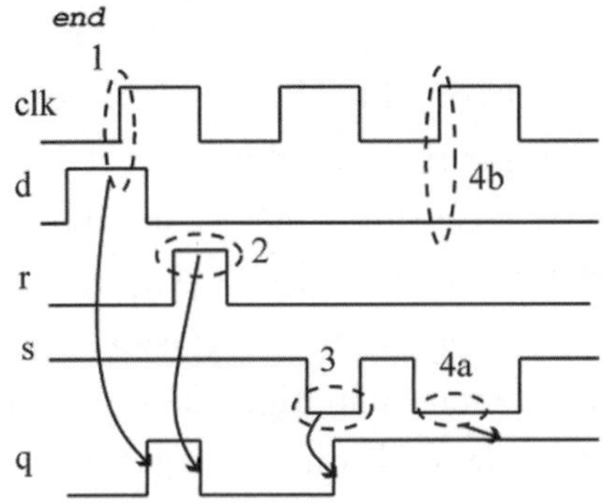

1. *d* input is captured on rising clock edge as *r*, *s* are negated; *q* becomes '1'.
2. *r* input is asserted; *q* becomes '0'.
3. *s* input is asserted; *q* becomes '1'.
4. assertion of *s* input (a) overrides clock input (b), so *q* output remains as '1'.

(b) DFF with synchronous preset/clear

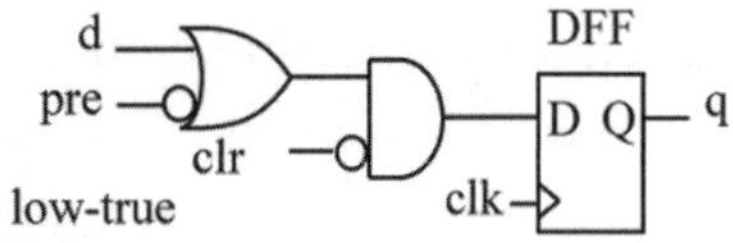

pre is low-true
clr is high-true
clr has priority over *pre*

```
always @(posedge clk)
begin
  q <= d;  //lowest priority
  if (!pre) q <= 1'b1;
  if (clr) q <= 0'b0; //highest priority
end
```

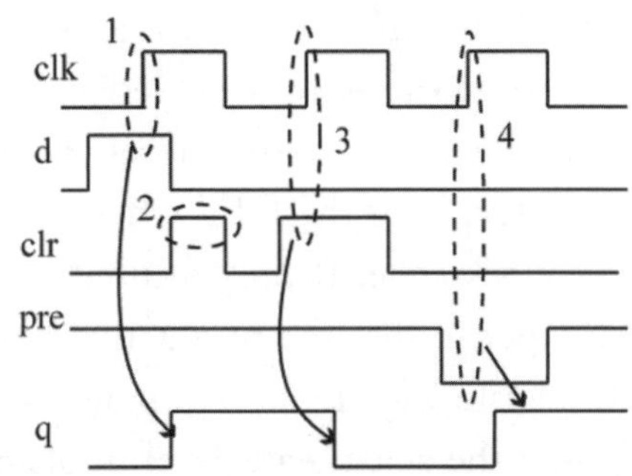

1. *d* input is captured on rising clock edge as *clr*, *pre* are negated; *q* becomes '1'.
2. assertion of synchronous input has no effect if does not occur on the active clock edge
3. *clr* input is asserted on rising clock edge; *q* becomes '0'.
4. *pre* input is asserted on rising clock edge; *q* becomes '1'.

FIGURE 1.26: Asynchronous versus synchronous inputs to a DFF

input that appears in the event list of the `always` block is the clock (`clk`) input as this DFF has no asynchronous inputs. The logic shown is a conceptual gate-level implementation of the synchronous clear and preset operations that are specified in the `always` block; be aware that the target implementation technology determines the actual synthesized logic. The `always` block is written such that the lowest priority assignment to the `q` output is done first, and the highest priority assignment to `q` is made last. An *if-else* chain similar to that of Fig. 1.26(a) could have also been written; unlike the asynchronous `always` block, there is no particular format for specification of synchronous logic. The timing diagram of Fig 1.26(b) indicates that assertion of a synchronous input does not affect the DFF's internal state until the next active clock edge.

In general, asynchronous inputs to DFFs are reserved for power-on logic, while synchronous inputs are used during normal operation.

1.5.4 Registers, Counters, and Shift Registers

The combinational building blocks of Section 1.4 and DFFs can be combined to form commonly used sequential building blocks such as registers, counters, and shift registers. A common need in a sequential system is the ability to store an N-bit value over several clock periods. Paralleling N-DFFs forms a block that can store an N-bit value, but the problem with DFFs is that they can change value on every active clock edge. Fig. 1.27 shows that a *register* is built by combining a DFF with a 2-to-1 multiplexer to form a device that only changes its value when the load (`ld`) input is asserted on the active clock edge. The register contained in Fig. 1.27 is an 8-bit register with an asynchronous low-true reset (`r`). The multiplexer steers the external `d` input of the register to the D input of the DFF when `ld` is asserted, causing a new value to be loaded on the next active clock edge. If the `ld` input is negated, then the register's output is steered back to the DFF input so that the next active clock edge simply loads the register's old value, thus retaining the register's contents. The gate-level implementation in Fig. 1.27 is provided for conceptual purposes; the actual gate-level implementation depends on the target technology. The Verilog implementation is an `always` block of sequential logic whose synchronous section

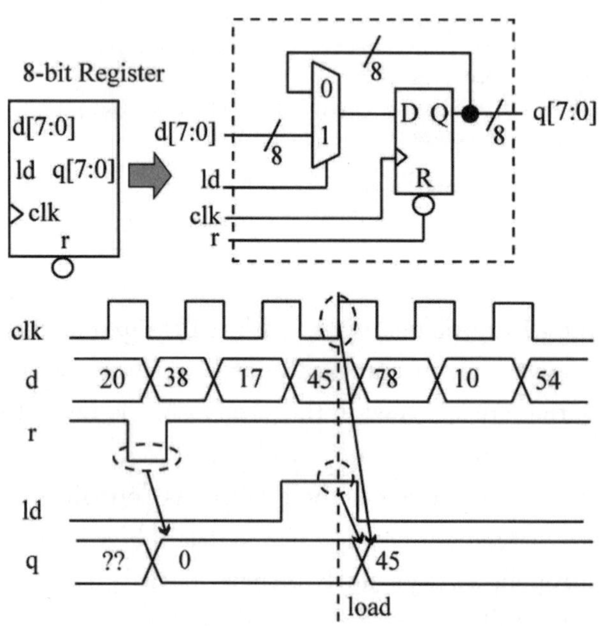


```
module reg8bit(clk,r,d,ld,q);

input clk,r,ld;
input [7:0] d;
output [7:0] q;

reg [7:0] q;

//register
always @(posedge clk
         or negedge r)
begin
  //async reset
  if (!r) q <= 8'b0;
  else begin
  //synchronous inputs
   if (ld) q <= d;
  end
end

endmodule
```

FIGURE 1.27: Register with low-true asynchronous reset

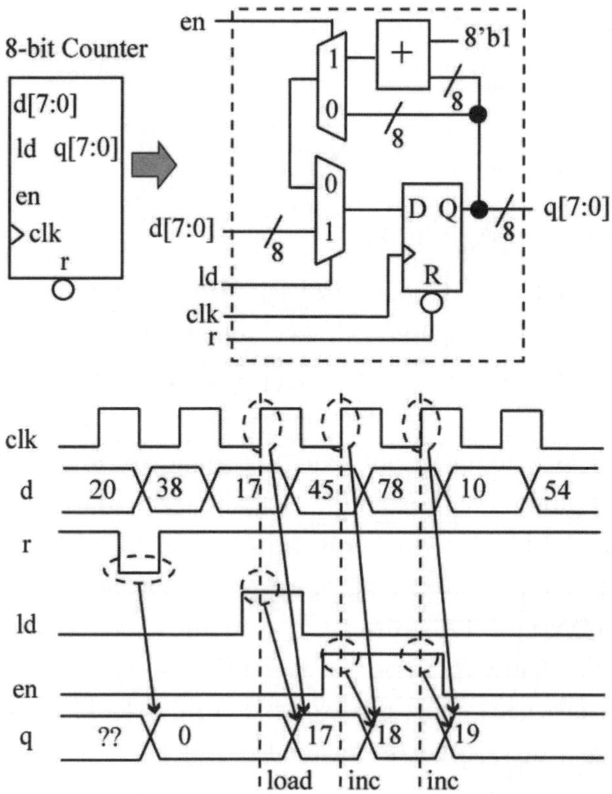


```
module cnt8bit (clk,r,d,
                ld,en,q);

input clk,r,ld,en;
input [7:0] d;
output [7:0] q;

reg [7:0] q;

//register
always @(posedge clk
          or negedge r) begin
 //async reset
 if (!r) q <= 8'b0;
 else begin
 //synchronous inputs
  if (en) q <= q + 8'b1;
  if (ld) q <= d;
 end
end

endmodule
```

FIGURE 1.28: Binary up-counter with low-true asynchronous reset

is an *if* statement that assigns the `d` input to the `q` output whenever the `ld` input is nonzero. If the `ld` input is zero, then no assignment is made to `q`, causing it to retain its last value.

Fig. 1.28 gives the logic symbol, conceptual gate-level logic diagram, timing diagram, and Verilog implementation for an 8-bit up-counter. In comparison to the register, the up-counter logic symbol has one extra control signal named `en`; the counter increases by one when `en` is asserted on an active clock edge. The timing diagram shows the counter loaded with the value 17 by asserting `ld` on an active clock edge, followed by assertion of `en` for two consecutive clock cycles causing the counter to increase to 18, and then to 19. The counter then holds the value 19 stable as both `ld` and `en` are negated for the remaining clock cycles. The Verilog implementation is the same as for the register except that the `always` block contains an extra *if* statement that performs the increment of `q<=q + 8'b1` when `en` is nonzero. The *if* statement for the load capability follows the *if* statement for the increment operation, giving load precedence over increment if both `ld` and `en` are asserted in the same clock cycle. This matches the multiplexer structure that is shown in the conceptual gate-level logic diagram. As one might expect, counters are useful for counting the number of operations performed in a digital system. However, another

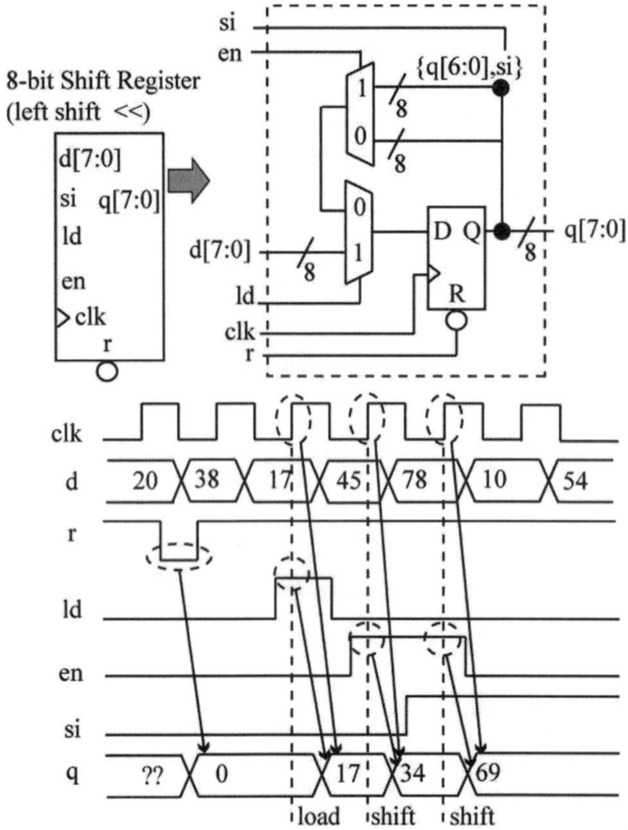


```
module sreg8bit (clk,r,d,ld,
                 en,si,q);

input clk,r,ld,en,si;
input [7:0] d;
output [7:0] q;

reg [7:0] q;

//shift register
always @(posedge clk
          or negedge r) begin
 //async reset
 if (!r) q <= 8'b0;
 else begin
 //synchronous inputs
  if (en) q <= {q[6:0],si};
  if (ld) q <= d;
 end
end

endmodule
```

FIGURE 1.29: Shift register (left shift) with low-true asynchronous reset

common use of counters is for providing addresses to a memory system, as memory locations are typically accessed in a sequential fashion.

Fig. 1.29 gives the logic symbol, conceptual gate-level logic diagram, timing diagram, and Verilog implementation for an 8-bit shift register (left shift). This has the same structure as the counter except that the incrementer in the multiplexer feedback path has been replaced by a 1-bit wired shift-left using the s_i input as shift-in for the least significant bit. The timing diagram loads the shift register with the value 17, and then shifts this value to the left in the next consecutive clock cycle by asserting the en input. The result of the first shift left is 34 (17*2+0) as the s_i input is "0," while the result of the following left shift is 69 (34*2 +1) because s_i is now "1." The Verilog implementation is the same as for the counter, except that the *if* statement that implemented the increment operation has been replaced by an *if* statement that implements the shift-left capability. One use of shift registers is for fast multiplication or division by 2. Another more common usage is for serial communication, which sends or receives an *N*-bit datum 1 bit at a time over a serial link.

1.6 EVENT-DRIVEN SIMULATION AND VERILOG

In previous sections, combinational and sequential building blocks have been presented along with their Verilog representations; these building blocks are used to form digital systems. To verify the functionality of a digital system, both the Verilog code and the synthesized gate-level netlist can be simulated in a *digital logic simulator*. This section covers the basics of how digital logic simulators operate, and the differences between Verilog simulation (presynthesis) and gate-level simulation (postsynthesis).

1.6.1 Event-Driven Simulation Basics

Various approaches have been taken in the past in creating digital logic simulators. One of the first approaches is that of a so-called Levelized Compiled Code (LCC) simulation. The advantage of this approach is speed since each circuit is transformed into a computer program that is compiled. The disadvantage of this approach is that all timing information is lost and thus these types of simulations are often referred to as 0-delay (zero-delay) simulations.

In order to generate timing information, an event-based simulation model is typically used. An event is defined as a logic-value change in a net at some instant in time. Event-driven (ED) simulations are quite different from LCC since events can occur simultaneously in time in real circuits and thus the parallelism inherent in logic circuits is more accurately modeled. To illustrate how a basic ED simulator operates, consider the example circuit in Fig. 1.30.

In the example circuit, all of the nets are given labels (A–F, H, K–N, P). Events are defined for a net when a logic level changes. In a basic ED simulator, a list or a queue is maintained that contains every net for which an event occurred at some instant of time. After this event list is built, it is traversed and a new list is built, called the gate queue. Whenever an event occurs on

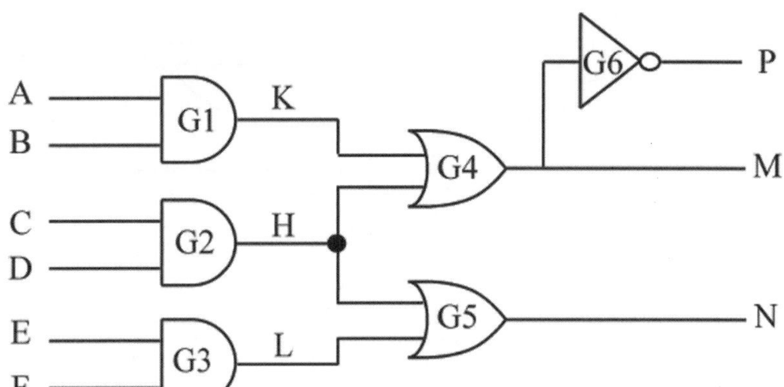

FIGURE 1.30: Example circuit for ED simulation

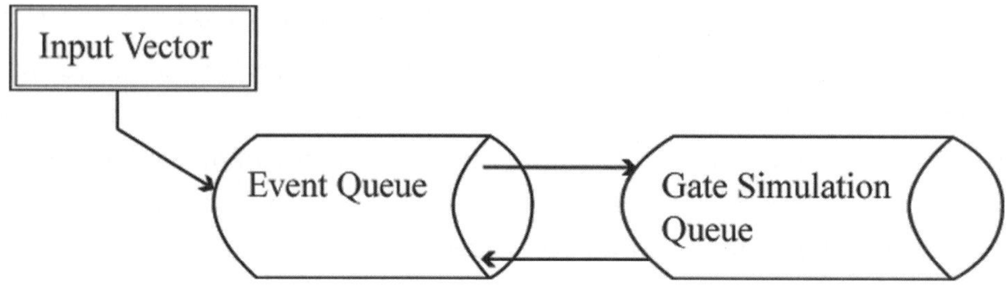

FIGURE 1.31: Timing wheel diagram

an input net to a gate, the simulator must simulate the gate to determine if the gate output net undergoes a corresponding event. The event queue and the gate queue are alternately formed and processed and the simulation is over when the queues are empty. Because the event and gate queues are alternately filled, processed, and emptied, this structure is sometimes referred to as the "timing wheel" and a diagram of a timing wheel is depicted in Fig. 1.31. The smaller box in the upper-left corner represents a new set of input stimulus values that initiate events on the input nets and cause a new simulation to occur. These are typically referred to as "test vectors" and the choice of an appropriate set of test vectors is a very important factor in effective simulation as well as using an efficient and timing accurate simulator. Here, we will focus on the anatomy of the timing accurate simulator only.

Using the timing wheel structure and the concept of an event-driven simulation, the example circuit is used to describe how the ED simulation occurs. Consider that the example circuit has been fully simulated for a test vector of A=1, B=1, C=0, D=0, E=1, F=0 and that at some instant in time the input test vector changes to A=0, B=0, C=0, D=0, E=1, F=1. This will cause events to occur on nets A, B, and F as shown in Fig. 1.32. In the figure, affected nets are shown with a previous logic value and the current logic value separated by a

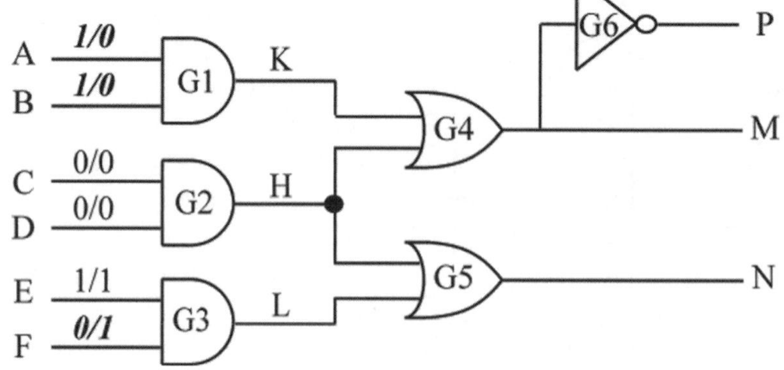

FIGURE 1.32: ED simulation example

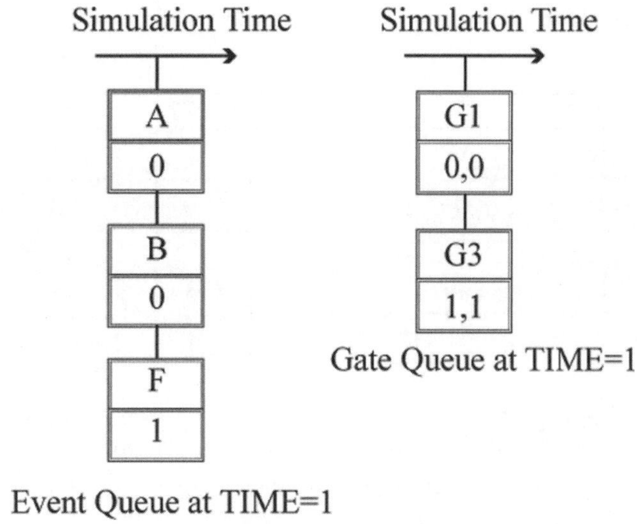

FIGURE 1.33: Event and gate queue content

slash. Whenever these two values are different, an event has occurred and these are denoted by the use of a ***bold-italic*** font.

The first step of the simulation is to build a list of events in the event queue. Each event may have other information included in the queue entry such as the net identifier, the new logic value, and other information. A very simplified figure illustrating the event content is shown in Fig. 1.33.

At this point in the simulation, the name of each net and the current logic value are stored in the event queue for all detected events. The simulator engine will next process this list of events and together with a graph depicting the circuit structure, the gate queue will be filled with a list of gates for which each event is driving (or fanning in to). As an example, event entries for nets A and B both serve as inputs to gate G1; thus this gate is placed in the gate queue along with all current input values. Note that G1 is placed in the gate queue only once.

As each event is processed, it is actually removed from the queue. However to better describe these concepts, we will leave previous events in the event queue diagram and draw the newer events to the left. As sets of events progress from left to right, a notion of time can be inferred from the simulation. For now we are assuming all gates have an equal delay and this is referred to as a "unit-time" simulation model. To complete this part of the simulation, each gate in the gate queue is simulated and if an event occurs on the gate output a corresponding entry is placed into the now empty event queue. Two time units later, the queue diagrams are shown in Fig. 1.34 and the corresponding circuit (or netlist) is shown in schematic form in Fig. 1.35. At this point one more gate will be scheduled for simulation (the inverter G6) and no more gates

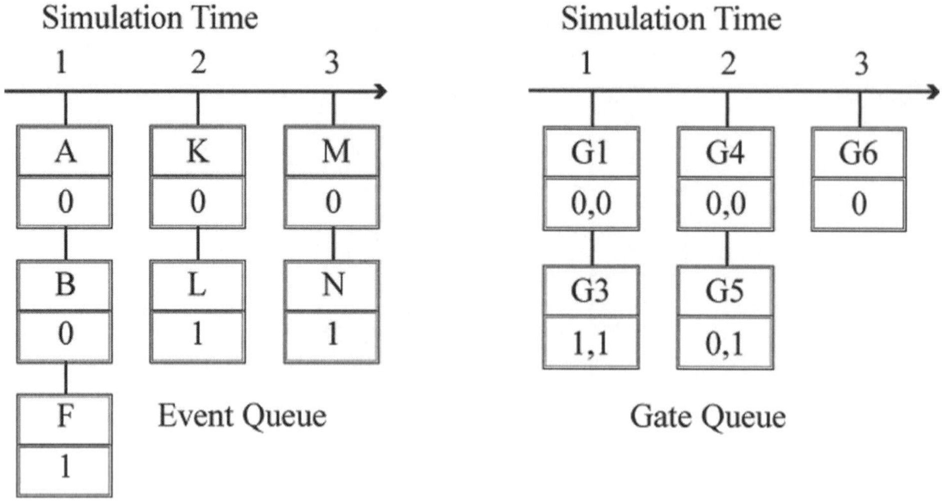

FIGURE 1.34: Event and gate queues at simulation time 3

will be scheduled afterward, yielding a total simulation time of four units. **Bold** font is used for events (previous/current logic values that differ) that have been processed during this time, with ***bold-italic*** font used for events that are yet to be processed. Since actual gate-level simulations are being performed, only those gates in the circuit that require simulation are simulated. As an example, gate G2 was never simulated for this test vector. Although only gates that require simulation are simulated, it is possible that the same gate can be simulated more than once for a single simulation run. Another important aspect of ED simulation is that the order in which the gates are simulated inside the gate queue does not matter; the scheduling mechanism ensures that all gates present in the gate queue for a given time instant are operating in parallel.

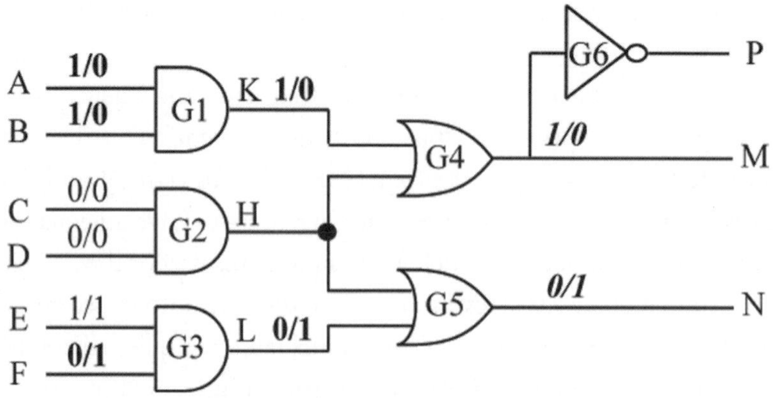

FIGURE 1.35: Example circuit with event annotations at simulation time 3

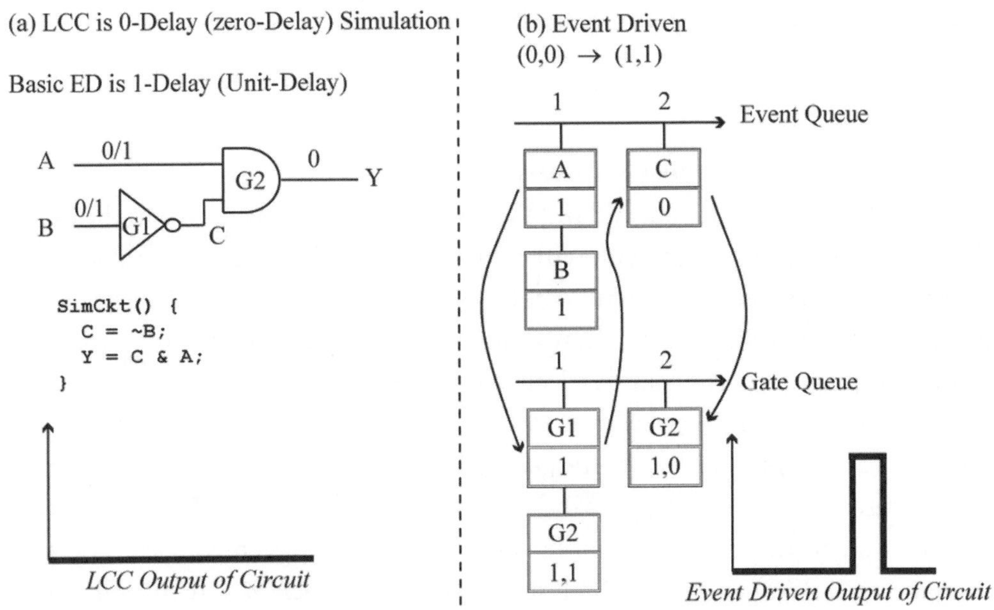

FIGURE 1.36: Example showing two simulations for same gate

1.6.2 Timing Considerations

As mentioned previously, a gate may be scheduled for simulation more than once for a particular test vector. This occurs when unequal delay paths are present in a circuit and actually allows for more accurate timing behavior in the simulated circuit. As an example, consider the simple circuit and the associated event and gate queue content in Fig. 1.36. In this example, the bottom-left corner of the figure contains the LCC code (in the C programming language) and the associated output waveform (all zero-valued) when the input vector changes from (0,0) to (1,1). On the right side of the figure, the queue content and the resulting output waveform are shown for a unit-delay ED simulation. Note that the AND gate is scheduled for simulation twice, once when the top-most input net undergoes an event, and then one time unit later when the output of the inverter causes an event to occur. This more accurately describes real circuit behavior and the output waveform shows that the AND gate outputs a logic-1 value briefly due to the delay in the inverter.

The unequal delays in the paths from the primary circuit inputs to the AND gate inputs cause the circuit output to go high for an amount of time equal to the unit-delay associated with the inverter. For more accurate timing, variable-delay nominal values could be used since a single time value is still inadequate to accurately capture detailed timing information. It is possible to also run three simulations using minimal, maximal, and nominal delays but this increases the overall design time and some timing errors can still be masked. Most event-driven simulators operate with worst-case delay models and thus produce pessimistic timing results.

1.6.3 Presynthesis Versus Postsynthesis Simulation

Verilog simulators are event driven, and the Verilog code that has been used in the examples to this point all use zero-delay assignments, either blocking, nonblocking, or continuous. Nonzero delays can be specified in Verilog assignments by using a # operator and a delay value. However, the examples in this book all use zero-delay assignments as the provided Verilog code is written for synthesis purposes, and the gate-level netlist synthesized from the Verilog code provides the delays. When simulating Verilog, one must be aware if the simulation is *presynthesis* (also known as *functional*) or *postsynthesis*, as the postsynthesis simulation reflects the timing delays of the implementation technology. Fig. 1.37(a) shows the C code of Fig. 1.36 written as a Verilog `always` block and synthesized to two different implementation technologies. The functional simulation in Fig. 1.37(a) shows no glitch on the output Y when A and B both transition to 1 simultaneously as it is a zero-delay simulation. The postsynthesis simulation in Fig. 1.37(b) has a timing glitch on the output Y as the implementation technology uses discrete gates with physical delays. Fig. 1.37(c) shows a different implementation technology in which the synthesis tool maps the `always` block logic to a 4 location by a 1-bit (4 × 1) memory device (a *lookup table*). This eliminates the intermediate C value, and thus there is no glitch in the Y output of the simulation even though the memory device has a nonzero delay. Presynthesis simulation

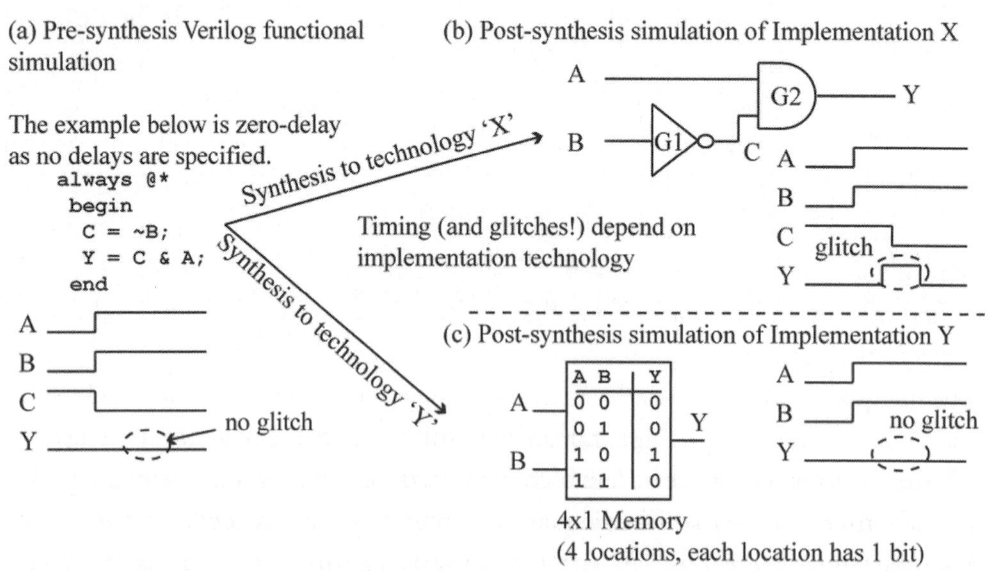

FIGURE 1.37: Presynthesis versus postsynthesis simulation

requires much less CPU time than postsynthesis simulation for complex systems that require a large number of test vectors for verification.

1.6.4 Blocking Versus Nonblocking Assignments and Synthesis

In the synthesis examples presented to this point, blocking assignments ("=") are used in `always` blocks that synthesize combinational logic, while nonblocking assignments ("<=") are used in `always` blocks that synthesize to sequential logic. This is due to the manner in which these two assignment types are handled by the Verilog event-driven simulation model, and its translation to hardware elements. What follows is a simplified explanation of the differences between these two assignment types; the reader is referred to [1] for a complete discussion. Zero-delay blocking assignments are so named because the assignment of the right-hand side (RHS) to the left-hand side (LHS) is completed without any intervening Verilog code allowed to execute, i.e., the assignment *blocks* the execution of the other Verilog code. For nonblocking assignments within an always block, all RHS expressions are evaluated, and are only assigned to the LHS targets after the `always` block completes. This causes a logic synthesis tool to treat blocking assignments differently from nonblocking assignments in terms of the synthesized logic. Fig. 1.38(a) shows an `always` block with two blocking assignments whose event list contains "`posedge clk`," inferring that the logic to be synthesized is edge triggered. The first blocking assignment "`q1 = d`" is triggered on the rising clock edge, and thus a DFF is synthesized to represent this assignment. Because this is a blocking assignment, the assignment is assumed to complete before the next assignment. Thus, the `q1` on the RHS of the second assignment "`q2 = q1`" is the `q1` after the clock edge has occurred, or is the `q1` that represents the output of the synthesized

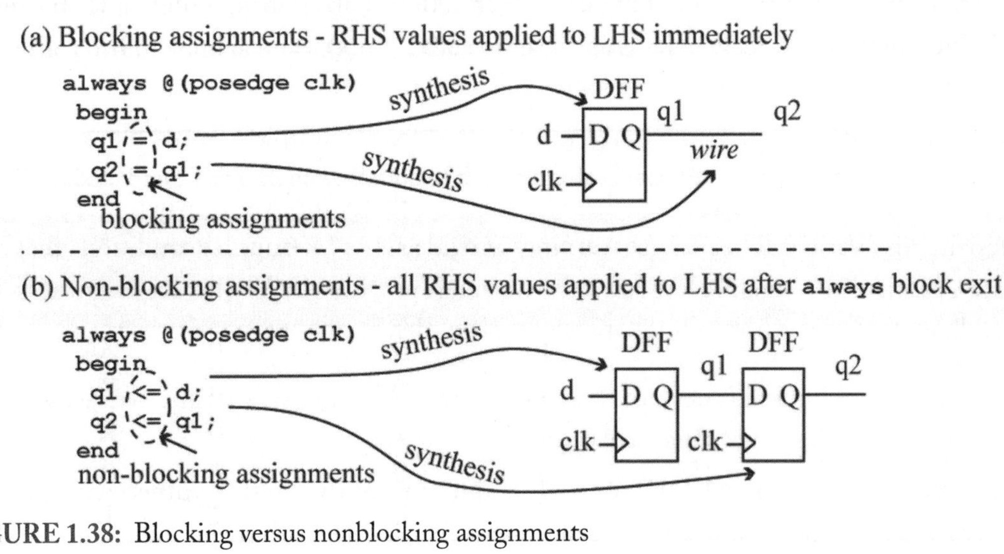

FIGURE 1.38: Blocking versus nonblocking assignments

DFF. The second blocking assignment "`q2 = q1`" copies this `q1` to the `q2`, which synthesizes to a wire as previously seen in Fig. 1.25(b).

Contrast this with Fig. 1.38(b), which is the same `always` block except that the blocking assignments have been replaced by nonblocking assignments. In this case, the RHS values of `d` and `q1` are their values at the time the process is triggered by the rising clock edge as all RHS values of all nonblocking assignments are evaluated before any nonblocking LHS assignments are made. This causes the synthesis tool to infer a DFF for each of these assignments, forming the DFF chain as shown in Fig. 1.38(b).

1.7 VERILOG CODING GUIDELINES

The following list contains a few guidelines adopted from [3] for the Verilog HDL that will help users who are new to logic synthesis to write the Verilog RTL code that will synthesize to the user's expected hardware realization.

1. Use blocking assignments ("=") in `always` blocks that are meant to represent combinational logic.
2. Use nonblocking assignments ("<=") in `always` blocks that are meant to represent sequential logic.
3. Do not mix blocking and nonblocking assignments in the same `always` block. If an `always` block contains a significant amount of combinational logic that requires intermediate wires (and thus, intermediate assignments), then place this logic in a separate `always` block.
4. If an `always` block for combinational logic contains complicated logic pathways due to *if-else* branching or other logic constructs, then assign every output a default value at the beginning of the block. This ensures that all outputs are assigned a value regardless of the path taken through the logic, avoiding inferred latches on outputs.
5. Do not make assignments to the same output from multiple `always` blocks.

Also, pay attention to any warnings that the logic synthesis tool issues when compiling and synthesizing the Verilog RTL. These warnings provide alerts for unusual conditions that often indicate a mistake in coding. A few common warnings provided by Verilog logic synthesis tools are as follows (exact wording is simulation tool dependent):

1. "Input X is unused (does not drive any logic)." This means that the synthesized logic does not make use of a particular input.
2. "Output X is stuck at VDD (or GND)." This means that in the synthesized logic there is an output that has been reduced to a fixed "1" or "0" with no gating.

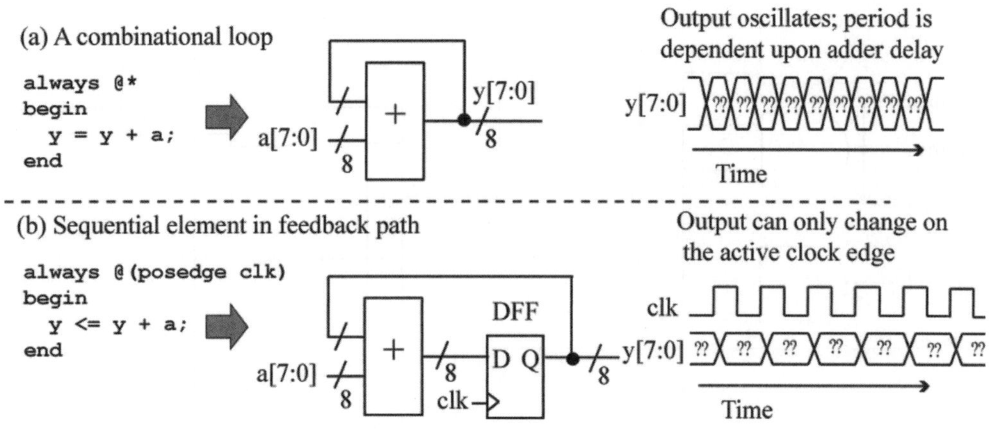

FIGURE 1.39: Combinational loop

3. "Outputs X and Y share the same net." This means that the logic for outputs *X* and *Y* is the same, and that outputs *X* and *Y* are driven by the same gate.
4. "Output X has no driver." This means that an output has never been assigned, and thus no logic has been synthesized to drive it.
5. "Combinational loop detected on net X." This means that the synthesis tool has found a feedback path from a combinational gate output back to one of its inputs without an intervening latch or DFF. Unless an *asynchronous* (i.e., no clock is used) digital system is being designed, this is an error in coding as all feedback paths should be broken by a sequential element. 1.39(a) shows an example of a combinational loop formed by feeding the output of an adder back to one of its inputs. For a nonzero value on the input a, this causes the adder to oscillate with a period equal to the delay path through the adder. To correctly sum the previous adder output with a new input value, a sequential element such as a DFF must be placed in the feedback path. This allows the clock signal to control the sequencing of new output values from the adder.

1.8 SUMMARY

Hardware description languages and logic synthesis offer a powerful mechanism for the specification and implementation of digital systems. This chapter has reviewed digital logic fundamentals in a Verilog HDL context, which is our HDL of choice in this book.

during state transitions. One way to do this is to convert a Mealy machine description to that of a Moore machine so that the circuit outputs are actually portions of the state encodings, or that they at least depend only on the state encodings. Conversely, transformation of a Moore machine to a Mealy machine can lead to circuits that require fewer states and hence memory storage elements that may result in a savings in area or component usage. Although the examples of Mealy to Moore conversion and vice versa are shown using state diagrams, the methodology for this conversion can also be accomplished directly with ASM charts.

[illegible]

[illegible] are replaced with edges [illegible] the next states have the [illegible] appended to the original state encodings. As an example, consider the portion of the state transition diagram in Fig. 2.12 that is originally in Mealy form on the left and is transformed to Moore form on the right.

From the example in Fig. 2.12, it is seen that the transformation from a Mealy to a Moore state transition diagram is accomplished by splitting states with more than one incoming edge into separate states and then by renaming all destination states. The output values in the Mealy machine are then put into additional bits in the state encodings. In the example these new bits were appended to the end of the state vectors of the Mealy machine encodings but in general they may be added anywhere within the state vector as long as each resulting bit [illegible]

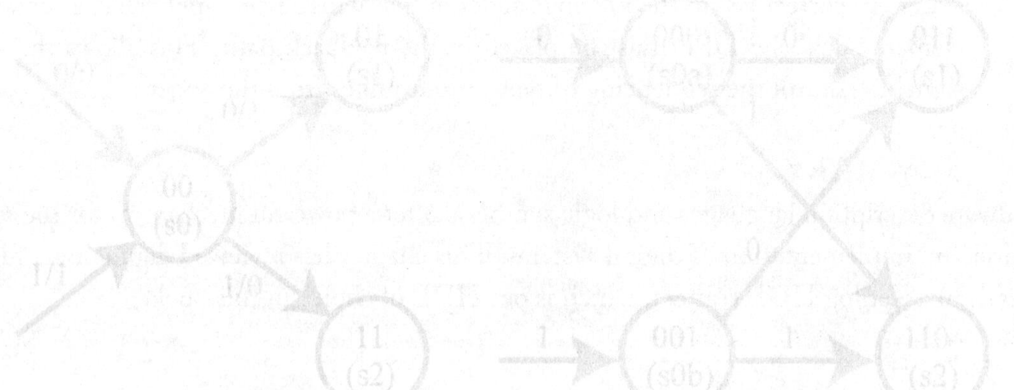

FIGURE 2.12: Portion of a Mealy state machine transformed to a Moore machine

CHAPTER 2

Synchronous Sequential Circuit Design

This chapter will review the underlying theory of sequential circuits and will heavily emphasize synchronous sequential circuits. Topics such as ASM charts, state encoding, Verilog descriptions, and their effect on resulting automatically synthesized circuits are included. These types of circuits are very common and useful when they are used as controllers for guiding input data through a data processing circuit.

2.1 LEARNING OBJECTIVES

After reading this chapter, you will be able to perform the following tasks:

- Describe a synchronous sequential circuit in contrast to other types of circuits.
- Express the operation of a synchronous sequential circuit in terms of an *Algorithmic State Machine* (ASM) chart and other common models.
- Understand how a synchronous sequential circuit can be used as a controller for a datapath.
- Describe what state encoding is and what the effects are in terms of the resulting synchronous circuit.
- Develop a *Register Transfer Level* (RTL) Verilog description of a synchronous sequential circuit.
- Be familiar with different Verilog coding styles for a synchronous sequential circuit.
- Develop either Mealy- or Moore-type synchronous sequential circuits and understand their differences in output signal timing.

2.2 SEQUENTIAL CIRCUITS

Sequential circuits differ from combinational logic circuits in that the outputs of these circuits depend on both the input signals and the value of an internal state. The state value is present by way of a memory capability in the sequential circuit. Memory circuits can be constructed in a variety of ways including the 1-bit storage elements described in Chapter 1. Sequential circuits can be broadly classified as being synchronous or asynchronous referring to the presence or absence of a periodic signal as an input.

2.2.1 Sequential Circuit Motivation

The common sequential devices described in Chapter 1 are used so commonly that we refer to them as sequential logic building blocks. In general, a sequential circuit may have any set of arbitrary states and these types of circuits are extremely useful in the design of digital circuits in the role of a controller. We will use the terms "controller" and "sequential circuit" interchangeably in the following text, although we will always assume that we are dealing with synchronous controllers, those that depend on a synchronizing clock signal for state changes.

There is interest in synchronous controllers since they are very commonly used as part of an entire digital system, including processors, bus bridges, physical layer protocol devices, and other digital circuits. These types of digital systems can be viewed as being composed of two main subcircuits: a datapath and a controller. Although this chapter is not devoted entirely to datapath + controller circuit design, a brief description of the architecture of these types of circuits is provided here to give context for the remainder of the discussion on controllers.

A general block diagram of a digital system composed of a datapath and a controller is shown in Fig. 2.1. The datapath is the portion of the circuit that contains components that transform input data signals into output data signals. Datapaths may be purely combinational or they may also contain synchronous components such as dedicated counters. The controller

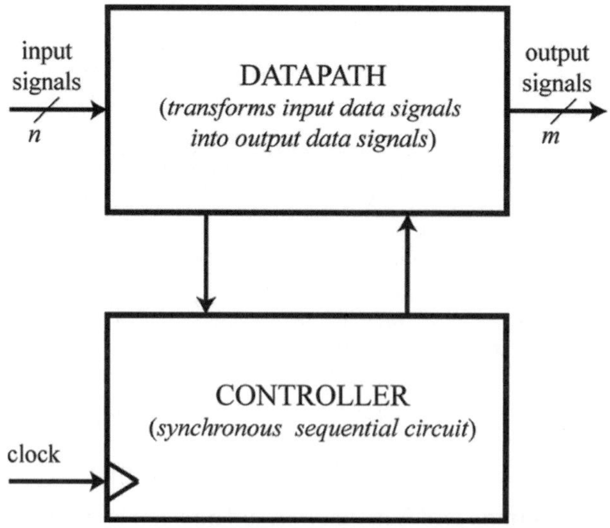

FIGURE 2.1: Diagram of digital circuit composed of a datapath and a controller

subcircuit is responsible for guiding the input signals through various components in the datapath and can allow the "sharing" of components in a datapath. For example, if a digital system needs to apply a multiplication operation to two different sets of input signals, but only one multiplier circuit is present in the datapath, the controller can generate signals that sequences and allows different input signals to be present at the multiplier circuit inputs at different times. Common types of signals that the controller circuit provides to a datapath include device reset signals, multiplexer select signals, register load and clear signals, and in general any signal that is considered to be a "control" versus a "data" signal. The controller may also receive input signals generated by the datapath. An example might be a completion signal generated by a datapath component that would in turn be used by the controller to change state.

2.2.2 Synchronizing Signals: The Clock

In this chapter, we focus on synchronous circuits, which are those that require a synchronizing periodic signal (or a clock signal) as an input. Fig. 2.2 illustrates a typical clock signal as a voltage versus time plot with various parameters such as period (τ), frequency (f), pulse width (P_W), pulse height (P_H), and duty cycle (D_C) as a percentage depicted. The clock signal in the figure is ideal in that it is depicted with zero-valued rise- and fall-times. In reality, the rise- and fall-times are finite valued and not necessarily equal; however, unless stated specifically, we can assume this ideal model for the purposes of our discussion. The P_H parameter is shown in units of voltage, which is the most common unit used for modern digital circuits since they are based on voltage mode circuits at the transistor level.

During the simulation of synchronous circuits using Verilog, it is necessary to generate a clock signal. This can be accomplished by a separate Verilog module for that purpose. Two Verilog modules are shown in Fig. 2.3 that can serve as clock signal generators. The first module `clk_gen1` produces an output clock signal that is composed of exactly 50 clock cycles. The

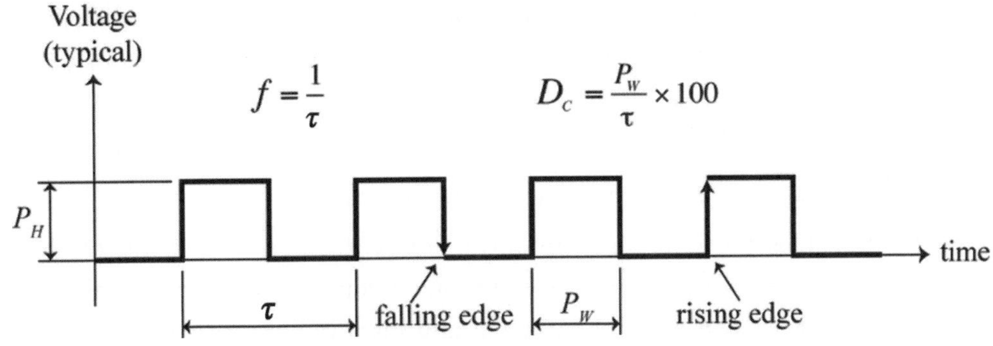

FIGURE 2.2: Typical synchronous sequential circuit clock signal

```
module clk_gen1(clk);
output clk;
reg clk;
initial
    begin
       clk = 1'b0;
       repeat (100)
       #5 clk = ~clk;
    end
endmodule
```

```
module clk_gen2(clk);
output clk;
reg clk;
initial  clk = 1'b0;
always     #5 clk = ~clk;
endmodule
```

FIGURE 2.3: Verilog modules for clock signal generation

`repeat (100)` statement causes the Verilog line `#5 clk = ~clk;` to be repeated 100 times. Every 5 simulation time units, the clock signal changes polarity so 10 time units are required for one complete period. The `clk_gen2` module generates a clock signal that continually oscillates until the simulation is halted, because the `#5 clk = ~clk;` statement is present in an `always` block. The clock signal produced by the Verilog modules in Fig. 2.3 have duty cycles of 50%, a pulse width of 5 simulation time units, and a period of 10 simulation time units. More sophisticated clock generators can be written that allow for varying the duty cycle of the clock generation circuit.

For practical reasons in terms of circuit synthesis, the clock generator module should be written as a separate module and not included as part of the module containing your circuit design description, since the purpose of the clock generator module is for simulation only. After your circuit has been simulated and you are ready to synthesize it, you will not want the synthesis tool to produce clock generation circuitry. By keeping the clock generator as a separate module, you can remove it from your file when you are ready to synthesize the circuit and provide the synthesis tool with only the module describing the circuit under design.

Many synthesis tools only support a synthesizable subset of the Verilog language. As an example, the clock signal generation modules in Fig. 2.3 cannot be simulated using the Altera Quartus tool since the timing delay parameters (e.g., #5) are not supported since the Quartus tool is designed for synthesis and supports only the synthesizable constructs of Verilog. Other simulation tools may require the insertion of simulator directives such as `$stop` or `$finish` in the clock modules for proper operation.

2.2.3 Synchronous Sequential Circuit Architectures

Just as combinational logic circuits are commonly modeled as binary-valued Boolean functions, it is also common and convenient to model controller circuits based on a mathematical model known as an automaton. The two most popular automata models for controller circuits are referred to as Moore or Mealy machines, named after researchers who published early papers on their structure [4, 5]. The distinguishing characteristic between Mealy and Moore machines is that the output of a Moore machine depends only upon the state of the sequential circuit whereas the output signals of a Mealy machine depend upon both the state and the input signals of the circuit. This has a practical effect in that the output signals of a Moore machine only change after output logic delays following a clock signal edge whereas the output signals of a Mealy machine may change at any time shortly after an input signal changes value. Because the outputs may change at any time for a Mealy machine, the outputs for a given state transition can only be considered to be valid shortly after a state transition that occurs in relation to a clock edge event.

Block diagrams of the architecture of synchronous Mealy and Moore machines are shown in Fig. 2.4. Each of these is shown with n input lines, m output lines, k present state values, and q excitation values. The k present state values are the content of the k 1-bit storage memory elements called the state of the circuit. The q excitation values are used as inputs to the storage devices so that their content may change during the next clock event. If the storage devices are simple registers or D flip-flops, then the $k = q$ and the q signals are referred to as the next state. The m output signals are produced by the lower combinational logic blocks in the figure. Note that the output combinational logic is optional for the Moore model. Many times Moore machines are designed such that some or all of the k state values are the desired circuit output signals. This is generally desirable in terms of timing since the output signals of a Moore machine that are actually the present state values will only change shortly after a clock event and will remain stable until the next clock event.

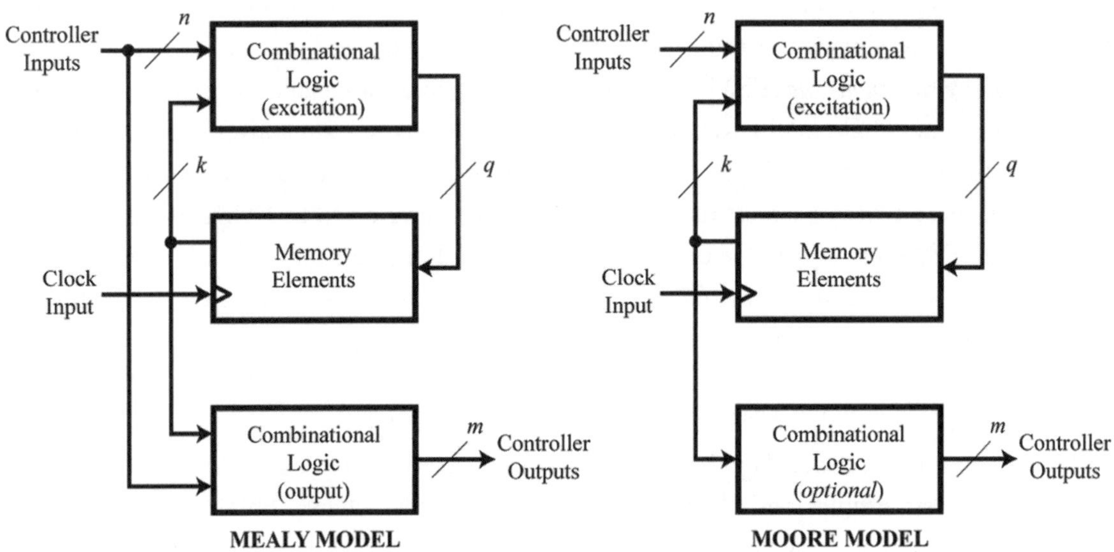

FIGURE 2.4: Block diagrams of Mealy and Moore machine digital circuitry

In comparing the two diagrams in Fig. 2.4, the Moore model has the same structure as the Mealy model with the exception of a missing bus connecting the input signals to the output combinational logic block. This has a practical effect in terms of building circuits that adhere to the Mealy- or Moore-type structure. In reality, it is usually the case that some outputs are derived directly from the present state values only and we will refer to these as "Moore-type" outputs. Other outputs may depend on both the input values and the present state; these are the "Mealy-type" outputs. Circuits containing both Moore-type and Mealy-type outputs are still technically Mealy machines. It is usually desirable from a design point of view to build circuits with as many Moore-type outputs as possible since those output signals have the desired property of changing only shortly after a clock event and they remain stable throughout the entire clock period.

Since Moore and Mealy machine models are mathematical structures, there is a rigorous theoretical background underlying the Mealy and Moore models referred to as the automata theory. The automata theory is the basis behind the traditional model of computation and is used for many purposes other than controller circuit design, including computer program compiler construction, proofs of algorithm complexity, and the specification and classification of computer programming languages [6]. While these subjects are not covered here, we mention this in order to point out that the synchronous sequential circuits we are studying here are a subset of the more general category of automata models.

Because automata are mathematical models that produce values dependent upon some internal state and possibly some dependent input values, they are also often referred to as *state machines*. An automaton may allow for a finite or an infinite set of possible states and furthermore, they may have deterministic or nondeterministic behavior. A deterministic state machine is one whose outputs are always the same for a given internal state and a set of dependent input values. A *finite state machine* (FSM) is one where all possible state values form a finite set. The synchronous sequential circuits that are the focus of this chapter are conveniently modeled as deterministic finite state machines that are modeled as either Mealy or Moore machines. For this reason, these types of circuit models are sometimes referred to as FSMs in the literature. Although an FSM refers to a type of mathematical model of an automaton, it is sometimes the case that designers refer to a controller as an FSM when technically a controller is a circuit that is a realization or embodiment of an FSM model.

2.3 MODELS OF FINITE STATE MACHINES

FSMs are commonly modeled by digital designers in a variety of ways, including state diagrams, state equations, state tables, and *algorithmic state machine* (ASM) charts. This section will provide an overview of each of these types of models for an example FSM. Our method for controller circuit design will involve specifying the behavior by first developing an FSM model and then translating the model into a Verilog description, which in turn will be synthesized yielding a circuit realization. Verilog descriptions are also models but we will describe the development of these models in a separate section since our preferred approach for controller design is to first develop an ASM chart and then to translate the chart into a Verilog description.

2.3.1 Basics of Algorithmic State Machine (ASM) Charts

Algorithmic State Machine (ASM) charts are a way to model a synchronous sequential circuit and these are our preferred model to use before the generation of a Verilog description commences. There are several reasons for this preference:

- ASM charts strictly adhere to a few rules and this allows them to be easily translated directly into the Verilog HDL description.
- ASM charts are usually much easier to comprehend by a human than a Verilog listing.
- The combination of ASM charts and a Verilog listing provides a very powerful and comprehensive form of documentation for a synchronous sequential circuit.
- ASM charts provide an easy way to distinguish between Moore- and Mealy-type outputs.

ASM charts are a rigorous form of flowcharts that were initially introduced for describing software programs. ASM charts consist of three different symbols connected by directed edges that depict the flow of events within a circuit. As shown in Fig. 2.5, a single rectangular box is present for each state, an oval is present for circuit outputs that depend on both input values and the current state (note that these dependent output ovals are only present in Mealy model descriptions), and the diamond shaped symbols are used for decisions in next-state transitions based on circuit inputs.

The rectangular state-representing symbol is sometimes optionally shown with a state encoding or a state name. If these are present in the ASM chart, they should be shown outside the box since everything inside the state-representing symbol refers to an unconditional (or Moore-type) output signal. It is also common to designate one of the state-representing signals as a "reset" or "initialization" state. While there is no strict convention for denoting this state, common practice is to have a flow direction symbol pointing to the rectangle with the other end labeled "RESET." State-representing symbols may have one or more flow direction symbols pointing to them but they should always have exactly one exiting flow direction symbol.

The diamond-shaped decision symbol is used to allow signal flow to vary in a conditional manner based on an input signal. Decision symbols should always contain an input signal name or a Boolean expression composed of input signals. Basic decision symbols are single-bit valued which implies that they have exactly two exiting flow direction symbols, one indicating that

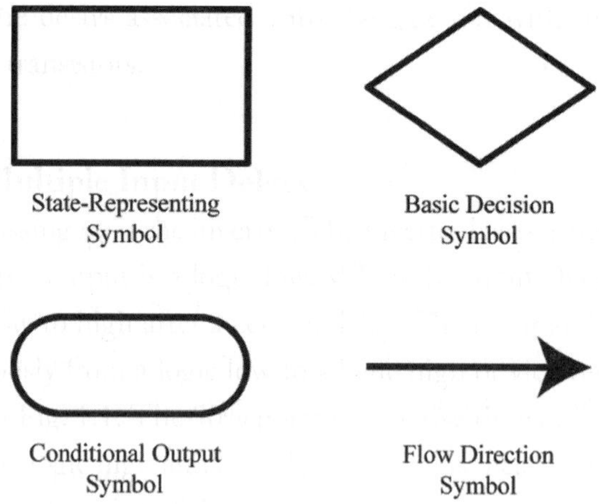

FIGURE 2.5: Symbols used in ASM charts

the contained expression or input signal is a “1” and the other indicating evaluation to a “0.” Each of the two exiting flow direction arrows is labeled by having either a 0 or a 1 next to them. A basic decision symbol always has a single input flow direction symbol originating from either a state-representing symbol or another basic decision symbol. The two outputs of a basic decision symbol should either point to a conditional output symbol or to a state-representation symbol.

A conditional output symbol is only used in Mealy machine ASM representations. These symbols always have exactly one entering and one exiting direction flow symbol. Output signals or equations defining output signals are contained within conditional flow boxes. In a correctly drawn ASM chart, the entering direction flow symbol to a conditional symbol must always originate from an output of a decision flow symbol. For this reason, the output designated by the conditional symbol depends (or is conditional upon) on the input signal in the decision flow symbol. These ASM chart symbols are sometimes referred to as Mealy output symbols since they denote outputs that depend on both the current state and an input signal value. The exiting flow direction arrow points to a state-representing symbol or another basic decision symbol.

We refer to the diamond-shaped symbol as a basic decision symbol since some designers use a more general decision symbol with more than two output flow direction arrows that depend on more than one input signal. The more complex generalized decision symbol can always be represented by a cascade of basic decision symbols, each labeled with a single input signal and having two (binary) exiting flow arrows. The order of the cascade of basic symbols does not matter since evaluation of a path through the decision boxes is assumed to occur simultaneously, shortly after a clock event in a synchronous sequential circuit; nevertheless, some people prefer

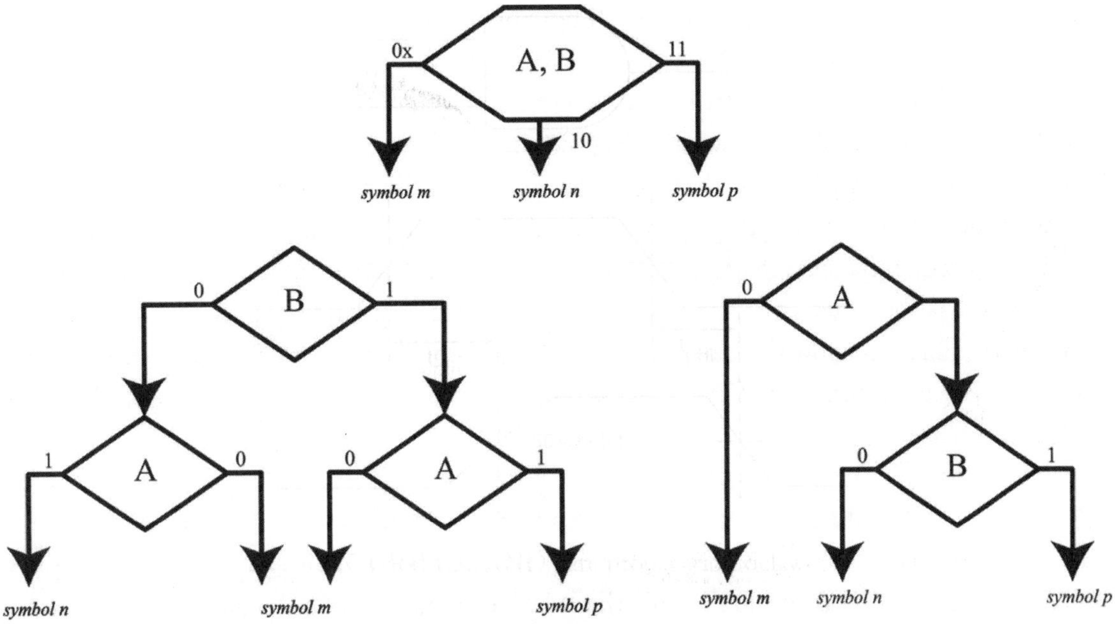

FIGURE 2.6: Complex and corresponding basic ASM charts

using these more general decision symbols since they avoid the issue of ordering in a cascade of basic decision symbols. A common way to draw a general decision symbol is shown in Fig. 2.6 on the top and two corresponding cascades of basic decision symbols are shown below.

All the three ASM chart portions shown in Fig. 2.6 are equivalent. The top decision symbol has three exiting flow symbols that correspond in total with all four different 2-bit values of the input signal pair A, B. In the worst case, the general decision symbol would have four different exiting flow symbols corresponding to the four different 2-bit valuations of the ordered pair of input signals A, B $= \{00,01,10,11\}$. For conciseness, the don't care *x* is used to indicate that if the input signal A $= 0$, the activated path in the chart is to the ASM symbol m. The two cascades of basic decision symbols are identical to the generalized symbol. There are advantages and disadvantages of each convention. The advantage of the generalized symbol is clearly the conciseness of the ASM chart. Also, if fully generalized decision boxes are used everywhere, the rule about the exiting flow symbol from conditional output symbols may be simplified to state that they must always point to a state-representing symbol. It is permissible to insert conditional output symbols in intermediate stages of cascades of flow symbols, provided that they follow all decision symbols whose input signals affect the conditional output signal. The disadvantage of the generalized decision symbol is that the designer must ensure that all possible combinations of input variables are uniquely covered. That is, there must be one and only one exiting decision symbol for all possible combinations of input signals inside the generalized decision symbol.

Finally, because there are two possible ways to express the generalized decision symbol using basic decision symbol cascades in Fig. 2.6, many designers who are just learning about ASM charts mistakenly assume that the order in which the decision boxes are encountered have some bearing on the underlying circuit. This is not the case since ASM charts represent synchronous sequential circuits. In synchronous sequential circuits, all decision boxes are evaluated as soon as the input signals change regardless of the order in which they are drawn in an ASM chart. This can allow the paths through the decision cascades to change as input signals change. However, in terms of effecting state transitions, only the activated path shortly after a clock event matters.

2.3.2 The ASM Chart Model and an Example Controller

The same example FSM is used for all the models described in this chapter. The example is intended to represent a controller circuit that receives inputs from a datapath that consists of a memory with extra circuitry that allows for writing a sequence of zero values over a range of addresses. The additional circuitry in the datapath consists of two registers for holding the lower and upper addresses, a counter for generating intermediate addresses, a comparator for determining when the upper address value has been reached, and multiplexers for switching the memory device input signals between the "zeroing" function and normal operation. A simplified block diagram of the datapath is shown in Fig. 2.7 with the controller circuit represented as a large box at the bottom of the diagram.

The datapath is designed to allow the memory to either function normally or in a *zeroing* mode. During normal functioning, addresses (addr[5:0]) are connected to the memory address input (addr) through the uppermost multiplexer (M1) and data to be written to the memory is passed through the lowermost multiplexer (M2) to the memory data input port. In the zeroing mode, the multiplexer select lines are changed to connect the output of the counter to the memory address input and the connection to a constant logic "0" to the memory data input port (din). One of the purposes of the controller circuit is to generate the appropriate multiplexer select line values.

In this example, it is assumed that the memory with a zeroing capability circuit is interfaced to other circuitry that is responsible for ensuring that the appropriate addresses are stored in registers R1 and R2 before a zeroing operation is asserted. The zeroing operation is asserted by external circuitry to indicate that the system should switch from normal operation to the zeroing mode. This external input signal, `zero`, serves as input to the controller which in turn will issue the appropriate signals to the datapath to initiate the zeroing mode. The overall circuit generates

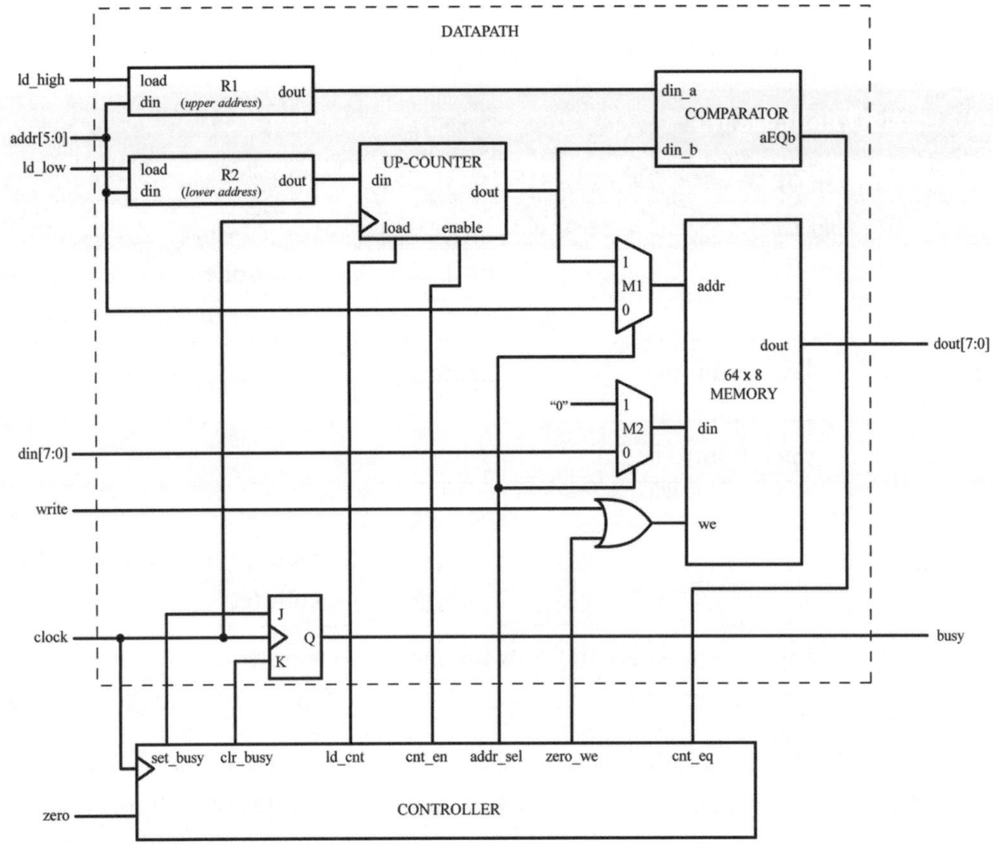

FIGURE 2.7: Block diagram of a datapath and controller for an example digital system

one output, `busy`, that is generated by the controller outputs `clr_busy` and `set_busy` via a JK flip-flop in the datapath. The output signal, `busy`, can be used by other external circuitry to determine when normal memory reads and writes can occur. The two-input OR gate in the datapath is needed to allow the single memory write enable (`we`) to be controlled externally during normal memory operations, or by the controller during the zeroing operation. The counter is responsible for generating the next incremental address to which a "0" value is to be written during each clock cycle of a zeroing operation. This counter has the capability to have an initial value loaded, in this case from R2, and is an up-counter. The output of the counter is fanned out to both the M1 multiplexer and an input of a comparator. The purpose of the comparator is to determine when the zeroing operation is complete. This is accomplished by comparing each generated address from the counter with the last address to be zeroed present in register R1. When the addresses are the same, the comparator outputs the signal `cnt_eq`

TABLE 2.1: Example Controller Input and Output Signal Names

INPUTS		OUTPUTS	
DESCRIPTIVE NAME	PURPOSE	DESCRIPTIVE NAME	PURPOSE
		`Set_busy`	Controller output indicating zeroing operation has begun
`zero`	External input causing zeroing operation to begin	`Clr_busy`	Controller output indicating zeroing operation has ended
		`load_cnt`	Controller output causing R2 content to be loaded into counter
`cnt_eq`	Input from datapath causing zeroing operation to halt	`addr_sel`	Controller output connected to select inputs of M1 and M2
		`zero_we`	Controller output asserting the we input of the memory
		`cnt_en`	Controller output asserting the enable input of the counter

that serves as an input to the controller, which in turn generates the control signals that allow the datapath to resume operation as a normal memory unit.

Examination of the Fig. 2.7 indicates that our example controller has two input signals from the datapath and produces six output signals that control the datapath. The inputs and outputs of the controller are summarized in Table 2.1.

Fig. 2.8 contains an ASM chart that models the example controller. When constructing the ASM chart, it is natural and easy to determine which outputs are Moore-type versus Mealy-type—this is an advantage of the ASM chart. State `s0` represented by the top state-representing symbol is the state at which our example circuit is behaving as a normal memory. The decision box below `s0` indicates that at each clock event, the input zero is checked to determine if an external request for a zeroing operation has been asserted, if so the controller transitions into

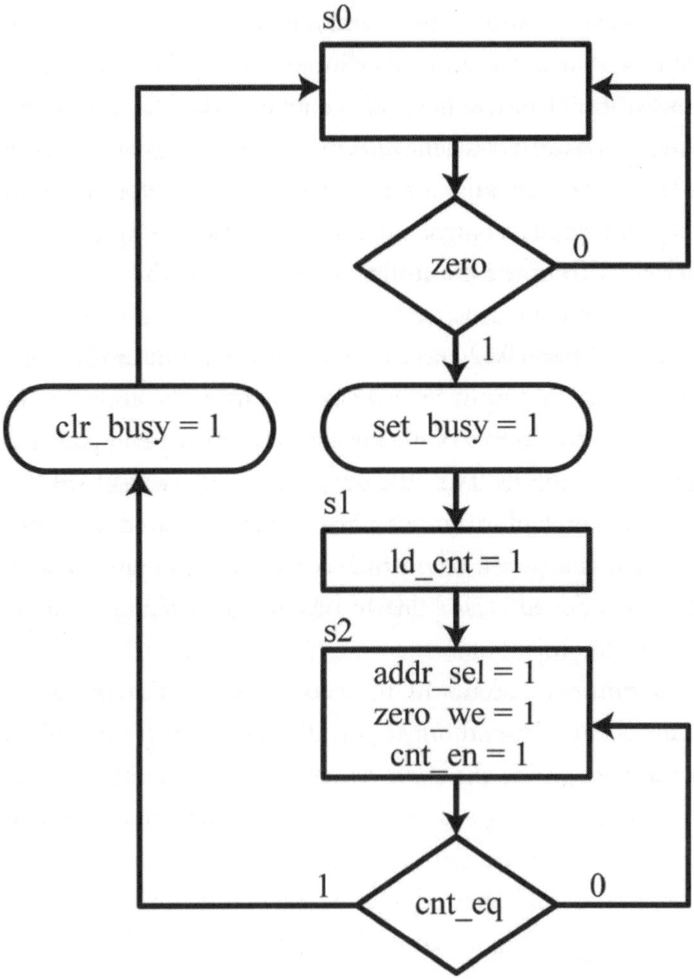

FIGURE 2.8: Example ASM chart for a controller

state `s1`, otherwise it remains in state `s0`. Note that whenever the controller is in state `s0` and `zero` is asserted, the conditional output `set_busy` is asserted. This implies that `set_busy` is a Mealy-type output since it depends upon the input signal `zero`.

After `zero` has been detected as asserted, and the controller has transitioned into state `s1`, it is necessary to load the counter in the datapath with a low-range value of the memory address where the zeroing operation is to begin. This is accomplished when the controller asserts the unconditional output `ld_cnt` signal that is connected to the load input of the counter. The purpose of state `s1` is to initialize the counter to begin at the address that was loaded into register R2. The controller stays in state `s1` for exactly one clock cycle since there is no decision block after the `s1` block in the ASM chart.

The controller transitions into state s2 and remains there as long as the zeroing operation is occurring. While in state s2, the controller input cnt_eq is checked at each clock cycle as indicated by the lowermost decision box. The cnt_eq signal is the output of the comparator in the datapath and is asserted whenever the counter is equal to the upper range address present in register R1. When cnt_eq is asserted, the controller can then transition back to state s0 and generate the appropriate control signals that cause the unit to resume operation as a normal memory device. While the controller is in state s2, the three unconditional outputs addr_sel, zero_we, and cnt_en are asserted. The addr_sel signal is connected to the select lines of the multiplexers M1 and M2 and causes the addr input of the memory to receive data from the counter and the din input of the memory device to be connected to a logic "0" value. The zero_we signal that is connected to the input of the OR gate is asserted, causing the memory device to be in the write mode. The cnt_en signal is the enable signal for the counter, and its assertion causes the counter to increase by one at each clock edge. Note that the clr_busy conditional output signal is asserted when the controller is in state s2 and the cnt_eq signal is asserted. The clr_busy signal is also clearly a Mealy output since it depends upon both the state (s2) and the controller input cnt_eq.

The four unconditional outputs of the controller are the signals ld_cnt, addr_sel, zero_we, and cnt_en. These depend only upon the state of the controller and for this reason they appear (when asserted) inside the state-representation symbols. Typically, when unconditional outputs such as these four signals are not asserted, they are not present in the state boxes of the ASM chart. For example, ld_cnt=0 is not written inside the s0 or s2 box in the ASM chart. The nonasserted values are referred to as default values and they have a direct translation into default Verilog statements when a controller is described with Verilog.

2.3.3 The State Diagram Model

The state diagram model for a deterministic FSM is a directed graph with a vertex set where each member of the set uniquely maps to each possible state and an edge set that represents all possible transitions from one state to another during a single clock event. The vertices are usually drawn as circles or ovals that are annotated with a label representing a state and the edges may be labeled with circuit input and/or output values.

It is easy to recognize whether a state diagram represents a Mealy or a Moore machine by examining the annotations on the graph edges. The state diagram for a Mealy machine contains edge annotations of the form *input/output* where the "/" symbol is used to delimit the input labels or values from the output labels or values. The inputs and outputs may be given in terms of symbolic or numerical values. Fig. 2.9 is a Mealy-type state diagram representation of our example controller with the numerically labeled edges. The edge labels should be interpreted as the input and output signal values in the order (zero cnt_eq/set_busy

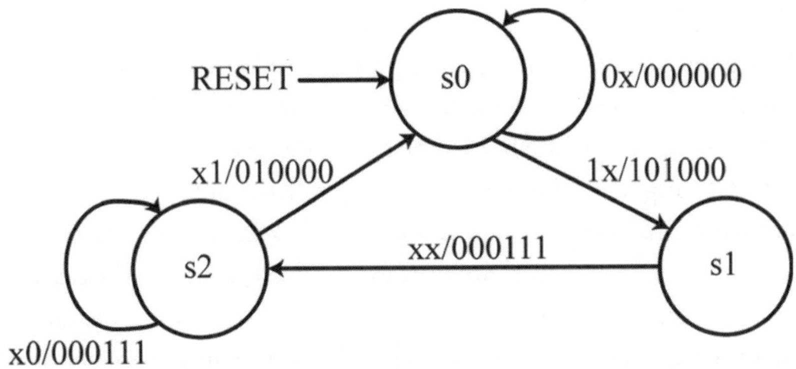

FIGURE 2.9: State diagram for a FSM representing a two-input/six-output Mealy machine

clr_busy ld_cnt addr_sel zero_we cnt_en). If symbolic edge labels were used then instead of 1x/101000 for the edge label indicating a transition from state s0 to s1, we would use (zero x/set_busy $\overline{\text{clr\_busy}}$ ld_cnt $\overline{\text{addr\_sel}}$ $\overline{\text{zero\_we}}$ $\overline{\text{cnt\_en}}$).

If the state diagram has no "input/output" designations on each transition arc, then the machine is a Moore model. Moore machines have outputs that are dependent only on current state values. In some cases the output signals may be equivalent to some or all of the state bits. In general, the outputs are specified values in the vertices of the state transition diagram, typically separated by the state values by using a "/". Alternatively, the particular output signals can directly be encoded into each state vector. Fig. 2.10 contains two state diagrams representing Moore machines that are equivalent to the Mealy machine model shown in Fig. 2.9.

While these two state diagrams look the same, there are differences in the underlying circuits that are synthesized from the two. The leftmost state diagram indicates the state

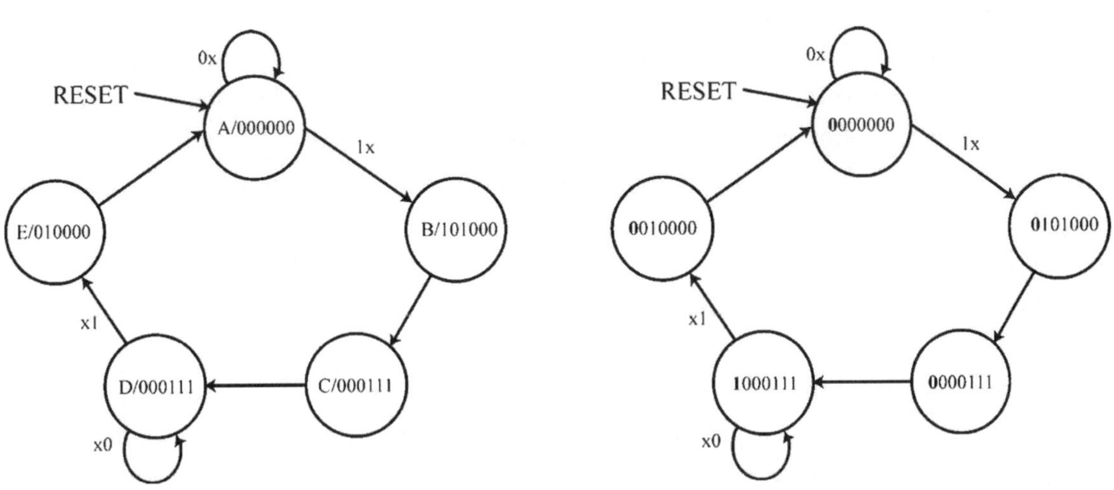

FIGURE 2.10: Two state diagram representations of the same Moore machine

symbolically {A, B, C, D, E} with the output signals shown to the right of the "/" symbol. The rightmost state diagram gives the state encoding values explicitly and the output signals are the least significant 6 bits of each state vector that are not present in bold. Note that an additional bit is needed in each state vector (the bolded bit) since the output signals are identical for states D and C. Referring to Fig. 2.4 where block diagrams of the architecture of controller circuits are shown, note that the Moore model has a lower block labeled optional output logic. This combinational logic is present when a circuit is synthesized from the state diagram on the left in Fig. 2.10. In this case, it is a combinational logic circuit whose inputs are only the current state values that are assigned to {A, B, C, D, E}. Because the outputs depend only on the state values, this satisfies the definition of a Moore machine; however in this case, the outputs are not registered. Conversely, when the circuit is synthesized from the state diagram on the right, outputs are directly particular state values and are thus registered.

These two types of Moore circuits have advantages and disadvantages. When the combinational output logic block is used, the number of bits representing each state may be minimized resulting in using fewer memory devices; however fewer memory devices can cause an increase in the amount excitation combinational logic as well as requiring an output logic block. Additionally, the production of the output signals by propagating current state values through an output logic block causes their stabilization to occur at different times and can cause the signals to exhibit static hazard behavior (bouncing or ringing) after each state transition. For these reasons, we recommend that Moore machine always be implemented where output values are state values directly when possible.

2.4 STATE ASSIGNMENT

State assignment is the process of assigning a unique bit string to each state represented in a description of a controller circuit. An example of state assignment was given in the rightmost state diagram of Fig. 2.10 where each state was labeled by a unique bit string rather than a symbolic symbol as is shown in the ASM chart example in Fig. 2.8, the Mealy state diagram in Fig. 2.9, and the leftmost Moore machine state diagram in Fig. 2.10. Before a controller can be implemented, the symbolic symbols representing each state must be transformed into a unique set of identifying bit strings. For each bit in these binary-valued state names, a memory element is synthesized.

While any arbitrary assignment can be made to a state machine as long as each state has a unique bit string, the exact assignment will affect the resulting synthesized logic in terms of both area and maximum clock frequency. The theory of optimal state assignment has been studied for many years and this problem has been proven to be intractable, which is to say that no existing method is known for finding the best assignment that does not require runtime proportional to trying all possibilities. Furthermore, the state assignment that yields the least amount of circuitry is likely not the same as the state assignment that allows the circuit to be clocked as fast as possible, so tradeoffs are present.

The minimum number of bits possible in a state encoding for a controller with N states is $\lceil \log_2(N) \rceil$; thus for the example ASM chart above we need $\lceil \log_2(3) \rceil = \lceil 1.5849 \rceil = 2$ bits for each state. Since there are 4! different possibilities, there are 24 different possible controller circuits possible for a minimum state-bit encoding. As the number of states increases, this combinatorial property increases dramatically, and this is only for minimal-bit encoding. Furthermore, studies have shown that fewer than 5% of all possible minimal-bit encodings are optimal in terms of area [7] for some controllers.

Because the problem is intractable, designers generally rely on using "rules of experience" or heuristic methods to perform state encoding. A heuristic is an observation of behavior that usually yields good results. As an example, for most controllers, if the state bits are mostly the same (but at least one bit must differ) between adjacent states, then the resulting circuit is generally a good tradeoff between speed and area. Adjacent states are any two states in an ASM chart that are connected either by a single direction arrow or by a path through decision and conditional outputs only. In the best case, we would have only a single bit changing in the state encoding of two adjacent states and we would have as few state bits as possible labeling each state. When this can be accomplished, it is referred to as a Gray code state encoding. However, Gray code state encoding is not always possible depending on the shape of the ASM chart. In our example ASM chart, Gray code state encoding is impossible since there are no possible ways to assign unique bit strings to the adjacent pairs {(`s0`, `s1`), (`s1`, `s2`), (`s2`, `s0`)}.

The next best alternative is to assign bit patterns that differ by only two bits for every adjacent pair of states. It turns out that this is always possible, and one way to achieve this is to use the so-called *one-hot* encoding method. In one-hot encoding, the number of bits in each state label is equal to the total number of states in the controller. The name one-hot comes from the fact that for each bit string only a single bit is a "1" and all others are a "0." As an example, a one-hot encoding for the example ASM chart in Fig. 2.8 is {`s0`=001, `s2`=010, `s3`=100}. Because only a single bit is a "1" in each encoding, we are guaranteed to have a distance of 2 between any two adjacent states regardless of the shape of the ASM chart.

Many times, state encoding simply does not matter and a simple arbitrary minimum assignment method works fine. When you develop your Verilog module, you will need to make an arbitrary state assignment before synthesis (note that newer HDLs such as SystemVerilog allow for symbolic enumerations that avoid this). We recommend that you make such an arbitrary assignment and validate that your controller is functionally correct first. Only in later stages of design iterations should you consider using a heuristic state assignment technique or an automated tool to optimize your design; however it is important to be aware of this issue. In highly optimized designs, such as modern microprocessors, good state assignment is essential.

It has also been our experience that one-hot encoding is generally only beneficial if your controller has more than approximately 10 states, and this benefit is generally in terms of increasing the maximum clock frequency. For small controllers, we recommend using a minimal-bit encoding strategy and assigning the codes such that as many of the adjacent state pairs as possible differ by a single bit. Finally, we note that some synthesis tools such as Synopsys Design Compiler allow for separate scripts to be employed where a designer can specify a type of state encoding to be used, such as one-hot. These tools analyze the Verilog code and automatically invoke an automated state assignment algorithm to determine the actual state assignments based on an area versus performance tradeoff constraint.

2.5 LOW-LEVEL MODELS OF CONTROLLERS

The preferred approach for controller design is to generate an ASM chart and then to immediately transform this description into a corresponding Verilog RTL-level description. This description can then be supplied as an input to a synthesis tool to produce a netlist file.

The synthesis task can be performed manually resulting in a low-level model of the controller; however, this is not practical for controllers of even moderate size. For the sake of completeness and pedagogy, we describe two low-level models of controllers in this section: state equation descriptions and a state table.

2.5.1 State Equations

A set of state equations describing a controller consists of a complete set of transition equations and output signal equations. State equations differ from binary Boolean logic equations in that the concept of discrete time is included within them. For synchronous sequential circuits, state equations are often written using a discrete time notation where t represents values in the current state and $t + 1$ represents the next state. Transition equations are those that give information about next state values, states at time $t + 1$, as a function of inputs and present state values, states at time t. Output equations are those that provide information about output values at time t as a function of input and state values at time t.

State equations may be written either before or after state assignment and memory element selection. When they are written before state assignment, there is exactly one transition equation for each state in the ASM chart and when they are written after state assignment and memory element selection, there is exactly one transition equation for each bit in the state encoding bit string. We refer to these forms as state-level and bit-level state equations, respectively.

2.5.1.1 Example of State-Level State Equations

State-level state equations can be directly derived from an ASM chart. Referring to the example ASM chart in Fig. 2.8, we examine each state box to determine each state transition equation.

The left-hand side of each transition equation is the state at time $t + 1$ and the right-hand side consists of an expression that has a product term for every path entering the state box with all products joined by the disjunctive (OR) operator. Consider state `s0` in the ASM chart; there are two paths possible to enter this state. One path occurs when the input zero is not asserted and the current state is `s0`. The other path to enter state `s0` occurs when `cnt_eq` is asserted and the current state is `s2`. Thus, the transition equation for state `s0` becomes

$$\texttt{s0}(t+1) = \overline{\texttt{zero}} \cdot \texttt{s0}(t) + \texttt{cnt\_eq} \cdot \texttt{s2}(t).$$

The state box for state `s1` has a single path entering it and that path occurs when the input `zero` is asserted and the current state is `s0`. Note that we do not include the information about the conditional output `set_busy`, which is also in the path since a separate output equation will be formed for this quantity. The product term formed for the transition into `s1` is the conjunction (AND) of `zero` and state `s0`:

$$\texttt{s1}(t+1) = \texttt{zero} \cdot \texttt{s0}(t).$$

The state box for `s2` also has a single entering path and this path originates directly from the state box of `s1`; thus the transition equation is very simple:

$$\texttt{s2}(t+1) = \texttt{s1}(t).$$

Next, we need to determine the output equations. The output equations for the unconditional outputs are very simple; they are equal to the state(s) in which they are asserted. If unconditional outputs are asserted in more than one state (not the case in our example), all the states in which they are asserted are joined together by the OR operation. The ASM chart in Fig. 2.8 has four unconditional outputs and the respective output equations are

$$\begin{aligned} \texttt{ld\_cnt}(t) &= \texttt{s1}(t), \\ \texttt{addr\_sel}(t) &= \texttt{s2}(t), \\ \texttt{zero\_we}(t) &= \texttt{s2}(t), \\ \texttt{cnt\_en}(t) &= \texttt{s2}(t). \end{aligned}$$

The final step is to develop the output equations for the conditional outputs. Conditional outputs depend on both the present state and the inputs. The conditional output `set_busy` is asserted only when the input `zero` is asserted and the present state is `s0`. Likewise, the conditional output `clr_busy` depends on the input `cnt_eq` and a present state value of `s2`. The two conditional output equations are

$$\begin{aligned} \texttt{set\_busy}(t) &= \texttt{zero} \cdot \texttt{s0}(t), \\ \texttt{clr\_busy}(t) &= \texttt{cnt\_eq} \cdot \texttt{s2}(t). \end{aligned}$$

2.5.1.2 Example of Bit-Level State Equations

Bit-level state equations depend on controller inputs and various output bits of the state holding devices rather than the entire state. This means that bit-level state equations can only be developed after state assignment has been performed. At the bit level, transition equations are formed for each bit in the state values rather than each state.

It is the case that if one-hot encoding is used, the bit-level state equations can be immediately determined from the state-level state equations. Assume that the one-hot state assignment {`s0=001`, `s1=010`, `s2=100`} is made with three flip-flops used as state holding elements. The current state of the controller, $\mathtt{s_i}$, is given as $\mathtt{s_i} = Q_C Q_B Q_A$ where each Q represents a flip-flop output. With this constraint, the bit-level state equations are the same as the state-level equations by substituting the flip-flop output value for each appropriate state value. This is possible for two reasons:

1. In one-hot encoding there are the same number of state-level and bit-level transition equations.
2. There is exactly one unique state bit asserted in each state of a one-hot encoded system.

In our example, Q_C replaces `s0`, Q_B replaces `s1` and Q_A replaces `s2`, and the bit-level state equations are

$$
\begin{aligned}
Q_C(t+1) &= \overline{\mathtt{zero}} \cdot Q_C(t) + \mathtt{cnt\_eq} \cdot Q_A(t), \\
Q_B(t+1) &= \mathtt{zero} \cdot Q_C(t), \\
Q_A(t+1) &= Q_B(t), \\
\mathtt{ld\_cnt}(t) &= \overline{Q_A}(t) Q_B(t) \overline{Q_C}(t), \\
\mathtt{addr\_sel}(t) &= Q_A(t) \overline{Q_B}(t) \overline{Q_C}(t), \\
\mathtt{zero\_we}(t) &= Q_A(t) \overline{Q_B}(t) \overline{Q_C}(t), \\
\mathtt{cnt\_en}(t) &= Q_A(t) \overline{Q_B}(t) \overline{Q_C}(t), \\
\mathtt{set\_busy}(t) &= \mathtt{zero} \cdot \overline{Q_A}(t) \overline{Q_B}(t) Q_C(t), \\
\mathtt{clr\_busy}(t) &= \mathtt{cnt\_eq} \cdot Q_A(t) \overline{Q_B}(t) \overline{Q_C}(t).
\end{aligned}
$$

After the bit-level state equations are determined, excitation equations may be formed from the transition equations. Recall that an excitation equation is one that describes the input to the state holding device while a transition equation is one that describes the output of a state-holding device in the next state as a function of controller inputs and other present state bit values. The transformation from transition to excitation equations can be performed by using the state-holding device characteristic. In this book we are only considering D flip-flops as state-holding elements since they are the simplest to use and are the most predominant in programmable logic. The characteristic input/output relation for the D flip-flop is very simple:

$$Q(t+1) = D,$$

where $Q(t+1)$ represents the output of the D flip-flop after a clock edge and D is the synchronous input. If we use three D flip-flops, D_A, D_B, and D_C, to represent the present state of the circuit, and use the D flip-flop characteristic equation and the bit-level transition equations, then the excitation equations become

$$\begin{aligned} D_C &= \overline{\texttt{zero}} \cdot Q_C(t) + \texttt{cnt\_eq} \cdot Q_A(t), \\ D_B &= \texttt{zero} \cdot Q_C(t), \\ D_A &= Q_B(t). \end{aligned}$$

2.5.2 State Tables

In the previous section we showed how bit-level state equations are directly obtained from state-level state equations when one-hot encoding is used. If one-hot encoding is not used, a direct translation from state-level to bit-level state equations is not possible. One convenient way to obtain the bit-level state equations is to use a state table.

State tables are tabular descriptions of synchronous state machines. Because these types of circuits can be described as ASM charts or state diagrams which are both directed graphs, the state table can be considered an adjacency matrix representation of these directed graphs with extra information related to the external inputs and outputs. State tables are usually written with a number of rows equivalent to the number of distinct states the corresponding circuit possesses and with rows for each external input, each present-state bit, each next-state bit, and each circuit output bit. If a state encoding of {s0=00, s1=01, s2=10} is used for the example controller where the current state is $\texttt{s}_i = Q_A Q_B$, the corresponding state table is shown in Table 2.2.

TABLE 2.2: State Table Representation of the Example Controller with Minimal State Encoding

		PRESENT STATE		NEXT STATE			
zero	Cnt_eq	$Q_A(t)$	$Q_B(t)$	$Q_A(t+1)$	$Q_B(t+1)$	set_busy	clr_busy
0	X	0	0	0	0	0	0
1	X	0	0	0	1	1	0
x	X	0	1	1	0	0	0
x	0	1	0	1	0	0	0
x	1	1	0	0	0	0	1

Each row in the state table corresponds to a path in the ASM chart. As an example the first in the table indicates that if the present state is $Q_A(t)Q_B(t)$=s0=00 and the input zero is not asserted (i.e., zero=0), then the next state is $Q_A(t+1)Q_B(t+1)$=s0=00 and the conditional outputs are set_busy=0 and clr_busy=0. From the information in the state table, the bit-level transition equations can be derived. $Q_A(\mathrm{t}+1)$ is asserted as shown in rows 5 and 6 of the table and the corresponding transition equation is

$$Q_A(t+1) = Q_B(t) + \overline{\texttt{cnt\_eq}} \cdot Q_A(t)\overline{Q_B}(t).$$

Likewise, the state table indicates that $Q_B(t+1)$ is asserted in the third row when zero=1, $Q_A(t)$=0, and $Q_B(t)$=0. This yields the transition equation

$$Q_B(t+1) = \texttt{zero} \cdot \overline{Q_A}(t)\overline{Q_B}(t).$$

Using the state table, a similar analysis as that for determining the transition equations is performed for the conditional outputs yielding

$$\begin{aligned} \texttt{set\_busy}(t) &= \texttt{zero} \cdot \overline{Q_A}(t)\overline{Q_B}(t), \\ \texttt{clr\_busy}(t) &= \texttt{cnt\_eq} \cdot Q_A(t)\overline{Q_B}(t). \end{aligned}$$

The unconditional output equations are equal to the state values in which they are asserted as before and they are given as

$$\begin{aligned} \texttt{ld\_cnt}(t) &= \texttt{s1}(t) = \overline{Q_A}(t)Q_B(t), \\ \texttt{addr\_sel}(t) &= \texttt{s2}(t) = Q_A(t)\overline{Q_B}(t), \\ \texttt{zero\_we}(t) &= \texttt{s2}(t) = Q_A(t)\overline{Q_B}(t), \\ \texttt{cnt\_en}(t) &= \texttt{s2}(t) = Q_A(t)\overline{Q_B}(t). \end{aligned}$$

Using the D flip-flop characteristic, the corresponding excitation equations are

$$\begin{aligned} D_A &= Q_B(t) + \overline{\texttt{cnt\_eq}} \cdot Q_A(t)\overline{Q_B}(t), \\ D_B &= \texttt{zero} \cdot \overline{Q_A}(t)\overline{Q_B}(t). \end{aligned}$$

Using the excitation equations and the output equations, a logic diagram can be drawn and is shown in Fig. 2.11. Although basic logic gates are shown in Fig. 2.11, the actual logic generated depends on the target technology. For example if a device in the Altera Flex10K is used, four-input lookup tables (i.e., 16×1 memory circuits) are used instead of discrete logic gates.

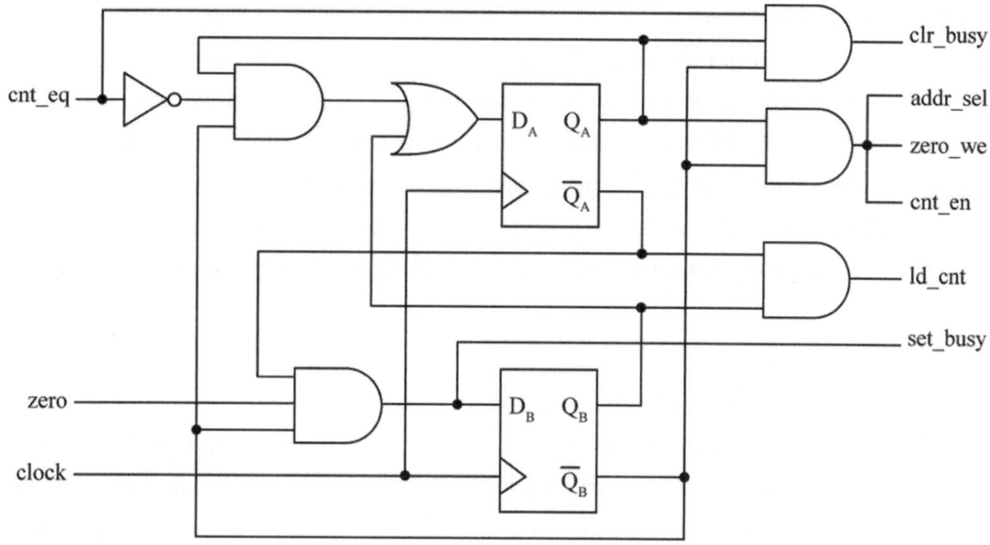

FIGURE 2.11: Logic diagram corresponding to bit-level state equations with minimal encoding

2.5.3 Controller Circuit Analysis

The inverse problem of logic synthesis is that of analysis. The analysis problem gives a logic diagram and requires the development of state equations. The procedure for determining a set of state equations for a given circuit diagram is as follows:

1. Determine the combinational logic expressions that represent the memory device synchronous input signals in terms of external inputs and present state variables.
2. Use the characteristic equations of the memory storage devices to find the next state equations in terms of the memory device inputs.
3. Substitute the expressions found in step 1 into those of step 2 yielding the next-state equations for each state bit.
4. Determine the output equations for each output bit by forming an equation based on the present state values and the external circuit inputs.

2.6 MEALY AND MOORE MACHINE CONVERSION

All synchronous sequential circuits may be implemented based on either Mealy or Moore machine models. Occasionally it is desirable to convert between Mealy and Moore machine models of synchronous state machines. As an example, it may be advantageous to realize a sequential circuit with outputs that are registered to avoid output signal changes in the middle of a clock period, or in other words, to ensure that the output is always valid except during state transitions. One way to do this is to convert a Mealy machine description to that of a Moore machine so that the circuit outputs are actually portions of the state encodings, or that they at least depend only on the state encodings. Conversely, transformation of a Moore machine to a Mealy machine can lead to circuits that require fewer states and hence memory storage elements that may result in a savings in area or component usage. Although the examples of Mealy to Moore conversion and vice versa are shown using state diagrams, the methodology for this conversion can also be accomplished directly with ASM charts.

2.6.1 Mealy to Moore Machine Conversion

The procedure for conversion from a Mealy to a Moore machine is to augment the state vectors with the output bits. All of the transition edges in a state diagram labeled with output bits are replaced with edges with the output bits omitted and the next states have the output bits appended to the original state vector encodings. As an example, consider the portion of the state transition diagram in Fig. 2.12 that is originally in Mealy form on the left and is transformed to Moore form on the right.

From the example in Fig. 2.12, it is seen that the transformation from a Mealy to a Moore state transition diagram is accomplished by splitting states with more than one incoming edge into separate states and then by preserving all destination states. The output values in the Mealy machine are changed into additional bits in the state encodings. In the example these new bits were appended to the end of the state vectors of the Mealy machine encodings but in general they may be added anywhere within the state vectors as long as each resulting bit

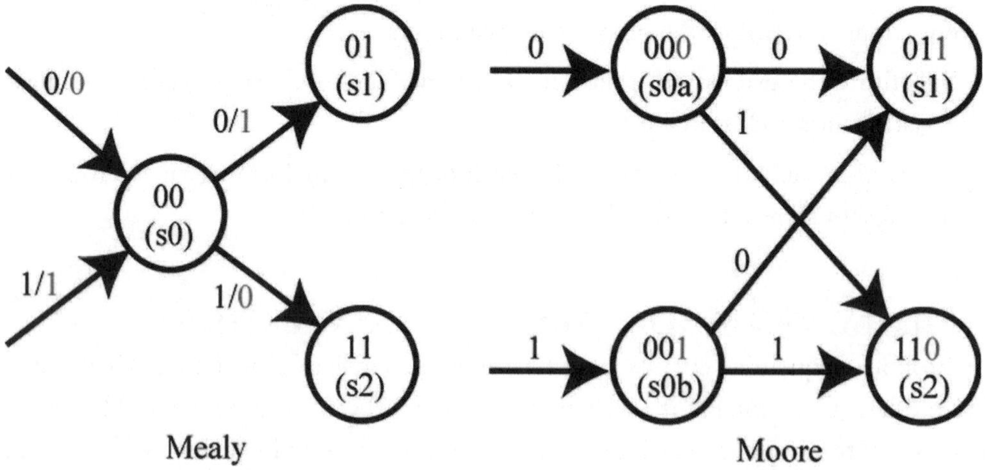

FIGURE 2.12: Portion of a Mealy state machine transformed to a Moore machine

string is unique. It is easy to see that the Moore equivalent of a Mealy machine will generally require more flip-flops since the number of states is increased and the state vectors grow in length.

2.6.2 Moore to Mealy Conversion

When a designer wishes to minimize the number of memory elements required and it is not necessary to have registered outputs of a synchronous sequential circuit, conversion from a Mealy to a Moore model of the circuit can be advantageous. Since sequential circuits realized from Mealy models generally require fewer states, the resulting circuit will usually require fewer state holding elements; however, the amount of combinational logic may also increase so a careful tradeoff should be performed if savings in area is the ultimate goal. The Moore to Mealy translation process is performed in two steps. The first step translates a Moore to Mealy machine and results in a state diagram with the same number of states. Clearly, there is no savings in terms of states after this translation is performed thus a second step should be performed that searches for equivalent states and collapses them into a single state. As an example of the first step, Fig. 2.13 contains a portion of a state diagram for a Moore machine on the left with the corresponding portion of the Mealy machine on the right. Essentially this transformation involves pushing the output values from the state vector to the outgoing edge of each vertex in the state diagram. The output values labeled A and B in Fig. 2.8 are those of the preceding states that each respective edge originates from.

2.6.3 State Machine Equivalence

Two state machines are said to be equivalent if for all possible inputs the same outputs are produced. In general, a given input/output behavior does not require a unique state machine.

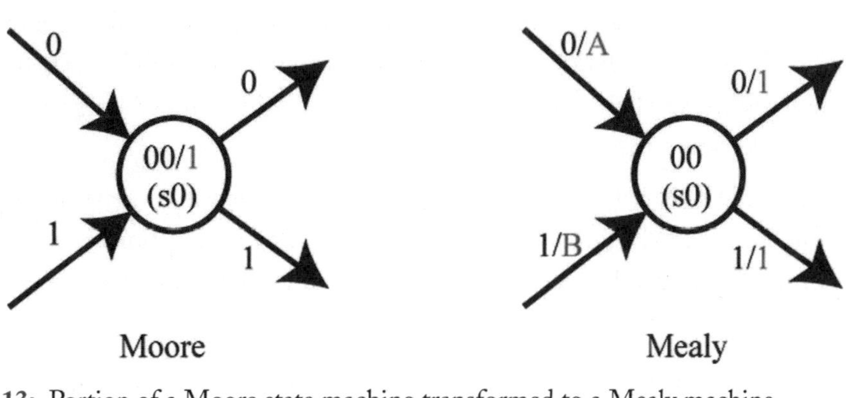

FIGURE 2.13: Portion of a Moore state machine transformed to a Mealy machine

Many different state machines can exist with the same input/output behavior with different numbers of internal states. As synchronous sequential circuit designers, we are interested in specifying state machines with as few states as possible to minimize the number of state holding elements.

During the process of transforming a Moore machine to a Mealy machine, it is always desirable to determine if the transformed Mealy machine can be replaced by another equivalent Mealy machine that is composed of fewer internal states. A methodology for minimizing the number of required states in a state machine is to generate the state table model of the circuit and then to search for equivalent states. Equivalent states are those for which the same input/output behavior occurs. When two equivalent states are found, one is discarded from the state table and all other occurrences of the discarded state are relabeled with the state that is retained.

The methodology for finding equivalent states in a Mealy machine is a classic problem in logic synthesis and design. One of the most common ways to find state equivalence involves the use of implication tables as described in [8] and many other texts on basic digital logic design and we refer the reader to these for more details.

2.7 VERILOG DESCRIPTIONS OF SYNCHRONOUS SEQUENTIAL CIRCUITS

The design of synchronous sequential circuits usually begins with a high-level description or notion of what the circuit is to accomplish in a behavioral sense. The first step in the design process is to transform this behavior into a state machine model. Given the model, the synchronous sequential circuit can be derived by a variety of means. In this text, we recommend the following procedure:

1. Transform the behavioral description of the circuit into an ASM chart. It may be easier for the designers who are more familiar with state diagrams to first generate another model such as a state diagram before finally deriving the ASM chart; however after experience with ASM chart generation has been obtained, most designers can generate the ASM chart as a first step.
2. Once the ASM chart has been developed, the next step is to generate a Verilog description of the ASM chart. Different styles of Verilog descriptions can be generated at this point, but all should synthesize into a controller with essentially the same timing characteristics.

Many designers accomplish this first step by drawing an ASM chart or a state diagram. Typically, the state encoding is unknown at this point and states are often given symbolic labels that may be letters of alphabets or descriptive words. After the first model of the state machine has been determined, an ASM chart is created if it has not already been produced. We strongly emphasize and suggest that ASM charts be generated before writing the Verilog code description. The "ASM first" approach is recommended because the rules of proper ASM chart production allow a designer to easily transform the chart into a Verilog description without performing low-level and tedious design work, such as state assignment and derivation of bit-level transition and excitation equations that are required for tabular or equation descriptions. The latter is to be avoided since simple changes in the circuit behavior result in more work to update the state table or state equations and then the resulting Verilog code.

It is very easy to fall into the trap of generating the initial ASM chart and corresponding Verilog description and to then modify only the Verilog code during design debugging. This should be avoided at all costs since the ASM chart provides a quick graphical depiction of the operation of the circuit and bugs can easily be detected by updating the ASM chart in each design iteration.

In the previous section on ASM charts we mentioned the rules of their proper construction as we described each symbol. We will group the rules together and restate them here for conciseness.

1. State-representation symbols
   a. Should only contain zero or more unconditional output expressions inside.
   b. May or may not have state names or state encodings appear outside but near the symbol.
   c. Must have one or more flow direction symbols (arrows) pointing to them.
   d. Must have a single flow direction symbol (arrow) exiting them.
2. Decision symbols
   a. Should contain input signals or expressions dependent upon input signals only.
   b. Must have a single flow direction symbol pointing to them that originates from a state-representation symbol.
   c. Must have two (single bit, basic decision symbol) or more (generalized decision symbol) outgoing flow direction symbols.
   d. Each outgoing flow direction symbol must be labeled with a unique and complete set of all possible values of the input signals or set of expressions contained within them.
   e. All outgoing signal flow arrows must point to another decision symbol, a conditional output symbol, or another state-representation symbol.

3. Conditional output symbols
   a. Must have exactly one incoming flow symbol in ASMs where generalized decision symbols are used.
   b. May have more than one incoming flow symbol if they all originate from decision symbols in ASM charts containing cascades of decision symbols.
   c. Must have exactly one outgoing flow symbol that points to another decision symbol or a state-representation symbol.
   d. The incoming flow signal must always originate from a decision symbol and never from a state-representation symbol.

Although the rules above allow conditional output symbols to be interspersed among cascades of decision symbols, we recommend that as a matter of practice, conditional output symbols always follow complete cascades of decision symbols, which implies that all outgoing flow arrows from conditional symbols will always point to state-representation symbols.

2.7.1 Example Verilog Descriptions

There are a variety of ways to describe a controller given an ASM chart description. One way to classify these different styles is through the number of `always` blocks used in the implementation. We will show three different Verilog listings that implement the example controller described by the ASM chart in Fig. 2.8. An easy way to classify these different approaches is to note that sections of the code represent the different blocks shown in the architecture models shown in Fig. 2.4. A fourth Verilog example is shown that synthesizes the Moore machine description of the controller modeled by the rightmost state diagram in Fig. 2.10.

2.7.2 Verilog Descriptions for the Mealy Machine Model of an Example Controller

In each of the three following Verilog modules we adhere to the coding guideline rule of using blocking assignments inside `always` blocks that synthesize combinational logic and nonblocking assignments for those always blocks that synthesize registered logic. Verilog `always` blocks that synthesize into registered logic are those that utilize the Verilog keywords `posedge` or `negedge` in the signal activation lists (i.e., the arguments of the `always` statement). It is possible to use blocking assignments in `always` blocks that produce registered logic but this can be tricky since the sequential order of the assignments matters and an incorrect order can produce the wrong results. For this reason, we promote and recommend using nonblocking assignments in `always` blocks that produce registered logic as this allows designers to not be concerned about the order of the assignment statements. The three examples that follow were constructed with a correct ordering of the assignments meaning that it would be safe to use blocking assignments in the following code listings as well as nonblocking.

In each of the following Verilog examples, we have included output ports for the current state values as indicated by the comments in the modules. These output ports are for debugging purposes only and are convenient during simulation. After the controller is verified to be working properly, these output ports should be removed prior to synthesis.

Fig. 2.14 contains a Verilog code listing that reflects the behavior of the ASM chart in Fig. 2.8.

The Verilog listing in Fig. 2.14 contains a single `always` block. When synthesized, the `always` block in Fig. 2.14 is responsible for generating the logic present in the top two

```
module ramfsm_ex1 (state, addr_sel, cnt_en, ld_cnt, zero_we, set_busy,
clr_busy, clk, reset, zero, cnt_eq);
   input        clk, reset, zero, cnt_eq;
   output [1:0] state;  //state output for debugging
   output       addr_sel, cnt_en, ld_cnt, zero_we;
   output       set_busy, clr_busy;

   reg    [1:0] state;

//  State Encoding Here
   parameter S0=2'b00, S1=2'b01, S2=2'b10;

// Register and Combinational Transition logic here
   always @(posedge clk or posedge reset)
       begin
           if (reset == 1'b1)
                state<=S0;
           else
                case (state)
                S0: if (zero == 1'b1) state <= S1;
                S1: state <= S2;
                S2: if (cnt_eq == 1'b1) state <= S0;
                default: state <= S0;
                endcase
        end
// Combinational output logic here
assign set_busy  = (state==S0 && zero==1'b1) ? 1'b1 : 1'b0;
assign ld_cnt    = (state==S1) ? 1'b1 : 1'b0;
assign addr_sel  = (state==S2) ? 1'b1 : 1'b0;
assign zero_we   = (state==S2) ? 1'b1 : 1'b0;
assign cnt_en    = (state==S2) ? 1'b1 : 1'b0;
assign clr_busy  = (state==S2 && cnt_eq==1'b1) ? 1'b1 : 1'b0;

endmodule
```

FIGURE 2.14: Verilog code listing of an example controller with a single `always` block

rectangles in Fig. 2.4 of the Mealy machine model. These are the memory storage elements and the combinational logic representing the excitation equations. The six concurrent statements that occur just before the `endmodule` are responsible for generating the output logic as shown in the bottommost rectangle in the Mealy model of Fig. 2.4. Referring back to the discussion on discrete-event simulation, we know that these are asynchronous statements that may be scheduled for simulation at any time an event occurs, meaning a signal on the right-hand side of the statement changes value. This in turn represents unregistered outputs in terms of synthesized combinational logic whose output changes shortly after any of the dependent signals on the right-hand side of the statements toggle. It is also interesting to note that the signals `ld_cnt`, `addr_sel`, `zero_we`, and `cnt_en` depend only on the present state. Hence, they will change value only after a short time that the state changes and are Moore-type outputs although they are not registered outputs. Although not shown in this example, it is possible to have outputs that depend only on the present state but with more complicated logic combining various bits of the present state in other controller examples. In this case output signals will all change shortly after a state change but not necessarily at the exact same time due to propagation through the output logic. The other two output signals, `set_busy` and `clr_busy`, depend not only upon the present state, but also the input signals `zero` and `cnt_eq`. This means that these output signals may change either shortly after a state change or shortly after one of these input signals toggles in value.

The next example, in Fig. 2.15, contains a Verilog code listing that has all the output signals and transition logic generated from within an `always` block. Since this logic is generated from within an `always` block that is not protected by a clock edge, the synthesized outputs are not registered and combinational logic is produced. The first `always` block is responsible for producing the memory elements represented by the middle block in the Mealy architecture diagram in Fig. 2.4 while the second `always` block produces the combinational logic in the top and bottom blocks of Fig. 2.4. Note that nonblocking (<=) assignments are used in the first `always` block while blocking (=) assignments are used in the second `always` block. Also note the inclusion of the default assignments for the outputs at the beginning of the combinational `always` block. These are very important to remember to include to prevent the outputs from being "stuck" in an asserted state. These assignments correspond to the nonasserted output values that are generally not shown in the ASM chart.

The next example code listing is shown in Fig. 2.16 and contains sections of Verilog code that each correspond to one of the blocks shown in the Mealy architecture model in Fig. 2.4. The first `always` block is responsible for synthesizing the memory elements in the center block, the second `always` block produces combinational logic that synthesizes the excitation logic block at the top of the diagram, and the bottom set of six continuous assignment statements synthesizes into the output combinational logic block.

```
module ramfsm_ex2  (pstate, addr_sel, cnt_en, ld_cnt, zero_we,
                set_busy, clr_busy, clk, reset, zero, cnt_eq);
   input        clk, reset, zero, cnt_eq;
   output [1:0] pstate;  //state output for debugging
   output       addr_sel, cnt_en, ld_cnt, zero_we;
   output       set_busy, clr_busy;
   reg          addr_sel, cnt_en, ld_cnt, zero_we, set_busy, clr_busy;
   reg    [1:0] pstate, nstate;

   //  State Encoding Here
   parameter S0=2'b00, S1=2'b01, S2=2'b10;

   // Register logic here
   always @(posedge clk or posedge reset)
       begin
           if (reset == 1'b1) pstate <= S0;
             else pstate <= nstate;
       end

// Combinational transition and output logic here
   always @(pstate)
      begin
         // We must include default values here
         // to avoid inferred latches
         set_busy = 1'b0;
         ld_cnt   = 1'b0;
         clr_busy = 1'b0;
         addr_sel = 1'b0;
         zero_we  = 1'b0;
         cnt_en   = 1'b0;
// Transition and output logic in case statement
         case (pstate)
                S0: if (zero == 1'b1)
                      begin
                       nstate = S1;
                       set_busy = 1'b1;
                      end
                    else nstate = S0;
                S1: begin
                       nstate = S2;
                       ld_cnt = 1'b1;
                    end
                S2: begin
                     if (cnt_eq == 1'b1)
                       begin
                        nstate = S0;
                        clr_busy = 1'b1;
                       end
                      else nstate = S2;
                     addr_sel = 1'b1;
                     zero_we = 1'b1;
                     cnt_en = 1'b1;
                    end
                default: nstate = S0;
          endcase
      end
endmodule
```

FIGURE 2.15: Verilog listing of an example controller with two `always` blocks

```
module ramfsm_ex3 (pstate, addr_sel, cnt_en, ld_cnt, zero_we,
                 set_busy, clr_busy, clk, reset, zero, cnt_eq);
   input        clk, reset, zero, cnt_eq;
   output [1:0] pstate;  //state output for debugging
   output       addr_sel, cnt_en, ld_cnt, zero_we;
   output       set_busy, clr_busy;
   reg    [1:0] pstate, nstate;

   //  State Encoding Here
   parameter S0=2'b00, S1=2'b01, S2=2'b10;

   // Register logic here
   always @(posedge clk or posedge reset)
       begin
           if (reset == 1'b1) pstate<=S0;
             else pstate <= nstate;
       end

// Combinational transition logic here
   always @(pstate)
      begin
         case (pstate)
                S0: if (zero == 1'b1)
                        nstate = S1;
                    else
                        nstate = S0;
                S1: nstate = S2;
                S2: if (cnt_eq == 1'b1)
                        nstate = S0;
                    else
                        nstate = S2;
                default: nstate = S0;
          endcase
      end

// Combinational output logic here
assign set_busy  = (pstate==S0 && zero==1'b1) ? 1'b1 : 1'b0;
assign ld_cnt    = (pstate==S1) ? 1'b1 : 1'b0;
assign addr_sel  = (pstate==S2) ? 1'b1 : 1'b0;
assign zero_we   = (pstate==S2) ? 1'b1 : 1'b0;
assign cnt_en    = (pstate==S2) ? 1'b1 : 1'b0;
assign clr_busy  = (pstate==S2 && cnt_eq==1'b1) ? 1'b1 : 1'b0;

endmodule
```

FIGURE 2.16: Verilog listing of an example controller with two `always` blocks and concurrent assignments for output logic

2.7.3 Verilog Descriptions for the Moore Machine Model of an Example Controller

The example controller described by the ASM chart in Fig. 2.8 was transformed to a Moore machine represented by the rightmost state diagram in Fig. 2.10. As discussed previously, the timing characteristics for the outputs of a Moore type controller differ in that they change value only slightly after a clock event since the output signals are actually state bits.

```
module ramfsm_ex4 (state, addr_sel, cnt_en, ld_cnt, zero_we, set_busy,
clr_busy, clk, reset, zero, cnt_eq);
   input        clk, reset, zero, cnt_eq;
   output [6:0] state;  //state output for debugging
   output       addr_sel, cnt_en, ld_cnt, zero_we;
   output       set_busy, clr_busy;

   reg    [6:0] state;

//  State Encoding Here
   parameter A=7'b0000000, B=7'b0101000, C=7'b0000111, D=7'b1000111 ,
E=7'b0010000;

// Register and Combinational Transition logic here
   always @(posedge clk or posedge reset)
       begin
           if (reset == 1'b1)
                state<=A;
           else
                case (state)
                A: if (zero == 1'b1)
                       state <= B;
                   else
                       state <= A;
                B: state <= C;
                C: state <= D;
                D: if (cnt_eq == 1'b1)
                       state <= E;
                   else
                       state <= D;
                D: state <= E;
                E: state <= A;
                default: state <= A;
                endcase
         end

// Outputs are directly the state encoding signals
assign set_busy  = state[5];
assign clr_busy  = state[4];
assign ld_cnt    = state[3];
assign addr_sel  = state[2];
assign zero_we   = state[1];
assign cnt_en    = state[0];
endmodule
```

FIGURE 2.17: Verilog listing of an example controller with two `always` blocks and concurrent assignments for output logic

Fig. 2.17 contains the Verilog description of the Moore-type controller and consists of a single registered `always` block followed by a set of continuous assignments statements that simply map certain state bits to the various output signals. This type of controller requires more preliminary work to be performed since our preferred approach is to generate an ASM chart that usually contains outputs that are both Mealy-type (i.e., conditional outputs) and

Moore-type (unconditional outputs). After generation of this type of ASM chart, the chart must be converted into a chart with unconditional outputs only, and this involves performing state assignments that map to the output signals. In certain cases, where it is desired to have outputs that only change at clock edges, this is one way to generate such controllers.

2.8 SUMMARY

This chapter has provided a definition of sequential digital logic systems and focused specifically on synchronous or clocked sequential logic. We have described how this class of circuits is very useful for serving as controllers in a system containing a datapath. Models at various levels of abstraction were described including ASM charts, state diagrams, state equations, and state tables. Several design issues were discussed including conversions from Mealy- to Moore-type models and vice versa, timing aspects of output signals from Moore versus Mealy machine based controllers, and state assignment. We concluded this chapter with a discussion of various styles for describing a controller in synthesizable Verilog and emphasized the design approach of first creating an ASM chart and then translating it into a Verilog module.

References

[1] IEEE Std. 1364-2001, IEEE Standard Verilog® Hardware Description Language, p. 856.

[2] IEEE Std. 1364.1-2002, IEEE Standard for Verilog® Register Transfer Level Synthesis, p. 100.

[3] C. E. Cummings, "Nonblocking assignments in verilog synthesis, coding styles that kill!," *SNUG-2000*, San Jose, CA, 2000.

[4] G. H. Mealy, "A method for synthesizing sequential circuits," *Bell System Tech. J.*, Vol. 34, No. 5, pp. 1045–1079, 1955.

[5] E. F. Moore, "Gedanken experiments on sequential machines," *Automata Studies*. Princeton, NJ: Princeton University Press, pp. 129–153, 1956.

[6] J. E. Hopcroft and J. D. Ullman, *Introduction to Automata Theory, Languages, and Computation*. Reading, MA: Addison-Wesley, 1979.

[7] C. Maxfield, *Bebop to the Boolean Boogie*. Solana Beach, CA: Hightext Publications, 1995.

[8] M. M. Mano, *Digital Design*, 3rd edition, Upper Saddle River, NJ: Prentice Hall, 2002.

Biography

Robert B. Reese received the B.S. degree from Louisiana Tech University, Ruston, in 1979 and the M.S. and Ph.D. degrees from Texas A&M University, College Station, in 1982 and 1985, respectively, all in electrical engineering. He served as a Member of the Technical Staff of the Microelectronics and Computer Technology Corporation (MCC), Austin, TX, from 1985 to 1988. Since 1988, he has been with the Department of Electrical and Computer Engineering at Mississippi State University, Mississippi State, where he is an Associate Professor. Courses that he teaches include VLSI systems, Digital System design, and Microprocessors. His research interests include self-timed digital systems and computer architecture.

Mitchell A. Thornton received the BSEE degree from Oklahoma State University in 1985, the MSEE degree from the University of Texas in Arlington in 1990, and the MSCS in 1993 and Ph.D. in computer engineering in 1995 from Southern Methodist University in Dallas, Texas. His industrial experience includes full-time employment at E-Systems (now L-3 communications) in Greenville, Texas and the Cyrix Corporation in Richardson, Texas where he served in a variety of engineering positions between 1985 through 1992. From 1995 through 1999, he was a faculty member in the Department of Computer Systems Engineering at the University of Arkansas and from 1999 through 2002 in the Department of Electrical and Computer Engineering at Mississippi State University. Currently, he is a Professor of Computer Science and Engineering and, by courtesy, Electrical Engineering at Southern Methodist University. His research and teaching interests are in the general area of digital circuits and systems design with specific emphasis in EDA/CAD methods including asynchronous circuit and computer arithmetic circuit synthesis, formal verification/validation and simulation of digital systems, multiple-valued logic, and spectral techniques.

Finite State Machine Datapath Design, Optimization, and Implementation

Justin Davis
Raytheon Missile Systems

Robert Reese
Mississippi State University

MORGAN & CLAYPOOL PUBLISHERS

ABSTRACT

Finite State Machine Datapath Design, Optimization, and Implementation explores the design space of combined FSM/Datapath implementations. The lecture starts by examining performance issues in digital systems such as clock skew and its effect on setup and hold time constraints, and the use of pipelining for increasing system clock frequency. This is followed by definitions for latency and throughput, with associated resource tradeoffs explored in detail through the use of dataflow graphs and scheduling tables applied to examples taken from digital signal processing applications. Also, design issues relating to functionality, interfacing, and performance for different types of memories commonly found in ASICs and FPGAs such as FIFOs, single-ports, and dual-ports are examined. Selected design examples are presented in implementation-neutral Verilog code and block diagrams, with associated design files available as downloads for both Altera Quartus and Xilinx Virtex FPGA platforms. A working knowledge of Verilog, logic synthesis, and basic digital design techniques is required. This lecture is suitable as a companion to the synthesis lecture titled Introduction to Logic Synthesis using Verilog HDL.

KEYWORDS:

Verilog, datapath, scheduling, latency, throughput, timing, pipelining, memories, FPGA, flowgraph

Table of Contents

Table of Figures

CHAPTER 1

Calculating Maximum Clock Frequency

The purpose of this chapter is to find the maximum clock frequency and adjusted setup and hold times based on propagation delays for circuits with combinational and sequential gates. This chapter assumes the reader is familiar with digital gates and memory elements such as latches and flip-flops.

1.1 LEARNING OBJECTIVES

After reading this chapter, you will be able to perform the following tasks:

- Discover the longest combinational delay path through a circuit
- Calculate the three types of delays in sequential circuits
- Calculate chip-level setup and hold time based on internal registers
- Calculate board-level clock frequencies

1.2 GATE PROPAGATION DELAY

The simplest metric of performance of a digital device is computation time. Often this is measured in computations per second and depends on the type of computation. For general-purpose processors, it may be measured in millions of instructions per second (MIPS). For arithmetic processors, it may be measured in millions of floating point operations per second (MFLOPS). Computation time is based partly on the speed of the clock and partly on the number of clocks per operation. This chapter will focus on computing the maximum clock speed to enable the minimum computation time.

A digital logic gate is constructed from transistors arranged in a specific way to perform a mathematical operation. These transistors are operated like on/off switches. Ideally the transistors can switch on to off or off to on instantly; however, realistic transistors have a finite switching time. A leading factor in transistor switching time is their physical size. Smaller transistors will usually switch faster than large transistors. As transistor size is further miniaturized through emerging technologies, this delay continues to decrease. Modern transistors can switch exceptionally fast, but the delay must still be accounted for.

Specific types of transistors in a logic gate are not as important as their effect. The switching delay of the transistors creates a delay in the logic gate. The latter can be measured from the time an input changes to the time an output changes. This delay is called the ***propagation delay***(t_{pd}). This book will only consider the delays associated with the gate but with the understanding that it is defined by the underlying transistors.

1.2.1 Single Input/Multiple Input Delays

The simplest gate for discussing t_{pd} is the inverter. The inverter has one input and one output. While the input is a logic high, the output is a logic low. When the input changes from high to low, the output will change from low to high after a certain delay. The input and the output of the inverter do not change instantaneously from a logic low to a logic high or vice versa. These finite rise times and fall times are shown in Fig. 1.1. The 50% point on the rise time or fall time is when the voltage level is halfway between the logic high and logic low. The t_{pd} is measured between the 50% point of the input rise time and the 50% point of the fall time of the output.

The t_{pd} can be different for the output rise time and fall time. If the rise time is longer than the fall time, then the 50% point will be shifted, which results in a larger t_{pd}. Since the propagation delay can be different, each is denoted differently. When the output is changing from high to low, the delay associated with it is denoted t_{phl}. When the output is changing from low to high, the delay associated with it is denoted t_{plh}. For simplicity, the worst case is taken for the two propagation delays and is considered to be the total t_{pd} for the entire gate.

Even though each type of logic gate is constructed differently, the delay through the gates are measured the same. A multiple input gate has many more propagation delays. For example, an AND gate has at least two inputs as shown in Fig. 1.2. The t_{pd} must be measured from low to high and high to low for each input.

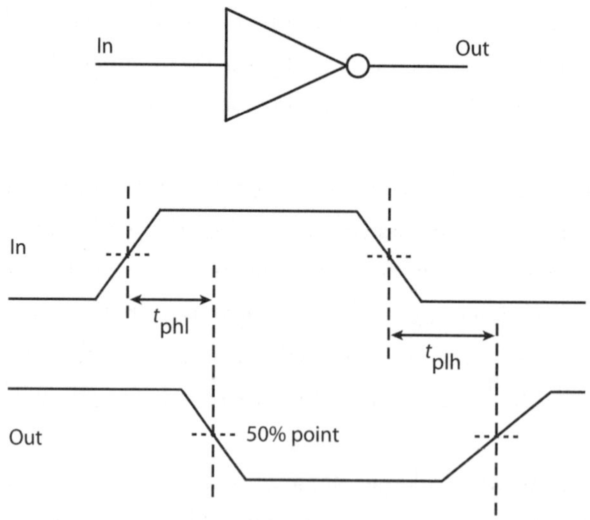

FIGURE 1.1: Inverter propagation delay.

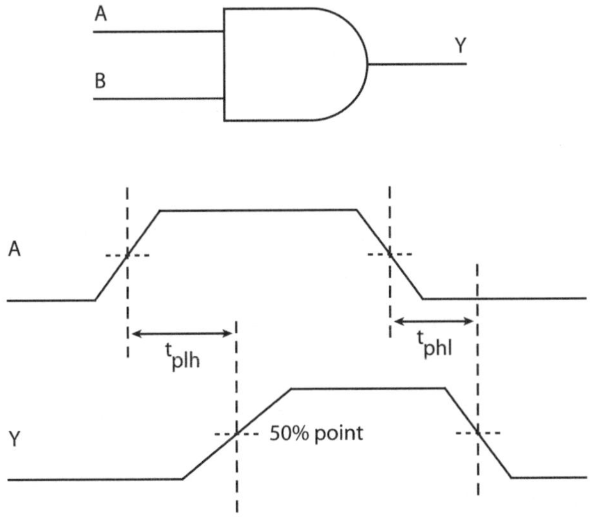

FIGURE 1.2: AND gate propagation delay.

For a two-input gate, four propagation delays are found: A2Y_t_{plh}, A2Y_t_{phl}, B2Y_t_{plh}, A2Y_t_{phl}. For simplicity, the worst case is taken for the four propagation delays and is considered to be the total t_{pd} for the entire gate (Y_t_{pd}). This is true for any number of inputs for a combinational gate. Typically, datasheets for a logic device contains the worst-case t_{pd} along with the typical t_{pd}.

1.2.2 Propagation Delay Effects

When multiple gates are connected together, the propagation delays on the individual gates can produce unwanted and incorrect results in the output called ***glitches***. The glitches can cause output values that are logically impossible with ideal logic gates. For example, an AND gate only outputs a logic high when both inputs are logic high. When the inputs to an AND gate are always opposite as in Fig. 1.3, then the output will never be logic high. If the inverter has a finite t_{pd}, then the output of the AND gate can become a logic high while the signal is propagating through the inverter. When the input X is a logic low, the output of the inverter is a logic high. When the input switches to a logic high, both the inputs to the AND gate are logic high because the change has not propagated through the inverter yet.

Because of propagation delays, whenever multiple gates are combined, the output could have glitches until after all the signals have propagated through all the gates. The output cannot be considered valid until after this delay. This is the reason why digital systems are usually clocked. The rising edge of the clock signifies when all the input signals are sent to the circuit. If the clock period is set correctly, by the time the next rising edge occurs, the glitches end and the output is considered valid. The clock period is set by analyzing all the propagation delays in the circuit.

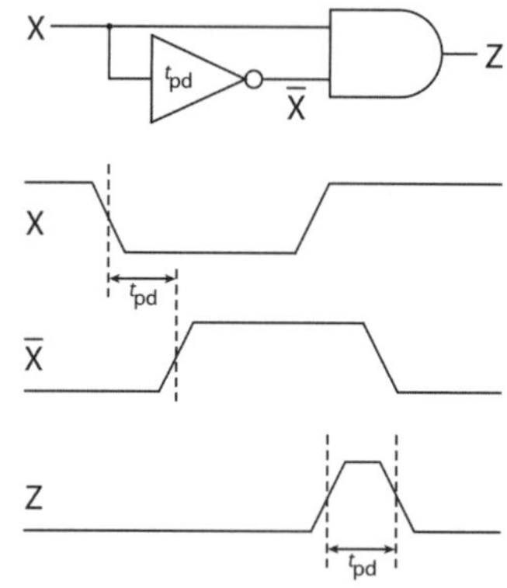

FIGURE 1.3: Glitches caused by propagation delay.

1.2.3 Calculating Longest Delay Path

The t_{pd} for a circuit is found by tracing a path from one input to the output. The propagation delay of each gate is added to the total delay for that path. This procedure is repeated for every path from each input to the output. After a set of all delays is constructed, t_{pd} for the circuit is chosen to be the largest delay in the set.

1.2.4 Example 1.1

An XOR gate can be constructed using AND, OR, and NOT gates as in Fig. 1.4. Using the circuit in Fig. 1.4 and the delays of the AND, OR, and NOT gates in Table 1.1, what is the worst-case t_{pd} for the entire circuit?

For the XOR gate, there are four individual paths from the input to the output. The first path starts at the X input and progresses through the A1 AND gate and the O2 OR gate. The total delay is $25 + 20 = 45$ ns. The second path from the X input progresses through the O1 OR gate, the N3 NOT gate, and the O2 OR gate for $20 + 10 + 20 = 50$ ns delay.

The Y input also has two paths. The first is through the N2 NOT gate, the A1 AND gate, and the O2 OR gate for a $10 + 25 + 20 = 55$ ns delay. The last path is through the N1 NOT gate, the O1 OR gate, the N3 NOT gate, and the O2 OR gate for a $10 + 20 + 10 + 20 = 60$ ns delay. All paths are listed in Table 1.2 .

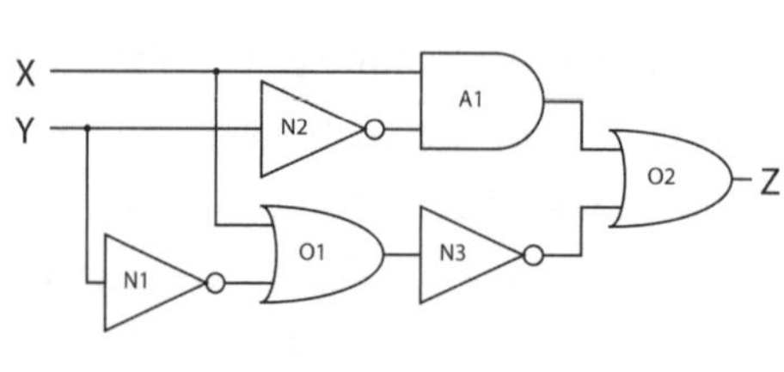

FIGURE 1.4: XOR gate architecture.

TABLE 1.1: Propagation delays for individual gates

Gate	Propagation Delay
NOT	10 ns
AND	25 ns
OR	20 ns

TABLE 1.2: Total set of all propagation delays

Starting Input	Path	Delay
X	A1 + O2	45 ns
X	O1 + N3 + O2	50 ns
Y	N2 + A1 + O2	55 ns
Y	N1 + O1 + N3 + O2	60 ns

The worst-case delay path is 60 ns. On the datasheet, the maximum t_{pd} would be listed as 60 ns. This is also the minimum period of the clock if the XOR gate is used in a real circuit.

1.2.5 Propagation Delays for Modern Integrated Circuits

Delay values for an integrated circuit are dependent upon the technology used to fabricate the integrated circuit, and the environment that the integrated circuit functions within (voltage supply level, temperature). The delays used in this chapter and the next are not meant to reflect actual delays found in modern integrated circuits since those delays are moving targets. Instead, the delay values used in these examples are chosen primarily for ease of hand calculation. The ns unit (nanoseconds,1.0e–9 s) was chosen because nanoseconds is convenient for describing off-chip delays as well as on-chip delays. Furthermore, using a real time unit such as ns instead of unit-less delays allows frequency calculations with real units. See Section 1.6 for a short discussion of how propagation delays for integration circuits have varied as integrated circuit fabrication technology has improved.

1.3 FLIP-FLOP PROPAGATION DELAY

Flip-flops and latches are considered memory elements because they can output a set value without an input. This value can be changed as needed. The input is transferred to the output when the device is enabled. In this book, a flip-flop will be defined by the enable (usually a clock) being an

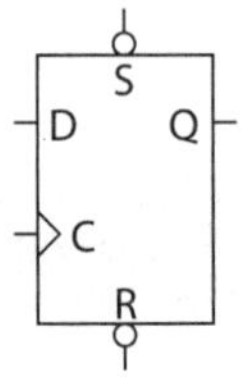

FIGURE 1.5: D-type flip-flop input options.

edge-triggered signal. For a latch, the enable is a level-sensitive signal. This book uses flip-flops in its examples since this is the most commonly-used design style. While many types of flip-flops exist such as SR flip-flops, D flip-flops, T flip-flops, or JK flip-flops, this book will only discuss D flip-flops since they are the simplest and most straight-forward. The other types of flip-flops can be analyzed using the same techniques as the D flip-flop. In D flip-flops, the input is copied to the output at the clock edge. The D flip-flop can have a variety of input options as shown in Fig. 1.5.

A specialized type of flip-flop is called a register. Registers have an enable input which prevents the latter from being transferred to the output in every clock cycle. The input will only be copied when the enable is set high. Registers can come in arrays, which all have the same control signals, but have different data inputs/outputs. Sometimes the term register is used synonymously with the term flip-flop.

The output for a memory element has a t_{pd} like a combinational gate; however, it is measured differently. Since the output for a register only changes on a clock transition, t_{pd} is measured from the time the clock changes to the time the input is copied to the output. Since the data output does not change when the data input changes, t_{pd} is not measured from the data input to the data output. However, the clock-to-output propagation delay (t_{C2Q}) is not the only delay associated with a register.

1.3.1 Asynchronous Delay

Other inputs are available for different types of registers. Some registers have the ability to be set to a logic high or reset to a logic zero from independent inputs. These set/reset inputs can take effect either on a clock edge or independent of the clock altogether. When an input is dependant on the clock edge, it is called a ***synchronous*** input. When an input is not dependant on the clock, it is called an ***asynchronous*** input. The data input to a register is always a synchronous input. An asynchronous set-to-output delay is labeled (t_{S2Q}) and an asynchronous reset-to-output delay is labeled (t_{R2Q}). If the set/reset inputs are synchronous, then there are no individual delays associated with them since the clock-to-output delay covers their delay. Other inputs are available for registers such as an enable input, but again any input, which is dependant on the clock, will not have a separate propagation delay.

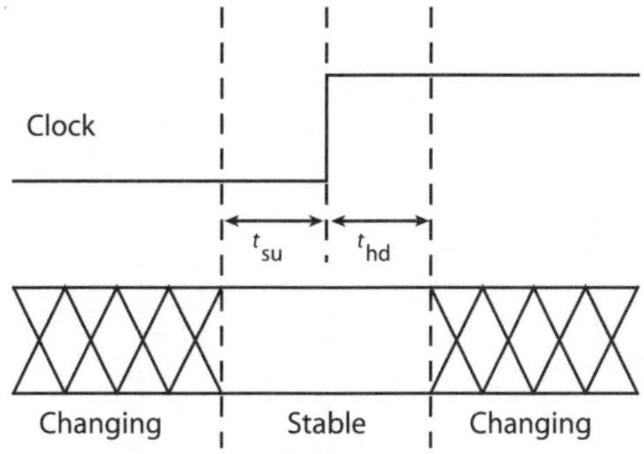

FIGURE 1.6: Relative setup and hold time timing.

1.3.2 Setup and Hold Time

Registers have an additional constraint to ensure that the input is correctly transferred to the output. For every synchronous input, the signal must remain at a stable logic level for a set amount of time before the clock edge occurs. This is called the ***setup*** (tsu) time for the register. Additionally, the input signal must remain stable for a set amount of time after the clock edge occurs. This is called the ***hold*** (thd) time for the register. If the input changes within the setup or hold time, then the output cannot be guaranteed to be correct. This specification is indicated on the datasheet for the register and is set by the characteristic of the internal transistors. Fig. 1.6 illustrates setup and hold time concepts.

1.4 SEQUENTIAL SYSTEM DELAY

Most digital systems contain both sequential and combinational circuits. These circuits can be more difficult to analyze for the longest delay path. Three different types of delay paths occur in the circuit. Each delay path is analyzed differently depending on the origin and destination of the path. The first type of path starts at the data or control inputs to the circuit and is traced through to the outputs of the circuit passing through only combinational gates. This is called a pin-to-pin propagation delay. The next type of path starts at the clock input and is traced to the outputs of the circuit passing through at most one register. This is called t_{C2Q}. The last type of path starts at a register and is traced to another register. This is called the register-to-register delay.

1.4.1 Pin-to-Pin Propagation Delay

A pin-to-pin propagation delay path (t_{P2P}) is defined by any path from an input to an output that passes through only combinational gates, which means ***it cannot pass through any registers***. This is similar to Section 1.2.3 when the longest delay path was found through multiple combinational gates. A path is formed from the input to the output and all of the gate delays are added together. This is repeated for all possible combinational paths. It is possible there are no paths from the input to the output that contain only combinational gates. In this case, t_{P2P} does not contribute to finding the minimum clock period.

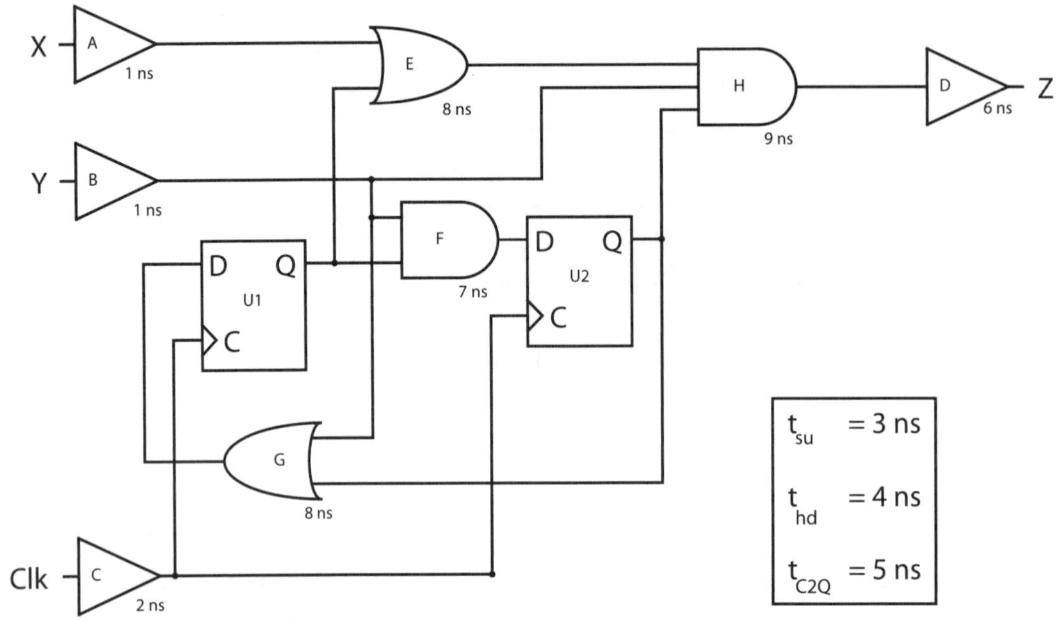

FIGURE 1.7: Sequential circuit for propagation delay.

1.4.2 Example 1.2

The circuit in Fig. 1.7 is the internal layout of a custom built chip. The t_{pd} for each gate is listed below it. The delays for the register are all the same and listed in the lower right corner. Input protection circuits and output fan-out circuitry can slow down the signal transmission on and off the chip. These delays will be represented as simple buffers on the schematic. Find t_{P2P}.

There are multiple pin-to-pin combinational paths for this circuit. The inputs X and Y both have combinational-only paths to the output. The clock (Clk) input does not have a combinational-only path to the output because any path would pass through one of the two registers.

For input X, the path starts at the input buffer A and proceeds through the OR gate E, the AND gate H, and the output buffer D. The propagation delays for these gates are added together to get $1 + 8 + 9 + 6 = 24$ ns.

$$A\_t_{pd} + E\_t_{pd} + H\_t_{pd} + D\_t_{pd} = t_{P2P} \tag{1.1}$$

$$1 + 8 + 9 + 6 = 24\,\text{ns} \tag{1.2}$$

For the input Y, the path starts at the input buffer B and proceeds through the AND gate H, and the output buffer D. The propagation delays for these gates are added together to get $1 + 9 + 6 = 16$ ns.

$$B\_t_{pd} + H\_t_{pd} + D\_t_{pd} = t_{P2P} \tag{1.3}$$

TABLE 1.3: Total set of all pin-to-pin propagation delays

Starting Input	Path	Delay
X	A + E + H + D	24 ns
Y	B + H + D	16 ns

$$1 + 9 + 6 = 16\,\text{ns} \quad (1.4)$$

The larger of these two delays is the worst-case t_{P2P} for this circuit. The path "A + E + H + D" is the worst-case with a delay of 24 ns. The list of delays is in Table 1.3.

1.4.3 Clock-to-Output Delay

The second type of t_{pd} path is the clock-to-output path (t_{C2Q}). These paths pass through ***exactly one register***. The clock input is routed to the registers in the circuit. A path is traced from the clock input of the system to the clock input of a register. Then the path continues through that register to the output of the circuit. The delays of the combinational gates along the path and the clock-to-output delay of the register are added to the total delay of the path.

Often two clock-to-output delays exist when analyzing a circuit. One is for the internal registers, and the other is for the entire circuit. The register C2Q will be a part of the system C2Q, so the register C2Q will always be the smaller of the two. The combinational delay before the register is listed as t_{comb_I2C}, and the combinational delay after the register is listed as t_{comb_Q2O}.

$$t_{comb\_I2C} + t_{C2Q\_FF} + t_{comb\_Q2O} = t_{C2Q\_SYS} \quad (1.5)$$

Some circuit analysis programs treat the clock-to-output delay the same as the pin-to-pin combinational delay, so sometimes on the analysis report there will be no clock-to-output delay listed. The clock input is counted as a regular input. Often these reports will list the worst-case delays for each input, so the clock-to-output delay can be found by searching this list.

1.4.4 Example 1.3

Using the same circuit in Fig. 1.7, find the worst-case t_{C2Q}.

There are two clock-to-output paths through the circuit. Both paths pass through the input buffer C. One path then proceeds through the first register U1, through the OR gate E, through the 3-input AND gate H, and finally to the output buffer D.

$$C\_t_{pd} + U1\_t_{C2Q} + E\_t_{pd} + H\_t_{pd} + D\_t_{pd} = t_{C2Q\_SYS} \quad (1.6)$$

$$2 + 5 + 8 + 9 + 6 = 30\,\text{ns} \quad (1.7)$$

TABLE 1.4: Total Set of all clock-to-output propagation delays

Starting Input	Path	Delay
Clk	C + U1 + E + H + D	30 ns
Clk	C + U2 + H + D	22 ns

The second path proceeds through the second register U2, through the 3-input AND gate H, and finally to the output buffer D.

$$C\_t_{\text{pd}} + U2\_t_{\text{C2Q}} + H\_t_{\text{pd}} + D\_t_{\text{pd}} = t_{\text{C2Q\_SYS}} \quad (1.8)$$

$$2 + 5 + 9 + 6 = 22\,\text{ns} \quad (1.9)$$

The larger of these two delays is the worst-case t_{C2Q} for this circuit. The path "C + U1 + E + H + D" is the worst-case with a delay of 30 ns. The list of delays is in Table 1.4.

1.4.5 Register-to-Register Delay

The last type of propagation delay is the register-to-register delay (t_{R2R}). This is usually the largest of the three types of delays in modern circuit designs. Consequently, it is usually the delay that sets the minimum clock period. As the name of this delay path suggests, this delay path starts at the output of a register and is traced to the input of another register. The path could even be traced back to the input of the starting register, but the route always involves ***at most two registers.*** The number of register-to-register paths in a circuit is proportional to the number of registers in the design. Specifically, the number of paths will be at most 2^N where N is the number of registers. Therefore, the number of paths that must be checked can increase very quickly as a design grows.

The t_{R2R} must be equal to or larger than the clock period. At the beginning of the clock period, the clock transitions from low to a high. This change propagates through the register for a fixed amount of time before the input is transferred to the output. This is the clock-to-output delay of the register. Once the input is present on the output, the combinational gates after the output will begin to switch. After the changes propagate through the combinational gates, the new signals will be ready at the inputs to the registers for transfer to the outputs of the registers. Furthermore, the new signals must satisfy the setup time of the register to ensure they will be transferred correctly to the output.

$$t_{\text{C2Q\_FF}} + t_{\text{comb\_R2R}} + t_{\text{su\_FF}} = t_{\text{R2R}} \quad (1.10)$$

TABLE 1.5: Total set of all register-to-register propagation delays

Starting Input	Path	Delay
U1	U1 + F + U2	15 ns
U2	U2 + G + U1	16 ns

1.4.6 Example 1.3

Using the same circuit in Fig. 1.7, find the worst-case t_{R2R}

There are two registers in this design. Starting with register U1, there is only one path from the output of this register to another register. This path passes through gate F to the input of register U2. Therefore, computing this register-to-register path is easy.

$$U1\_t_{\text{C2Q}} + F\_t_{\text{pd}} + U2\_t_{\text{su}} = t_{\text{R2R}} \tag{1.11}$$

$$5 + 7 + 3 = 15\,\text{ns} \tag{1.12}$$

Starting with register U2, there is only one path from the output to another register. This path passes through gate G to the input of register U1.

$$U2\_t_{\text{C2Q}} + G\_t_{\text{pd}} + U1\_t_{\text{su}} = t_{\text{R2R}} \tag{1.13}$$

$$5 + 8 + 3 = 16\,\text{ns} \tag{1.14}$$

The two register-to-register paths in Table 1.5 above are 15 ns and 16 ns. The worst-case t_{R2R} is therefore 16 ns through the path "U2 + G + U1". If all the registers have the same clock-to-output delay and t_{su} (as is often the case), the only difference between the paths is the combinational circuits between the registers. This can make computing t_{R2R} much easier.

1.4.7 Overall worst-case delay

Now that the maximum delays for the three types of paths have been found, the overall maximum delay of the sequential system can be found. The worst case is the largest delay of the three path types. For the example circuit in Fig. 1.7, the three worst cases are listed in Table 1.6.

The worst-case delay for this system is the clock-to-output delay at 30 ns. Therefore, for this sequential system, the minimum clock period is 30 ns in order to allow all gate outputs to reach stable values. This corresponds to a maximum clock frequency of 33.3 MHz.

1.4.8 Setup and hold adjustments

An additional requirement for sequential circuits is to ensure that t_{su} and t_{hd} requirements of the internal registers have been met. Signals external to the circuit must not violate t_{su} before the clock

TABLE 1.6: Total set of worst-case propagation delays

Path Type	Path	Delay
P2P	A + E + H + D	24 ns
C2Q	C + U1 + E + H + D	30 ns
R2R	U2 + G + U1	16 ns

and t_{hd} after the clock ***at the inputs to the internal register***. If the sequential circuit was going to be packaged into a chip and sold to a customer, the customer may not know how to check if the internal register setup and hold requirements have been met. Therefore t_{su} and t_{hd} requirements are recomputed for the entire sequential circuit and that information is passed to the customer.

For setup time, the data signal must not change for a given time before the clock edge. If the input signal is delayed, such as, through a combinational gate or input buffer as in Fig. 1.8, the input may violate the t_{su} requirement. Therefore, any delay added between the input pin and the register input must be added to the setup time requirement. The delay between the clock input pin and the clock input to the register must also be subtracted from t_{su} . This means if the delays between the pins to the register are the same, there will be no change in t_{su}. Only when there is a difference in the delays will the setup time change.

This procedure must be repeated for each register in the design that has an external input routed to its input through any combinational path. The longest delay from the data input to the registers is used as the worst case. The shortest delay from the clock input to the registers is used as the worst case. The difference between these two paths is the adjustment to the setup time.

$$(t_{pd\_data(MAX)} - t_{pd\_clk(MIN)}) + t_{su\_FF} = t_{su\_TOTAL} \tag{1.15}$$

For hold time, if the clock signal is delayed, such as through an input buffer, the input may violate the t_{hd} requirement. The worst case for t_{hd} is the opposite worst case for t_{su} : the longest delay from the clock input of the circuit to the register, and the shortest delay from the data input to the register. The difference between these two paths is the adjustment to the hold time.

$$(t_{pd\_clk(MAX)} - t_{pd\_data(MIN)}) + t_{hd\_FF} = t_{hd\_TOTAL} \tag{1.16}$$

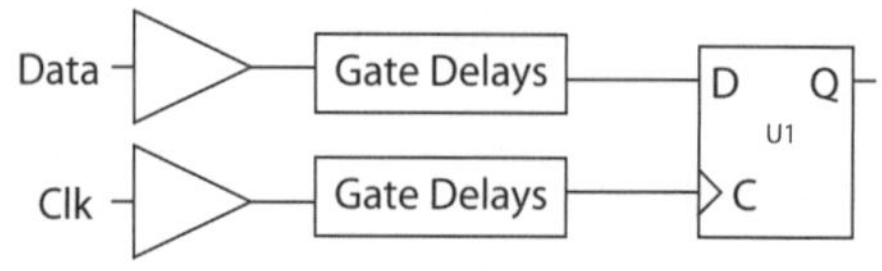

FIGURE 1.8: Calculating adjusted setup/hold times.

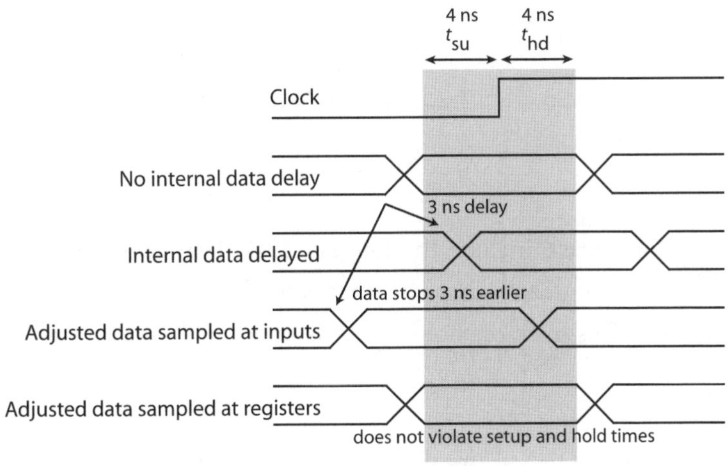

FIGURE 1.9: Adjusted setup and hold timings.

When t_{su} and t_{hd} have been adjusted correctly for the external inputs, the internal t_{su} and t_{su} at the register inputs will not be violated. The timing diagram in Fig. 1.9 shows the behavior of internal delays, which can cause changes in the setup and hold requirement.

1.4.9 Example 1.4

Using the same circuit in Fig. 1.7, find the adjustments to the t_{su} and t_{hd} for the circuit.

In this design, the data input is delivered to the input to two registers. The first path is routed from the Y input through the input buffer, through the OR gate G, and then to the input of the U1 register. The second path passes through the input buffer, through the AND gate F, and then to the U2 register. Note there are no paths from the X input to the inputs of any registers. Table 1.7 provides the set of all input to register delays.

The calculation for t_{su} will include the longest data delay and the shortest clock delay. For this example, the longest data delay is $t_{pd\_data\_U1}$ that will add 9 ns to t_{su}. The shortest clock delay is

TABLE 1.7: Total set of all input to register delays

Delay Path	Path	Delay	Path Name
Y to U1	B + G + U1	9 ns	$t_{pd\_data\_U1}$
Y to U2	B + F + U2	8 ns	$t_{pd\_data\_U2}$
Clk to U1	C + U1	2 ns	$t_{pd\_clk\_U1}$
Clk to U2	C + U2	2 ns	$t_{pd\_clk\_U2}$

$t_{pd\_clk\_U1}$ that will subtract 2 ns from t_{su}. Given t_{su} of 3 ns, the external t_{su} for this circuit is 10 ns.

$$(t_{pd\_data\_U1} - t_{pd\_clk\_U1}) + t_{su\_FF} = t_{su\_TOTAL} \tag{1.17}$$

$$(9 - 2) + 3 = 10\,\text{ns} \tag{1.18}$$

The calculation for t_{hd} will include the longest clock delay and the shortest data delay. For this example, the longest clock delay is $t_{pd\_clk\_U1}$ that will add 2 ns to t_{hd}. The shortest data delay is $t_{pd\_data\_U1}$ that will subtract 8 ns from the hold time. Given t_{hd} of 4 ns, the external t_{hd} for this circuit is -2 ns.

$$(t_{pd\_clk(MAX)} - t_{pd\_data(MIN)}) + t_{hd\_FF} = t_{hd\_TOTAL} \tag{1.19}$$

$$(2 - 8) + 4 = -2\,\text{ns} \tag{1.20}$$

The setup and hold window is 8 ns in which the data cannot change. The negative sign in the hold time calculation means the data input can actually start changing before the clock signal. This is not an intuitive behavior for a digital circuit, so often a negative t_{hd} will be specified as zero instead. By setting t_{hd} to zero, the effective setup and hold window has increased to 10 ns.

1.5 BOARD-LEVEL TIMING CALCULATION

A digital chip will usually be used in a larger system connected to other chips. Even if all chips in the system may be rated to operate at a specific clock frequency, the entire system may not.

1.5.1 Datasheet compilation

The datasheet of each chip should have all of the relevant timing information to compute the board-level maximum clock frequency. This data is similar to the gate delays when computing the chip-level maximum clock frequency. Six relevant pieces of data are needed to ensure the operation of the board-level system. The maximum clock frequency of each chip must be provided since the board-level system cannot operate faster than that. The t_{su} and t_{hd} must be provided to ensure no write violation to the registers internal to the chip. The combinational delay and clock-to-output delay must be known to compute the maximum clock frequency of the circuit. The needed information is presented in Table 1.8 along with the values for the example results.

Each chip can be treated as a sequential circuit with both synchronous and asynchronous delays much like a register. Each of the three worst-case delay path types can be computed with the above information to find the maximum clock frequency. The maximum clock frequency for the board will never exceed any individual chip's rating listed on the datasheet.

TABLE 1.8: Datasheet for the chapter example

Parameter	Description	Min	Max	Units
Tclk	Clock Period	26		ns
Fclk	Clock Frequency		33.3	MHz
t_{su_Y}	Y Setup Time	10		ns
t_{hd_Y}	Y Hold Time	0		ns
$X\_t_{pd\_P2P}$	Combinational delay	24		ns
t_{pd_C2Q}	Clock-to-output delay	30		ns

1.5.2 Board-level maximum frequency

The procedure to find the maximum clock frequency at the board-level is same as at the chip level. The worst-case delays must be found in three cases: the pin-to-pin combinational, the clock-to-output and the register-to-register delays. The minimum clock period is set to the largest of these three paths or the minimum clock period for each individual chip.

1.5.3 Example 1.5

Using the circuit in Fig. 1.10, find the maximum clock frequency. Each chip is the circuit in Fig. 1.7 and uses the timings in Table 1.6.

First, the pin-to-pin combinational delay is found for any path from the X input to the output. There is one pin-to-pin path from the input A to the X input of U1, to the X input of U2, to the

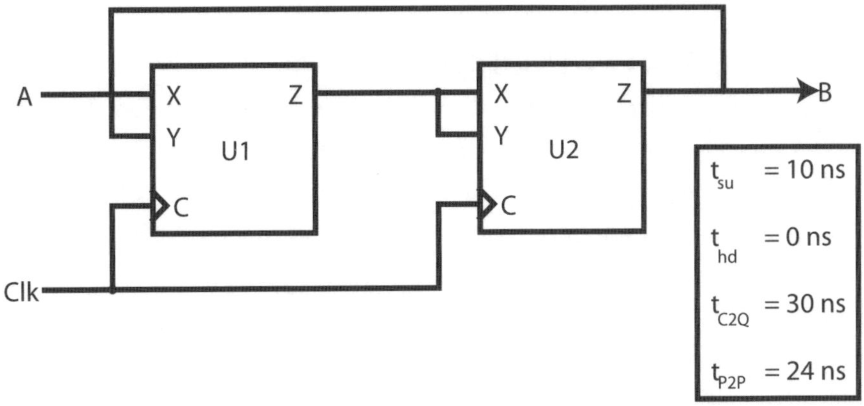

FIGURE 1.10: Board-level schematic to compute maximum clock frequency.

output B. The delay of this path adds the two pin-to-pin delays together $24 + 24 = 48$ ns.

$$X(U1)\_t_{pd} + X(U2)\_t_{pd} = t_{P2P} \tag{1.21}$$

$$24 + 24 = 48\text{ ns} \tag{1.22}$$

Two clock-to-output delays exist for this circuit. The first path passes through the clock input of U1, through the X input of U2. The second path passes only through the clock input of U2. Since the clock-to-output delays for each chip are the same, the first path will be longer since $30 + 24 = 54$ ns.

$$U1\_t_{C2Q} + X(U2)\_t_{pd} = t_{C2Q\_SYS} \tag{1.23}$$

$$30 + 24 = 54\text{ ns} \tag{1.24}$$

Three t_{R2R} exist for this circuit. The first path goes through the U1 clock-to-output, through the X input of U2, and then back to the Y input of U1. The second is through the U1 clock-to-output to the input of Y on U2. The third is through the U2 clock-to-output to the input of Y on U1. The longest path is the first since it passes through the combinational portion of U2 for $30 + 24 + 10 = 64$ ns.

$$U1\_t_{C2Q} + X(U2)\_t_{pd} + U1\_t_{su} = t_{C2Q\_SYS} \tag{1.25}$$

$$30 + 24 + 10 = 64\text{ ns} \tag{1.26}$$

The three worst-case paths and the chip minimum clock period limit the clock frequency for the board-level system. The largest of these values (48 ns, 54 ns, 64 ns, 30 ns) is 64 ns, which is the minimum clock period for the board which corresponds to 15.63 MHz. This frequency is much lower than the chip clock frequency. Note that the combinational delay of the chip contributes most of the slow-down to the circuit.

1.6 DELAYS AND TECHNOLOGY

As stated earlier, delay values for an integrated circuit are dependent upon the technology used to fabricate it, and the environment within which the integrated circuit functions (voltage supply level, temperature). Gate delays for complementary metal-oxide-semiconductor (CMOS) integrated circuits have become smaller over time because transistor channel lengths have become smaller, resulting in transistors that switch faster, and thus, smaller propagation delays for gates. Shrinking transistor sizes have allowed more transistors to be placed in the same integrated circuit, allowing for increased integrated circuit functionality. In programmable logic terms, this means that new generations of programmable logic are able to implement increasing numbers of logic gates in a single package.

TABLE 1.9: Xilinx Virtex FGPA delays over time (delays in picoseconds)

DELAY TYPE	VIRTEX 1 (2200 NM, 2.5 V, 1998)	VIRTEX-2 (1500 NM, 1.5 V, 2000)	VIRTEX-4 (90 NM, 1.2 V, 2004)	VIRTEX-5 (65 NM, 1.0V, 2006)
LUT Propagation Delay	700	390	170	90
DFF Tcq	1200	500	310	40
DFF setup	700	330	400	40
DFF hold	0	−80	−90	20
IOB out (LVTTL)	3200	1510	2020	1520
IOB in (LVTTL)	90	76	87	70

Notes: DFF T_{su}/T_{hd} for Virtex-5 are native setup/hold.
DFF T_{su}/T_{hd} for Virtex 1,2,4 include mux delay.
IOB out/in for Virtex 4,5 uses fast 24mA LVTTL.

Table 1.9 shows delay evolution for the Xilinx Virtex family of field programmable gate arrays (FPGAs) over time. The top row gives each FPGA family name as well as the CMOS technology, supply voltage, and date of first introduction. A CMOS technology designated as 2200 nm (nanometer = 1.0e–9 m) means that the shortest channel MOS transistors has a channel length of 2200 nm (the value 2200 nm is more commonly written as 0.22 μm, but nm is used for consistency purposes). The Xilinx Virtex FPGA family uses a static RAM lookup table (LUT) as the programmable logic element. A LUT is a small memory that is used to implement a boolean function; its contents are loaded from a non-volatile memory at power up. The Virtex 1, 2, and 4 families use a 16×1 LUT, which means that it can implement one boolean function of four variables; the Virtex-5 family uses a 64×2 LUT (two boolean functions of the same six variables). The LUT delays given in Table 1.9 are for a mid-range *speed grade* of these devices. CMOS integrated circuits being made on the same fabrication line can have a range of delays because of variations in the CMOS fabrication process. Thus, devices coming off a fabrication line are tested and separated into different speed grades, with the higher performing devices being sold at a premium price. The supply voltages of Table 1.9 have decreased over time because transistor-switching speeds reach a maximum at lower voltages as transistor channel lengths shrink. Lowering the supply voltage has the added benefit of reducing power consumption, which is important because excessive heating due to high power consumption has become a problem as increasing number of transistors are used in a single integrated circuit.

The delays of Table 1.9 are given in picoseconds (1 ps = 1.0e–12 s). Observe that the LUT propagation delays in Table 1.9 have decreased by almost an order of magnitude across the families (the Virtex-5 LUT t_{pd} would be even faster if it used the smaller LUT of the previous families). The D-flip-flops (DFF) Clock-to-Q propagation delay shows a similar improvement. The DFF t_{su} and t_{hd} are hard to compare because these times include a MUX delay on the D-input of the DFF for the Virtex 1, 2, and 4 families – the setup/hold times for the Virtex-5 DFF does not include this delay. However, in general, DFF t_{su} and t_{hd} also decrease as transistor channel lengths decrease. The Input/Output buffer (IOB) delays are relatively constant over this time because the bonding pad size used to connect the integrated circuit to the package does not shrink as transistor channel length shrinks. The delays associated with any digital logic within the IO pad decreases, but the IO pad delay is dominated by the off-chip load for an output pad, and by the input pad capacitive load for the input pad. Any changes in these delays over time are due to architectural changes in the pad design, such as providing different ranges of output drive strength current, or the need to accommodate different IO standards over time.

For modern programmable logic devices, the device delays are kept in a database that is included in the design toolkit being used to create the design. The timing analysis tool in the FPGA vendor's design toolkit uses these device delay times to calculate external setup and hold times, maximum operating frequency, and internal setup and hold constraints using the timing equations presented in this chapter.

1.7 SUMMARY

This chapter has discussed how to find the important timings of a circuit such as maximum clock frequency by analyzing the delay paths through the gates and registers. By categorizing the delay paths through the circuit, the total number of delay paths that need to be calculated can be minimized. These timings of the internal chip design can also be used to find the maximum clock frequency of the board-level system.

1.8 SAMPLE EXERCISES

For each of the following circuits:

a. Calculate the worst-case pin-to-pin combinational delay, clock-to-output delay, and register-to-register delay.

b. Use this data to find the maximum clock frequency.

c. Calculate t_{su} and t_{hd} for the external inputs.

1.

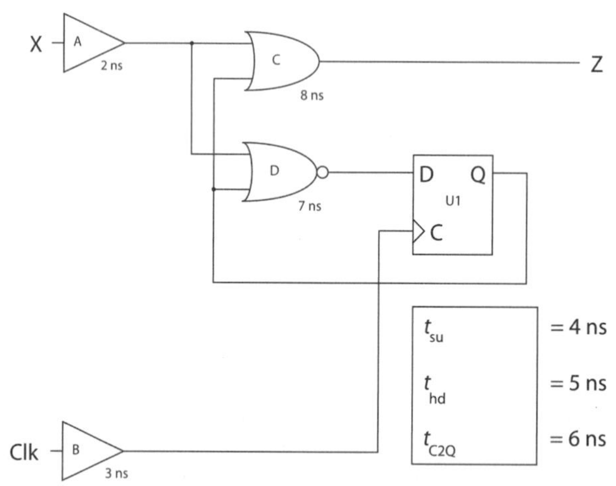

2.

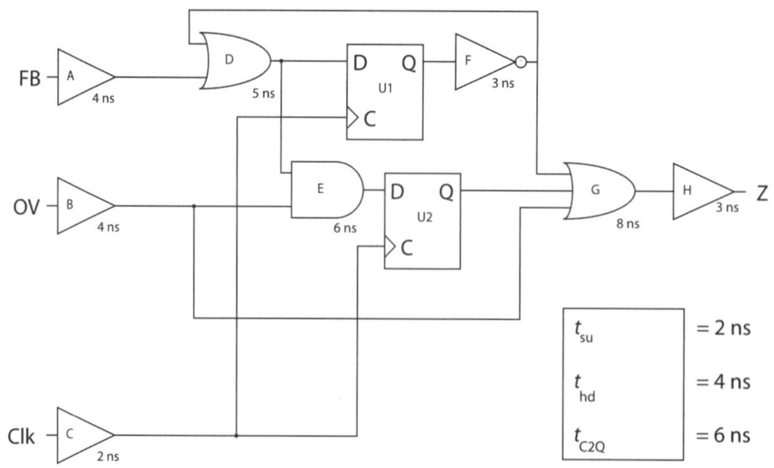

3. Caution, gate E adds a complicating factor!

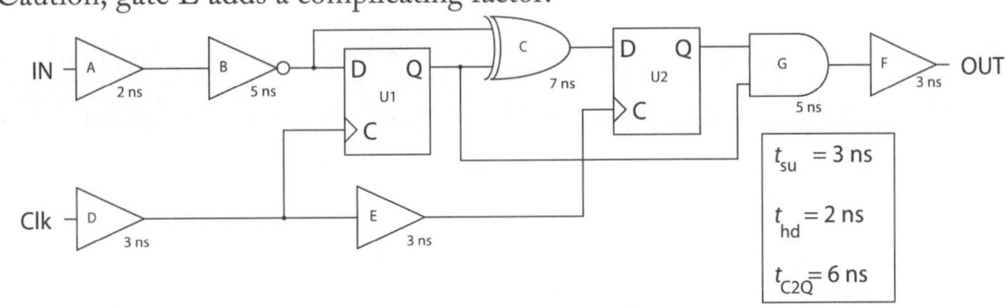

1.9 SAMPLE EXERCISE ANSWERS

1.

Parameter	Calculation	Min	Max	Units
$X\_t_{pd\_P2P}$	$2 + 8 = 10$	10		ns
t_{pd_C2Q}	$3 + 6 + 8 = 17$	17		ns
t_{pd_R2R}	$6 + 7 + 4 = 17$	17		ns
Tclk	max(10, 17, 17)	17		ns
Fclk	1/Tclk		58.8	MHz
t_{su_X}	$4 + (2 + 7) - 3 = 10$	10		ns
t_{hd_X}	$5 + 3 - (2 + 7) = -1$ or 0	0		ns

2.

Parameter	Calculation	Min	Max	Units
$OV\_t_{pd\_P2P}$	$4 + 8 + 3 = 15$	15		ns
t_{pd_C2Q}	$2 + 6 + 3 + 8 + 3 = 22$	22		ns
t_{pd_R2R}	$6 + 3 + 5 + 6 + 2 = 22$	22		ns
Tclk	max(15, 22, 22)	22		ns
Fclk	1/Tclk		45.5	MHz
t_{su_FB}	$2 + (4 + 5 + 6) - 2 = 15$	15		ns
t_{hd_FB}	$4 + 2 - (4 + 5) = -3$ or 0	0		ns

3.

Parameter	Calculation	Min	Max	Units
IN_t_{pd_P2P}	0	0		ns
$t_{pd\_C2Q\_}$	$3 + 3 + 6 + 5 + 3 = 20$	20		ns
t_{pd_R2R}	$6 + 7 + 3 + 3$ (gate E) $= 19$	16		ns
Tclk	max(0, 20, 19)	20		ns
Fclk	1/Tclk		50	MHz
t_{su_IN}	$3 + (2 + 5 + 7) - (3 + 3) = 11$	11		ns
t_{hd_IN}	$2 + 3 - (2 + 5) = -2$ or 0	0		ns

CHAPTER 2

Improving Design Performance

The purpose of this chapter is to increase the maximum clock frequency and improve the setup and hold timing by modifying the circuit design. This chapter assumes the reader is familiar with digital gates and memory elements such as latches and registers and can analyze a circuit to find the maximum clock frequency.

2.1 LEARNING OBJECTIVES

After reading this chapter, you will be able to perform the following tasks:

- Maximize the clock frequency by adding output registers
- Minimize the setup and hold window by adding input registers
- Adjust delay measurements when including a delay locked loop (DLL)
- Recalculate the timing of the board-level system after timing modification

2.2 INCREASING MAXIMUM CLOCK FREQUENCY

The three types of delays paths through a circuit set the maximum clock frequency for the design. The only way to increase the maximum clock frequency is to reduce the delay through these worst-case paths. Assuming the propagation delays of the gates and registers cannot be changed, only changing the circuit architecture can reduce the worst-case path delays.

Reducing the worst-case delays by adding circuit elements is not intuitive, but it is effective in increasing performance. For example, the pin-to-pin combinational delay through a circuit can be completely removed by ensuring there are no combinational paths from any input to any output. Likewise, t_{C2Q} can be minimized by reducing combinational paths between the clock input and the output. Both of these tasks can be accomplished by using the same method. Placing registers on all outputs of the circuit removes all combinational delay paths, and minimizes the combinational path of t_{C2Q}.

Adding registers to the design may seem like it would reduce the clock frequency, but in fact it can often increase it. Analyzing the worst-case paths is the only way to set maximum clock frequency. If the worst-case path delay is reduced, then the circuit naturally can be clocked faster. While the pin-to-pin combinational delay is inherently removed from the analysis, the clock-to-output is usually reduced to its minimum possible value. Since the registers are placed at the output of the circuit, there are no combinational circuits after this to add to the clock-to-output delay. The only clock-to-output delay paths possible are through these output registers, so the analysis is greatly simplified.

The output registers can only be added before the combinational output buffer delay because this is not an actual gate in the design. This delay represents the interface from the chip to the board. Often the output circuitry design has a significant delay because of the need for a high fan-out, larger voltage swing, and over-voltage protection. Therefore, placing the register immediately before this buffer is the optimum location.

One consequence of this approach is the impact of t_{R2R} through the circuit. Since there are more registers in the design, there are more register-to-register delays to be computed. Sometimes the worst-case t_{R2R} will increase because of this. If the clock frequency is being limited by the pin-to-pin delay or the clock-to-output delay, and then those delays are reduced, the clock frequency will still increase if t_{R2R} is not increased by a significant amount. If registers are added to the outputs, the worst-case t_{R2R} will usually become the largest delay path of the circuit.

Another consequence of this approach is the impact on latency. Latency is the time required for an input to propagate through a circuit to the output. If a circuit is all combinational, then the latency is in the same clock period in which the data input is applied. By adding registers to the output of the circuit, the latency increases into the next clock period. Adding a set of registers to all outputs of a device means the latency of each input will increase to the beginning of the next clock period. While this is a disadvantage, the impact on performance is usually not significant. The latency has increased, but the clock period has decreased as well (usually). Therefore, the combination of these two effects often cancels each other out.

While latency may have increased by one clock cycle, the rate at which data is being input and output is the same. New data is input and output every clock cycle. The throughput of the data is the same, even though the latency has increased. Therefore, the overall computing performance of the device will increase. This effect is called pipelining, which will be covered in much more detail in the next chapter.

2.2.1 Example 2.1 Add a register to the output of the circuit in Fig1.7 and recompute the maximum clock frequency. Compare the new computations with the computations before the circuit improvements. The new circuit is shown in Fig. 2.1.

The analysis for this circuit is the same as for all maximum clock frequency calculations. The worst-case pin-to-pin combinational delay, clock-to-output delay, and t_{R2R} must be found. Since the output is now registered, there is no pin-to-pin combinational delay. This measurement can be excluded from the analysis, or set to zero for continuity in the final comparison.

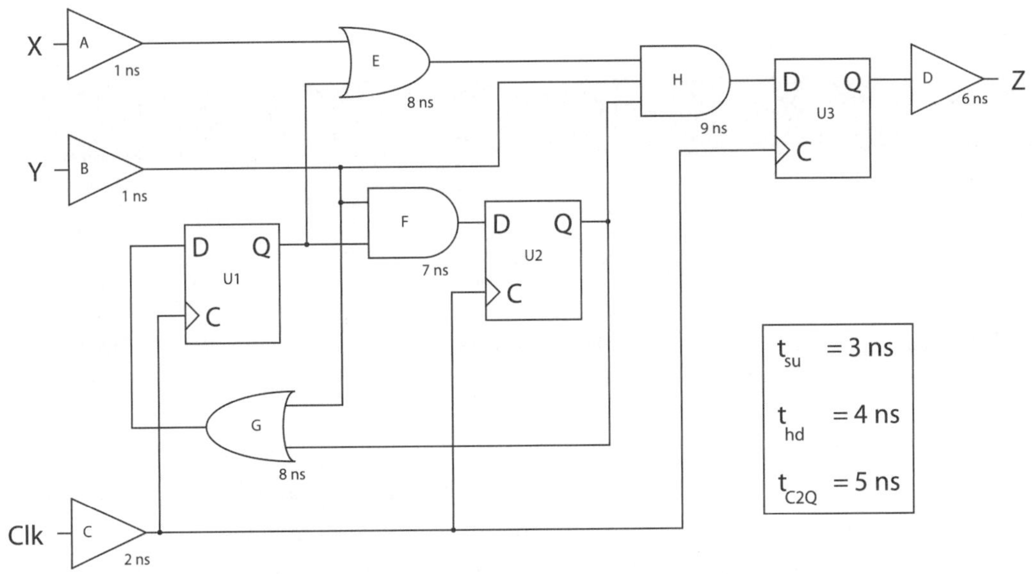

FIGURE 2.1: Adding an output register to the sequential circuit.

The clock-to-output delay only has one path to compute. Since this delay can pass through at most one register, the only register it can now pass through to the output is the new added register. This path proceeds from the clock buffer C, through the register U3, and through the output buffer D. The improved clock-to-output delay is 13 ns.

$$C\_t_{\text{pd}} + U3\_t_{\text{C2Q}} + D\_t_{\text{pd}} = t_{\text{C2Q\_SYS}} \tag{2.1}$$

$$2 + 5 + 6 = 13ns \tag{2.2}$$

The number of register-to-register paths has increased due to adding another register from two to four. The paths are listed in Table 2.1 . The worst-case path is from U1, through gates E and H, to the new output register U3 for a total delay of 25 ns.

TABLE 2.1: Total set of new register-to-register propagation delays

Starting input	Path	Delay
U1	U1 + F + U2	15 ns
U2	U2 + G + U1	16 ns
U1	U1 + E + H + U3	25 ns
U2	U2 + H + U3	17 ns

TABLE 2.2: Measured improvement of adding output registers

Measurement	Original delay	Improved delay
P2P	24 ns	0 ns
C2Q	30 ns	13 ns
R2R	16 ns	25 ns
Clock Period	30 ns	25 ns
Clock Frequency	33.3 MHz	40 MHz

The clock period is set by taking the largest of the three worst-case paths, zero ns for the pin-to-pin combinational delay, 13 ns for the clock-to-output delay, and 25 ns for t_{R2R}. Therefore, the minimum clock period is 25 ns, which corresponds to a maximum clock frequency of 40 MHz.

Before adding the register on the output, the minimum clock period was set by the clock-to-output delay. Since this delay decreased to 13 ns, it is no longer limiting the clock period. The t_{R2R} has increased, but is still less than the previous limiting value of 30 ns. This means the maximum clock frequency has significantly increased by adding a single register to the design. The total comparison of measured values is present in Table 2.2.

2.3 IMPROVING SETUP AND HOLD TIMES

Adding registers to the output of the circuit also changes t_{su} and t_{hd} for the circuit. If the circuit has a combinational path through the circuit and a register is added to the output, the longest combinational delay path from a circuit input to a register input could very likely be the newly added register. The setup and hold window could increase significantly because of the new output register. One way to minimize the effects of adding output registers is to place registers on the inputs of the circuit. This will reduce the combinational paths to the registers to minimize the setup and hold window. The input registers can only be placed after the input buffer delay since this is not an actual buffer much like the output buffer delay. Therefore, there will be an input buffer combinational delay to the register input.

2.3.1 Example 2.2

Recompute t_{su} and t_{hd} before and after adding registers to the inputs of the circuit as in Fig. 2.2. This circuit includes the output registers added in the previous example.

The t_{su} of the circuit before adding input registers is computed by finding the longest combinational path to any register in the design. The addition of the output register increases the worst-case delay to 18 ns from the circuit input X to the U3 register through gates A, E, and H. The minimum

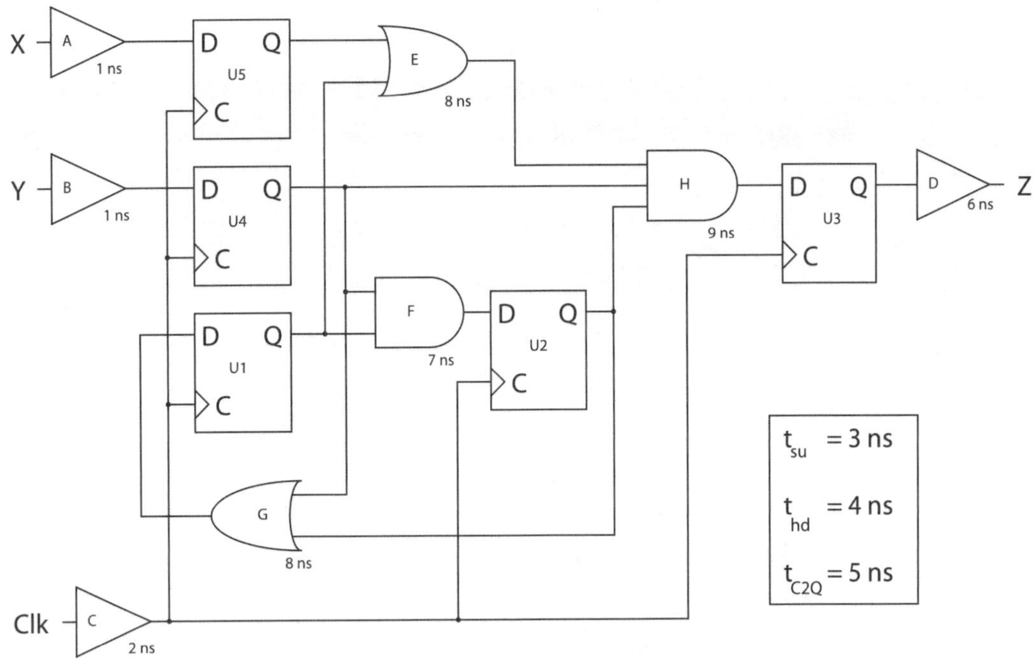

FIGURE 2.2: Adding input registers to the sequential circuit.

clock delay remains the same. Therefore, the new circuit t_{su} increases to 19 ns.

$$(t_{pd\_data\_U1} - t_{pd\_clk\_U1})) + t_{su\_FF} = t_{su\_TOTAL} \tag{2.3}$$

$$(18 - 2) + 3 = 19\,\text{ns} \tag{2.4}$$

The t_{hd} of the circuit before adding input registers is computed by finding the shortest combinational path to any register in the design. The addition of the output register does not increase this value. The shortest path is the same as the previous analysis at 8 ns. This means t_{hd} remains the same at –2 ns, which should be set to zero since it is negative. The setup and hold window is now 19 ns because of the addition of the output registers.

$$(t_{pd\_clk(MAX)} - t_{pd\_data(MIN)})) + t_{hd\_FF} = t_{hd\_TOTAL} \tag{2.5}$$

$$(2 - 8) + 4 = -2\text{ns} \tag{2.6}$$

Adding input registers after the input buffers simplifies the computations because the number of paths from each input is reduced to one per input. For this circuit, the combinational delay for each input is 1 ns, and the delay for the clock is 2 ns. This means the new t_{su} is 2 ns, and the new $t_{hd}$is 5 ns. This means the setup and hold window is now 7 ns. The comparison between t_{su} and t_{hd}

TABLE 2.3: Measured improvement of adding input registers

Measurement	Original	Added output registers	Added input registers
Setup Time	10 ns	19 ns	2 ns
Hold Time	0 ns	0 ns	5 ns
Setup and Hold Window	10 ns	19 ns	7 ns

is given in Table 2.3.

$$(t_{\text{pd\_data\_U1}} - t_{\text{pd\_clk\_U1)}}) + t_{\text{su\_FF}} = t_{\text{su\_TOTAL}} \tag{2.7}$$

$$(1 - 2) + 3 = 2\text{ns} \tag{2.8}$$

$$(t_{\text{pd\_clk(MAX)}} - t_{\text{pd\_data(MIN)}}) + t_{\text{hd\_FF}} = t_{\text{hd\_TOTAL}} \tag{2.9}$$

$$(2 - 1) + 4 = 5\text{ns} \tag{2.10}$$

The setup and hold window is nearly doubled when output registers were added to the design. When registers were added to the inputs, the setup and hold window decreased to the smallest possible window. The window cannot decrease below this because it is limited by the setup and hold window of the register, which is also 7 ns.

2.4 DELAY LOCKED LOOPS

Often modern designs that have internal clocks have some type of Phased Locked Loop (PLL) or Delay Locked Loop (DLL) to stabilize and adjust the clock. A PLL is a circuit that creates a completely new clock internal to the circuit, but based on the external clock provided to it. A DLL passes the external clock to the circuit, but adjusts its timing through a network of delays. There are significant differences between these two types of clock management schemes, but they are beyond the focus of this book. For this chapter, the term DLL will be used to describe both PLLs and DLLs. The relevant feature to this material is how DLLs can adjust the phase of the internal clock.

A clock signal can be easily manipulated because of its predictability. The clock will always have a repeating 1-0-1-0 pattern. Therefore, once the clock is active, the clock is the same from one clock period to the next. If the external clock signal is delayed by an input buffer, the internal clock will not be aligned with the external clock. A DLL can artificially make the clock appear to be aligned by inserting additional delay to the clock. For example, an external clock with a period of 8 ns passes through an input buffer that delays the signal by 1 ns as in Fig. 2.3. The DLL measures that the two clocks are not aligned, and then it inserts additional delay to the internal clock until they are aligned. In this example, the DLL would add a 7 ns delay to make the two clocks aligned.

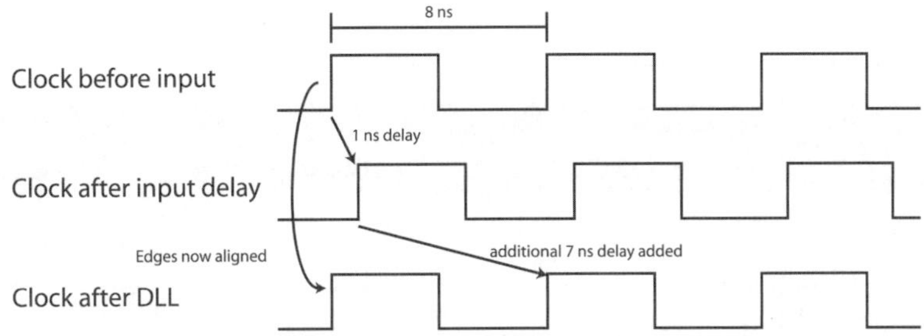

FIGURE 2.3: Operation of a delay locked loop.

A DLL can change the phase of the internal clock either manually or automatically. The advantage of this is that the active clock edge can be placed anywhere. This means the clock delay in the clock-to-output calculations and t_{su} and t_{hd} calculations can be set to whatever needed. Typically the DLL will align the internal clock with the external clock to remove any delays added by the input buffer for the clock signal. The input buffer will add a fixed delay to the clock signal, and the DLL will effectively reduce the delay by that same amount. Note that this technique is not possible to reduce the delays on the data signals because they don't have a predictable repeating pattern.

2.4.1 Example 2.3

Use a DLL to align the internal clock to the external clock in Fig. 2.2. Find any changes to the previous calculations.

Any equation that uses the delay of the input buffer C must be recalculated with that value set to zero. The first change is in the calculation of the clock-to-output delay for the circuit. There is only one clock-to-output path through the circuit through the output register. The new clock-to-output delay for this circuit is reduced by 2 ns to 11 ns.

$$C\_t_{pd} + U3\_t_{C2Q} + D\_t_{pd} = t_{C2Q\_SYS} \tag{2.11}$$

$$0 + 5 + 6 = 11\text{ns} \tag{2.12}$$

The pin-to-pin combinational delay and the register-to-register delay are not affected by the change to the clock because they do not include the clock buffer C. The maximum clock frequency must be checked because this change might affect it if the clock-to-output delay was the limiting factor. Typically t_{R2R} limits the maximum clock frequency, so often the clock frequency will not change when adding a DLL.

The t_{su} and t_{hd} also depend on the clock delay, so they will be affected by adding a DLL. The minimum and maximum clock delay is set to zero and t_{su} and $t_{hd}$are recalculated.

$$(t_{pd\_data\_U1} - t_{pd\_clk\_U1})) + t_{su\_FF} = t_{su\_TOTAL} \tag{2.13}$$

$$(1 - 0) + 3 = 4\text{ns} \tag{2.14}$$

TABLE 2.4: Datasheet for the improved circuit example

Parameter	Description	Old min	Old max	New min	New max	Units
Tclk	Clock Period	30		25		ns
Fclk	Clock Frequency		33.3		40	MHz
t_{su_Y}	Y Setup Time	10		3		ns
t_{hd_Y}	Y Hold Time	0		4		ns
$X\_t_{pd\_P2P}$	Combinational delay	24		N/A		ns
t_{pd_C2Q}	Clock-to-output delay	30		11		ns

$$(t_{pd\_clk(MAX)} - t_{pd\_data(MIN)}) + t_{hd\_FF} = t_{hd\_TOTAL} \tag{2.15}$$

$$(0 - 1) + 4 = 3\text{ns} \tag{2.16}$$

The new t_{su} is 4 ns, and the new t_{hd} is 3 ns. The setup and hold window has not changed from 7 ns.

2.5 BOARD-LEVEL TIMING IMPACT

The final calculation of the chip is to analyze how well the circuit will improve the board-level performance. The same circuit should be used as in last chapter's example even though the internal design is significantly different. The datasheet for the improved circuit is listed in Table 2.4 . The new calculations include both input and output registers and a DLL for clock adjustment.

2.5.1 Example 2.3

Using the circuit in Figure 2.4, find the maximum clock frequency. Each chip has the same circuit as in Figure 2.2 and uses the timings in Table 2.4.

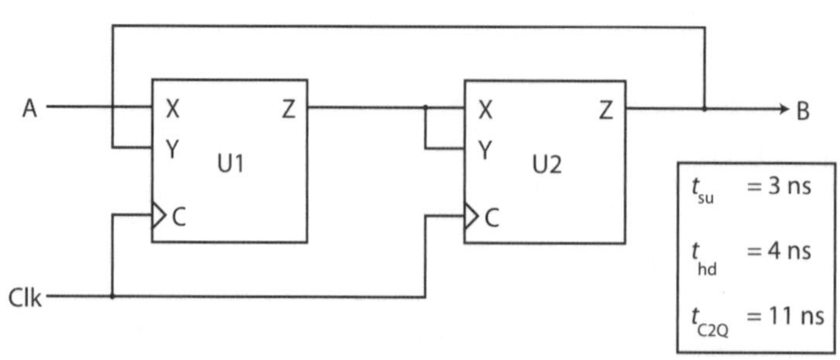

FIGURE 2.4: Board-level schematic to compute maximum clock frequency.

First, since there is no combinational path through the chip, there is no calculation for the pin-to-pin combinational path for the board. This value is excluded when computing maximum clock frequency.

One clock-to-output delay exists for this circuit. This path passes only through the clock input of U2. If there is no clock delay, the clock-to-output for the board is the same as the clock-to-output of the chip. This delay is 11 ns.

Two register-to-register delays exist for this circuit. The first is through the U1 clock-to-output to either input on U2. The third is through the U2 clock-to-output to the input of Y on U1. Both paths have the same delay of 11+ 4 = 15 ns.

$$U1\_t_{C2Q\_} + U2\_t_{su} = t_{C2Q\_SYS} \tag{2.17}$$

$$11 + 4 = 15\text{ns} \tag{2.18}$$

The three worst-case paths and the chip minimum clock period limit the clock frequency for the board-level system. The largest of these four values (0 ns, 11 ns, 15 ns, 25 ns) is 25 ns, which is also the minimum clock period for the chip. This means the board can operate at the same frequency as the chips on the board. Note the removal of the combinational paths greatly reduces the delays at the board level.

2.6 SUMMARY

By understanding the parameters that dictate the maximum clock frequency of a circuit, the design can be modified to reduce the longest delays to improve circuit performance. Reducing the combinational delay paths increases the maximum clock frequency by targeting the worst-case paths. By registering all inputs and outputs, the circuit can operate at its maximum frequency within a larger system. Using additional technologies like DLLs can further increase the circuit performance within a larger system.

2.7 SAMPLE EXERCISES

For each of the following circuits, place registers on all data inputs after the input buffer delay and place registers on all data outputs before the output buffer delay. Then,

a) calculate the worst-case pin-to-pin combinational delay, clock-to-output delay, t_{R2R},

b) use this data to find the maximum clock frequency,

c) calculate t_{su} and $t_{hd}$for the external inputs,

d) determine all effects on the circuit if a DLL was used to remove the clock input buffer delay.

1

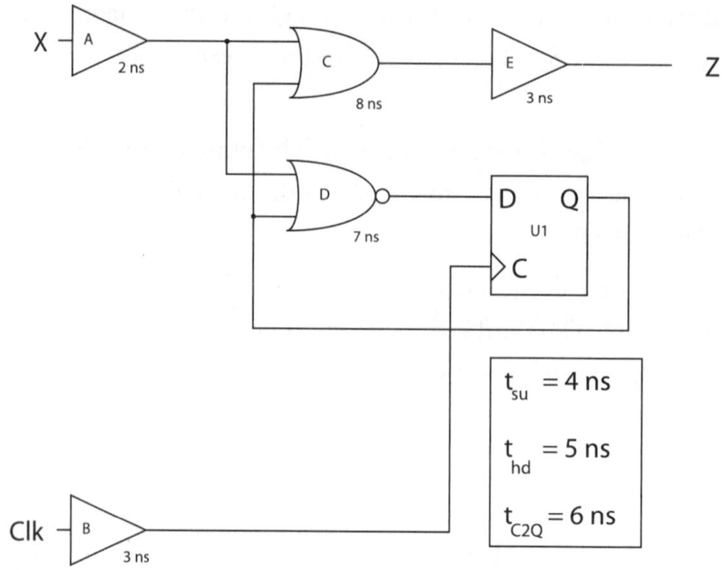

2

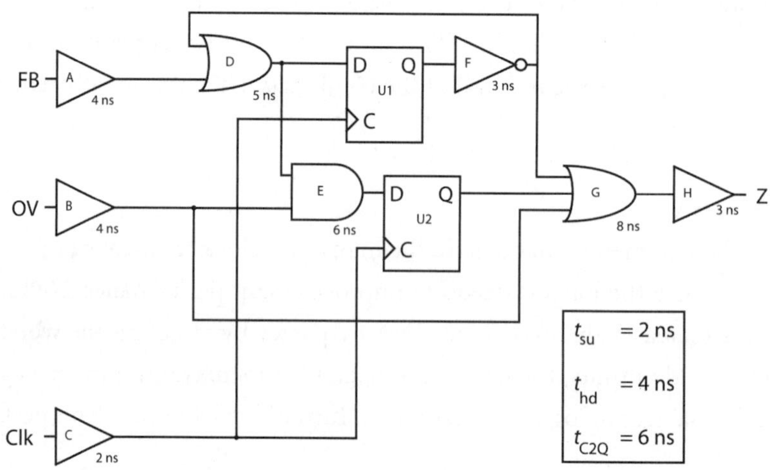

3

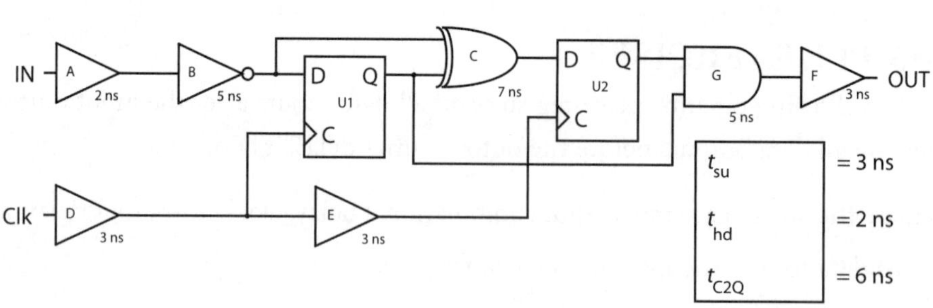

3. For this problem, assume the clock routed to the output register passes through both clock buffers, and the clock to the input register passes through only the first clock buffer, and the DLL only removes the delay in clock buffer D.

2.8 SAMPLE EXERCISE ANSWERS

1.

Parameter	Calculation	Min	Max	Units
X_t_{pd_P2P}	N/A	0		ns
t_{pd_C2Q}	$3+6+3=12$	12		ns
t_{pd_R2R}	$6+8+4=18$	18		ns
Tclk	max (0, 12, 18)	18		ns
Fclk	1/Tclk		55.6	MHz
t_{su_X}	$4+2-3=3$	3		ns
t_{hd_X}	$5+3-2=6$	6		ns

DLL effects:

Parameter	Calculation	Min	Max	Units
t_{pd_C2Q}	$0+6+3=9$	9		ns
t_{su_X}	$4+(2-0)=6$	6		ns
t_{hd_X}	$5+0-2=3$	3		ns

2.

Parameter	Calculation	Min	Max	Units
OV_t_{pd_P2P}	0	0		ns
t_{pd_C2Q}	$2+6+3=11$	11		ns
t_{pd_R2R}	$6+3+5+6+2=22$	22		ns
Tclk	max(0, 11, 22)	22		ns
Fclk	1/Tclk		45.5	MHz
t_{su_FB}	$2+4-2=4$	4		ns
t_{hd_FB}	$4+2-4=2$	2		ns

DLL effects:

Parameter	Calculation	Min	Max	Units
t_{pd_C2Q}	$0 + 6 + 3 = 9$	9		ns
t_{su_FB}	$2 + 4 - 0 = 6$	6		ns
t_{hd_FB}	$4 + 0 - 4 = 0$	0		ns

3.

Parameter	Calculation	Min	Max	Units
IN_t_{pd_P2P}	0	0		ns
t_{pd_C2Q}	$3 + 3 + 6 + 3 = 15$	15		ns
t_{pd_R2R}	$6 + 5 + 7 + 3 - 3 = 18$	18		ns
Tclk	max(0, 15, 18)	18		ns
Fclk	1/Tclk		55.6	MHz
t_{su_IN}	$3 + 2 - 3 = 2$	2		ns
t_{hd_IN}	$2 + 3 - 2 = 3$	3		ns

DLL effects:

Parameter	Calculation	Min	Max	Units
t_{pd_C2Q}	$0 + 3 + 6 + 3 = 12$	12		ns
t_{su_IN}	$3 + 2 - 0 = 5$	5		ns
t_{hd_IN}	$2 + 0 - 2 = 0$	0		ns

CHAPTER 3

Finite State Machine With Datapath Design

This chapter explores finite state machine with datapath (FSMD) design techniques for streaming data applications such as video or audio processing, which can require dedicated logic to meet throughput or latency requirements.

3.1 LEARNING OBJECTIVES

After reading this chapter, you will be able to perform the following tasks:

- Discuss fixed-point representation and saturating arithmetic.
- Transform a streaming data calculation expressed as an equation into a dataflow graph (DFG) format.
- Discuss speed and area tradeoffs in datapath design in relation to latency, throughput, initiation period, and clock period.
- Design datapaths using both non-overlapped/overlapped computations, and non-pipelined/pipelined execution units.
- Design a datapath to implement a DFG that meets target latency and initiation period requirements.

3.2 FSMD INTRODUCTION AND MOTIVATION

A *datapath* contains the components of a digital system that perform the numerical computations for the system. The datapaths described in this chapter perform addition and multiplication on fixed-point numbers with registers used to store intermediate calculations. In this chapter, the generic term of *execution unit* (EU) is used to refer to computation blocks such as adders and multipliers. This chapter uses execution units as black boxes; the reader is referred to a book such as [1] for detailed information on adder and multiplier design.

A finite state machine sequences the computations on the datapath's execution units, with the combined system referred to as FSMD. The FSMD designs in this chapter are tailored to execute a fixed sequence of computations on a dataset. Tradeoffs with regard to the number of required execution units versus the number of clock cycles to complete the computation are studied through

TABLE 3.1: Fixed-format examples

FORMAT	RANGE	EXAMPLES
8.0	0 to 255	143 = ‘b10001111; 37 = ‘b00100101
5.3	0 to 31.875	17.875 = ‘b10001111; 4.625 = ‘b00100101
0.8	0 to 0.99609375	0.55859375 = ‘b10001111, 0.14453125 = ‘b00100101

example implementations. An FSMD approach is used in an application if high performance is needed, as an FSMD implementation typically requires fewer clock cycles than a stored program (a computer) implementation. However, the FSMD logic is fixed, and can only perform its designed computation. A stored program implementation is more flexible, as altering the program that the computer executes modifies the target computation. This is the classic tradeoff of flexibility versus performance when choosing whether to use a stored program or FSMD approach for implementing a digital system. If an application is complex enough, its computations can be divided among cooperating digital systems, with an FSMD handling time-critical computations and a stored program system handling the remaining computations.

A good example of cooperating digital systems is found in a hand-held gaming system, whose task is to execute a game with three-dimensional (3D) graphics. The game application is handled by the microprocessor, while the 3D graphics is performed by a dedicated graphics processor whose core logic is an FSMD optimized for pixel processing. This chapter uses simplified equations from 3D graphics and digital signal processing to illustrate FSMD design tradeoffs.

3.3 FIXED-POINT REPRESENTATION

A *fixed-point* number is a binary number whose format is *X.Y*, where *X* and *Y* are the number of binary digits to the left and right of the decimal point, respectively. For unsigned numbers, the integer portion defined by *X* ranges from 0 to 2^X-1, while the fractional portion ranges from 0 to 1-2^{-Y}. Table 3.1 gives some examples of eight-bit fixed-point numbers for three different choices of *X* and *Y*.

To convert an unsigned decimal number to a *X.Y* fixed-point format, multiply the decimal number by 2^Y, drop any fractional remainder, and then convert this to its unsigned binary value using a *N*.0 format, where $N = X + Y$. From the 5.3 format example of Table 3.1, the multiplication 4.625 * 2^3 = 37, which is 0b00100101 as an eight-bit number.

To convert an *X.Y* unsigned binary number to its decimal representation, first convert the number to its decimal representation assuming an *N*.0 format, where $N = X + Y$. Then divide this number by 2^Y to produce the final decimal result. From the 5.3 format example of Table 3.1, the value 0b10001111 converted to its 8.0 value is 143, which is 17.875 when divided by 2^3.

The numbers in a fixed-point datapath are assumed to share a common *X.Y* format. The logic used to implement binary addition and multiplication works the same regardless of where the decimal point is located, as long as both numbers have the same X.Y format, i.e., the decimal points are *aligned*. This is in contrast to a *floating-point* datapath, which can perform computation on numbers whose decimal points do not align. Floating-point computation blocks require significantly more logic to implement than fixed-point logic blocks. Floating-point computation is used in applications that require an extended range for its numerical data. This chapter does not cover floating-point number encoding or implementation of floating-point computational elements. However, since this chapter treats computation elements as black boxes, the lessons learned in this chapter concerning clock-cycle versus execution unit tradeoffs in datapath design using fixed-point datapaths can easily be applied to floating-point datapaths.

3.4 FIXED-POINT REPRESENTATION IN 3D GRAPHICS

As mentioned previously, 3D graphics is a good example of an application that requires the performance of a dedicated FSMD engine. The *frame rate* of a 3D graphics processor is the number of times per second that a new image is generated for a 3D scene. Each frame is composed of pixels, with a typical resolution being 1280 × 1024 pixels, or 1,310,270 pixels. The color of each pixel is represented by three eight-bit values that specify the red, green, blue (RGB) color components. Many computations are performed on each RGB component of a pixel to determine the final RGB values of a pixel. Each eight-bit RGB component is a 0.8 fixed-point number. Thus, pixel computations can be thought of as computations on numbers whose range is [0–1.0), which means $0.0 < = c < 1.0$ if c is an RGB component value. From Table 3.1, it is seen that the maximum value of a 0.8 fixed-point number is 0.99609375, which is very close to 1.0. The advantage of the 0.8 fixed-point format is seen in the next section, which discusses saturating arithmetic for fixed point numbers.

3.5 UNSIGNED SATURATING ARITHMETIC AND FIXED-POINT NUMBERS FIXED-POINT REPRESENTATION

Overflow occurs in a computation when the numerical result is outside of the number range supported by a particular data format. A carry out of the most significant bit in an unsigned, fixed-point addition is an overflow indicator. Overflow indicates that the result is incorrect and typically this error condition is handled by the application. However, in real-time data computations such as 3D graphics, video, or audio processing there is no opportunity for the application to correct the error. In these cases, saturating arithmetic is used to saturate the result to the maximum or minimum number in the number range to produce a result that is closer to the correct answer than what overflow produces. Figure 3.1a shows an example of a fixed point addition using normal binary addition that

(a) unsaturating 8-bit addition

```
              8.0 format    0.8 format
  'h50           80          0.3125
+ 'hC0         + 192       + 0.75
  'h10           16          0.0625
```

8-bit result has overflowed

(b) saturating 8-bit addition

```
              8.0 format    0.8 format
  'h50           80          0.3125
+ 'hC0         + 192       + 0.75
  'hFF           255         0.99609375
```

8-bit result is saturated to maximum value

(c) unsaturating 8-bit subtraction

```
              8.0 format    0.8 format
  'h50           80          0.3125
- 'hC0         - 192       - 0.75
  'h90           144         0.5625
```

8-bit result has underflowed

(d) saturating 8-bit subtraction

```
              8.0 format    0.8 format
  'h50           80          0.3125
- 'hC0         - 192       - 0.75
  'h00           0           0.0
```

8-bit result is saturated to minimum value

FIGURE 3.1: Saturating addition.

overflows, as the result is greater than the maximum value of 255. A saturating adder that clips the result to its maximum value in the overflow case is shown for the same operation in Fig. 3.1b. While the results in Fig. 3.1a and Fig. 3.1b are both incorrect, the saturating operation produces a result that is closer to the correct answer, which is desirable in applications that cannot take any other corrective action on overflow. Figure 3.1c demonstrates an *underflow* case (a borrow into the most significant binary digit) for unsigned eight-bit subtraction. The same operation is performed in Fig. 3.1d using a saturating subtraction operation, which clips the result to its minimum value of zero.

An eight-bit unsigned saturating adder is shown in Fig. 3.2. The output is saturated to its maximum value of 'b11111111 when the eight-bit sum produces a carryout of '1'.

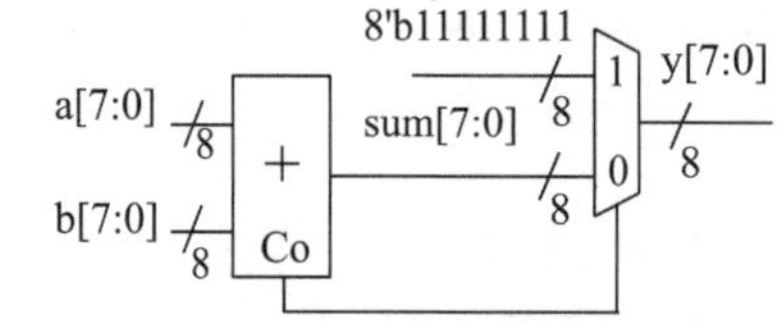

{1'b0,a}

This forms a 9-bit value whose most significant bit is '0', with the remaining 8-bits provided by **a**. The most significant bit of the 9-bit sum of **{1'b0,a} + {1'b0,b}** is the carry-out of the 8-bit sum **a + b**.

```
//saturating adder
module satadd (a, b, y);

input  [7:0] a,b;
output [7:0] y;

reg [7:0] y;
wire [8:0] sum;
wire cout;

//do 9-bit sum so that
//we have access to carry out
assign sum = {1'b0,a} + {1'b0,b};
assign cout = sum[8];

//saturate the result
always @(cout or sum) begin
 if (cout == 1) y = 8'b11111111;
  else y = sum[7:0];
end

endmodule
```

FIGURE 3.2: Unsigned saturating adder (8-bit).

3.6 MULTIPLICATION

A good question to ask at this point is "How does saturating arithmetic operate for multiplication?" To answer this, recall that the binary multiplication of two N-bit numbers, $N \times N$, requires a $2N$-bit result to contain all of the bits produced by the multiplication. However, it is usually not possible to retain these $2N$-bit in the datapath calculation, as successive multiplications would continually require the datapath size to double in order to prevent any data loss. Assuming that only N bits of an $N \times N$ bit multiplication is kept, then two strategies can be used for discarding half of the bits of the $2N$-bit product. If the fixed-number format used for the calculation is $N.0$ (integers), then a saturating multiplier can be built that saturates the result to the maximum value in case of overflow in the same manner as was done for addition. In this case, the upper N-bit of the $2N$ bit product is discarded and the lower eight-bit saturated to its maximum value.

Another approach is to encode the fixed-point numbers in a $0.N$ format, which means that the product of the $N \times N$ multiplication can never overflow, since the two N-bit numbers being multiplied are always less than one. Hardware saturation of the result is not required; instead, the lower eight-bit of the $2N$-bit product are discarded. The bits that are discarded are the least significant bits of the product, causing successive multiplications to automatically saturate towards a minimum value of zero, as precision is lost due to only retaining eight bits of the product. This will be the approach used in this chapter, as the multiplier design does not have to be modified and the examples used in this chapter assume a 0.8 fixed-point number format.

3.7 THE *BLEND* EQUATION

Equation. (3.1) gives the *blend* equation that is used to illustrate some basic datapath design concepts. The C_{new} value in the blend equation is a new color formed by blending two colors C_a and C_b via a blend factor F. The color values C_{new}, C_a, and C_b are 0.8 fixed-point values whose range is [0–1.0), i.e., $0 \leq C < 1.0$. However, the blend factor F is a nine-bit value encoded to allow the range [0.0–1.0], i.e., $0 \leq F \leq 1.0$. The inclusion of one in the range allows C_{new} to be equal to C_a if F is one, or C_{new} to be equal to C_b if F is zero.

$$C_{new} = C_a \times F + C_b \times (1 - F) \tag{3.1}$$

The nine-bit encoding of F is 'b100000000 if F is equal to one, and 0*dddddddd* for any other value of F, where *dddddddd* is the 0.8 fixed point equivalent of F. For computation speed purposes, the lower eight-bit of 1-F is computed as the one's complement value of the lower eight-bit of F when F is not equal to one or zero. The one's complement operation produces an error of one least significant bit (LSb), but this is deemed acceptable in pixel blend operations, in which computation speed is the most critical factor. The 1-F operation implementation is shown in Fig. 3.3. The *mxa* multiplexer and the zero detect logic handle the special case of $F = 0.0$ (0b000000000), in which case the output is 1.0 (0b100000000). The *mxb* multiplexer handles the case of $F = 1.0$, which is

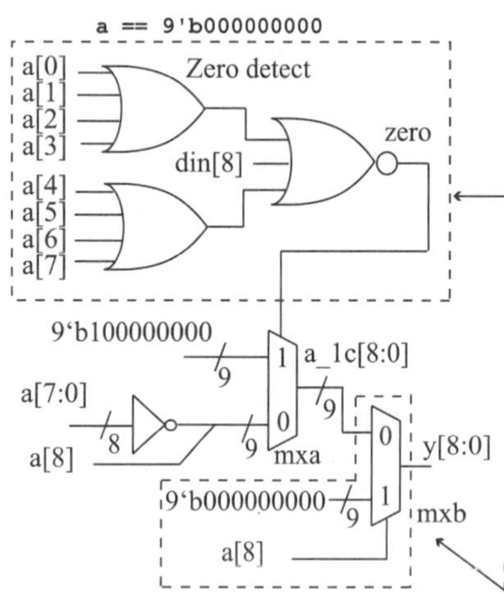


```
//do 1-F operation
module oneminus (a, y);

input  [8:0] a;
output [8:0] y;

reg [8:0] a_1c;

//handle '0' input case
always @(a) begin
 if ( a == 9'b000000000 )
  // input is zero, convert to '1.0'
  a_1c = 9'b100000000;
 else
  // do one's complement
  begin
    a_1c[8] = a[8];
    a_1c[7:0] = ~a[7:0];
  end
end

//handle '1.0' input case
//              a[8]==1          a[8]==0
assign y = a[8] ? 9'b000000000 : a_1c;

endmodule
```

FIGURE 3.3: Implementation for 1-*F* operation.

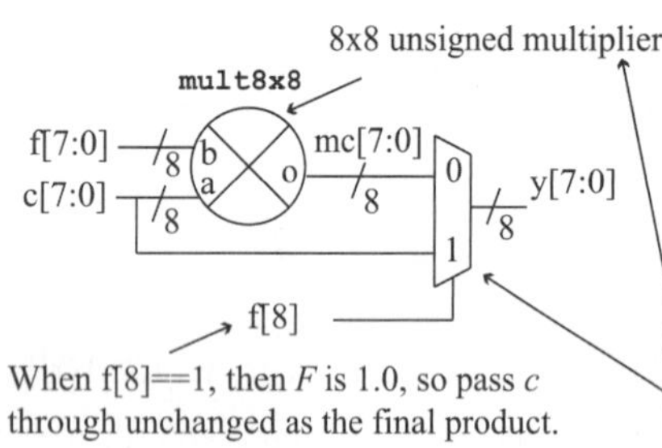


```
module bmult(c,f,y);
input   [7:0] c;
input   [8:0] f;
output  [7:0] y;

wire [7:0] mc;

mult8x8 m1 (.a(c),.b(f[7:0]),.o(mc));
//            f[8]==1   f[8]==0
assign y = f[8] ? c   :   mc;

endmodule;
```

FIGURE 3.4: Multiplication of an eight-bit color operand by nine-bit blend operand.

detected by examining the most significant bit (MSb) of *F*. If *F* is not equal to zero or one, then the output is the one's complement of the lower eight-bit. The most significant bit is not included in this one's complement operation, as this would make the output value equal to one.

The multiplication operations in the blend equation have an eight-bit color operand, either C_a or C_b, and a nine-bit blend operand, either *F* or 1-*F*. When the nine-bit blend operand is not equal to one, then the multiplication result is the product of the lower eight-bit of the nine-bit blend operand and the eight-bit color operand. When the nine-bit operand is equal to one, then the product of the multiplication should be exactly equal to the eight-bit operand, which is accomplished by using a multiplexer on the output of the multiplier and testing the most significant bit of the nine-bit blend operand. The multiplication implementation is shown in Fig. 3.4; the Verilog blendmult module assumes the availability of an 8×8 multiplier component named mult8×8.

TABLE 3.2: Example blend computations

		CASE A (CNEW = CA)	CASE B (CNEW = CB)	CASE C CNEW = 0.5 * CA + 0.5*CB
F	decimal	1.0	0.0	0.5
	binary	‘b100000000	‘b000000000	‘b010000000
1-F	decimal	0.0	1.0	0.49609375
	binary	‘b000000000	‘b100000000	‘b001111111
Ca	decimal	0.75	0.75	0.75
	binary	‘b11000000	‘b11000000	‘b11000000
Cb	decimal	0.25	0.25	0.25
	binary	‘b01000000	‘b01000000	‘b01000000
Ca*F	decimal	0.75	0.0	0.375
	binary	‘b11000000	‘b000000000	‘b01100000
Cb*(1-F)	decimal	0.0	0.25	0.12109375
	binary	‘b000000000	‘b01000000	‘b00011111
Cnew	dec	0.75	0.25	0.49609375
	bin	‘b11000000	‘b01000000	‘b01111111

Table 3.2 gives some example blend computations for three cases: A, B, and C. In Case A, the blend factor F is 1.0, causing C_{new} to be exactly equal to C_a. In Case B, the blend factor F is zero, causing C_{new} to be exactly equal to C_b. In Case C, the blend factor F is 0.5; note that the 1-F computation gives a value of 0.49609375 that is incorrect by one LSb due to the use of the one's complement to compute 1-F. This one LSb error is propagated to the final result of 0.49609375, which should be exactly equal to 0.5 if precise arithmetic is used for the computation of 0.75 * 0.5 + (1 – 0.5)* 0.25.

3.8 SIMPLE DATAPATHS AND THE BLEND EQUATION

Before designing an example datapath, some terms used in it are defined. The *input dataset* of a datapath contains the external values required by the datapath to perform the computation. The *output dataset* of a datapath contains the computational output of the datapath for a given input dataset. For example, the input dataset of the blend equation contains C_a, C_b, and F, while the output dataset contains C_{new}. The *latency* of a datapath measures the number of clock cycles required for a calculation on an input dataset and this number is from the first element of the input dataset to the last element of the output dataset. The *total computation time* of the datapath for an input dataset is the latency multiplied by the clock period. The *initiation period* measures how often a datapath can accept a new input dataset and is the number of clock cycles from the first element of the input dataset to the first element of the next input dataset. The *throughput* of a datapath is the number of input datasets processed per unit time; lowering the initiation period (providing input datasets more often) or decreasing the clock period increases the throughput of a datapath.

The *constraints* of a datapath determine how it is designed. Constraints are measured in both time and area (number of gates). One common constraint for datapath design is the minimum time constraint, i.e., design the datapath to perform computation in the least amount of time. Another common constraint is the minimum area constraint, i.e., design the datapath to use the minimum number of logic gates. These two constraints are contradictory to each other as performing a computation in a fewer number of clock cycles usually requires more execution units so that computations can be performed in parallel, which means more logic gates. In this chapter, we specify time constraints for a datapath as latency and initiation period values, which are measured in clock cycles. We do not specify a clock period constraint, as this is dependent upon the implementation technology such as the particular FPGA family used for the datapath.

Figure 3.5 shows the DFG of the blend equation. In a DFG, circles represent computations, with arrows linking circles to show the dataflow between computations. The operations (circles) of the DFG are labeled n1, n2, ...n*N* for referral purposes. DFGs are useful in high-level synthesis tools that synthesize a datapath solution given latency and initiation period constraints. Our DFG usage is very informal and is principally used to visualize dependencies between computations; the reader is referred to [2] for a complete discussion of DFGs.

While a DFG shows the data dependencies between computations, the *datapath diagram* shows an implementation of the DFG's computation. A datapath diagram shows the computation elements and registers that are used to perform the computation and how these elements interconnect. Figure 3.6 is a datapath diagram for a naïve implementation of the blend equation. This implementation is termed naïve as it is simply a one-to-one assignment of the nodes of the DFG to execution units. This is an undesirable implementation as the execution units are *chained* together,

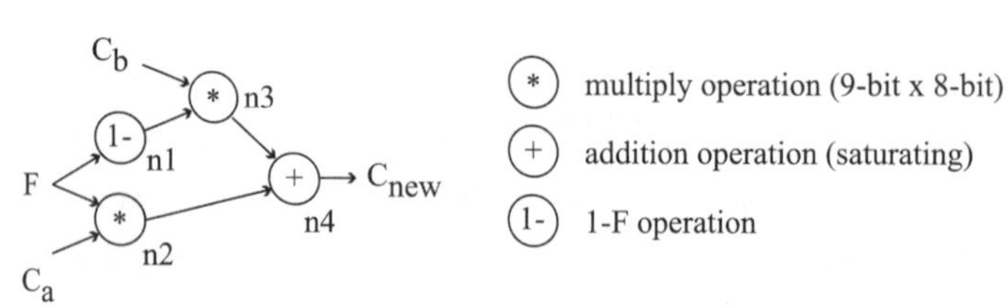

FIGURE 3.5: Dataflow graph of the blend equation.

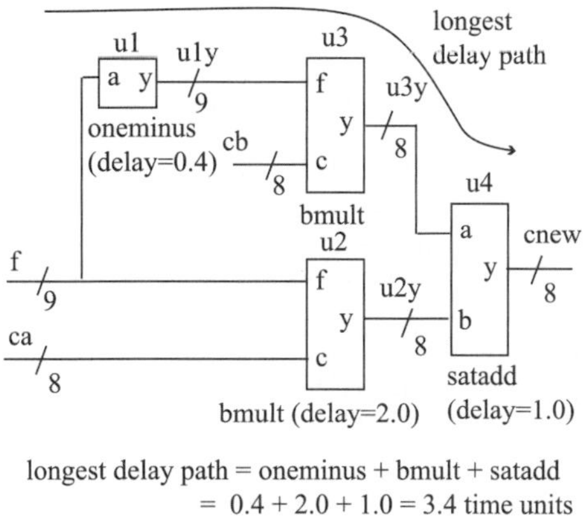


```
module blend1clk(ca,cb,f,cnew);
input  [7:0] ca,cb;
input  [8:0] f;
output [7:0] cnew;

wire [7:0] u2y,u3y;
wire [8:0] u1y;

bmult      u2 (.c(ca),.f(f),
               .y(u2y));
oneminus   u1 (.a(f),.y(u1y));
bmult      u3 (.c(cb),.f(u1y),
               .y(u3y));
satadd     u4 (.a(u3y),.b(u2y),
               .y(cnew));

endmodule
```

FIGURE 3.6: Naïve implementation of the blend equation.

creating a long delay path that results in a large clock period. For example purposes, relative delays of bmult = 2.0, satadd = 1.0, and oneminus = 0.4 are assumed with no time units specified. The longest combinational delay through this datapath is then 0.4 + 2.0 + 1.0 = 3.4 time units, which forces the clock period of the system to be at least T_{cq} (register clock-to-q delay) + 3.4 + T_{su} (register setup time) assuming the inputs and outputs of the datapath are registered. Assuming that T_{cq} and T_{su} are both 0.1, this gives a system clock period of 3.6 time units.

Figure 3.7 shows a better implementation of the blend equation where DFFs have been placed after the multipliers and after the adder to break the combinational delay path, assuming that the inputs originate from a registered source. This implementation still has the 1-F calculation chained with the n3 bmult execution unit, as the 1-F operation is designed for a low combinational delay by using the one's complement operation that allows it to be chained with another execution unit. Within the datapath's Verilog code, the DFFs are implemented by the always block and are synthesized as rising edge triggered via the posedge clk in the always block's sensitivity list. Observe that the longest t_{R2R} path of Fig. 3.13 is 2.6, which is shorter than the longest combinational path of Fig. 3.11, allowing for a higher clock frequency.

The cycle-by-cycle timing for the implementation of Fig. 3.7 is shown in Fig. 3.8 for the blend computations of Table 3.2. The latency of the datapath is two clock cycles due to the two DFFs in series for any path through the datapath. The initiation period as implemented in Fig. 3.8 is two clocks as new input values are only provided every two clock cycles. Observe that this datapath takes 2 * 2.6 = 5.2 time units to compute an output result for an input dataset, which is actually longer than the 3.6 clock period of Fig. 3.6. One reason for this is because dividing the combination delay by adding registers does not also divide the T_{cq} and setup times of the DFFs, which remain constant. Furthermore, the combinational delay path is not divided evenly when the registers are inserted. The delay of the register-to-register path that includes the adder is only 0.1 (T_{cq}) + 1.0

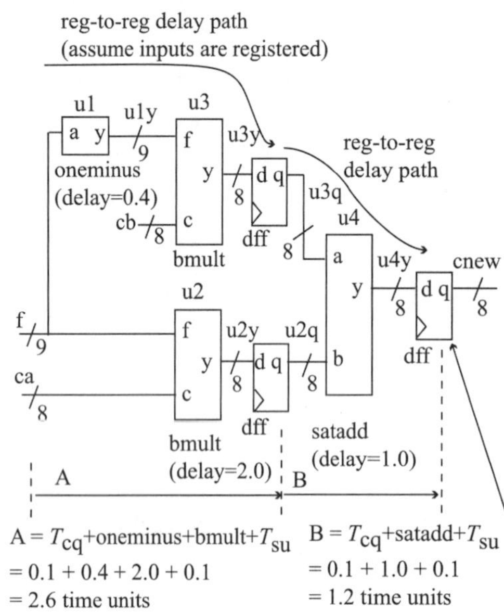


```
module blend2clk(clk,ca,cb,
                    f,cnew);
input  clk;
input  [7:0] ca,cb;
input  [8:0] f;
output [7:0] cnew;

wire [7:0] u2y,u3y,u4y;
wire [8:0] u1y;
reg [7:0] u3q, u2q, cnew;

bmult u2 (.c(ca),.f(f),
             .y(u2y));
oneminus  u1 (.a(f),.y(u1y));
bmult u3 (.c(cb),.f(u1y),
             .y(u3y));
satadd u4 (.a(u3q),.b(u2q),
             .y(u4y));

// always block that adds DFFs
// to datapath
always @(posedge clk)
 begin
  cnew  <= u4y; //dff on output
  u3q <= u3y; //dff on u3 output
  u2q <= u2y; //dff on u2 output
 end
endmodule
```

FIGURE 3.7: Blend equation implementation with latency = 2.

(satadd) + 0.1 (T_{su}) = 1.2 time units, as compared to the longest path of 2.6 time units. This is not a good division of work between the datapath stages; an optimium division of labor evenly divides the delay path between the datapath stages. However, this datapath's faster clock period of 2.6 time units allows computations outside of the datapath to execute faster than that possible with the datapath of Fig. 3.6.

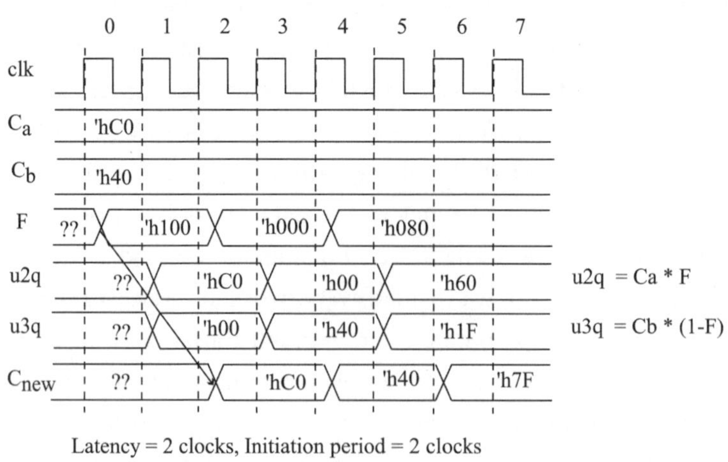

FIGURE 3.8: Cycle timing for latency = 2, initiation period = 2 clocks.

The timing diagram of Fig. 3.8 is one way to view a datapath's activity. A *scheduling table*, as shown in Table 3.3, provides another viewpoint of a datapath's activity. A scheduling table shows how DFG operations map to datapath resources such as input/output busses and execution units. Each row of the scheduling table shows the activity of the datapath resources for that clock cycle. A blank entry for a resource indicates that the resource is idle for that clock cycle. Indices such as '(0)', '(1)', etc., are used with input data values, output data values, and DFG node names to track the dataset computation that is being performed. The row entries in a schedule eventually repeat as the datapath performs the same operations on each input dataset. The last two rows in Table 3.3 form the *generalized* schedule, that is, the repeated operations on the datapath resources for each dataset. The percentage time that each datapath resource is busy during the generalized schedule is listed in the *%utilization* row of Table 3.3. Each of the datapath resources in Table 3.3 is only utilized 50% of the time as each resource is idle for one clock period of the two clock cycles that form the generalized schedule.

3.9 REGISTERING DATAPATH INPUTS VERSUS REGISTERING DATAPATH OUTPUTS

Our datapath examples place registers on the datapath outputs, and do not register the datapath inputs. The alternate choice of registering datapath inputs and leaving the outputs as unregistered is also valid, as long as consistency is followed in designing datapaths that are meant to connect together. If an external datapath with a registered output provides a value to a datapath with an unregistered input, then the communication delay from the external datapath is added to the execution unit delay that the input connects with. On large integrated circuits, the wire delay from one datapath to another can be significant if the datapaths are in different areas of the die. If the communication delay for an unregistered input value is large, then this input value should be registered in the destination datapath before being used as an input to an execution unit. The same can be said for an unregistered datapath output connected to a registered datapath input. If a datapath's input comes from off-chip or a datapath's output goes off-chip, then these signals should always be registered, as off-chip communication is slow compared to on-chip communication. Also, an unregistered datapath output should not be connected to an unregistered datapath input as the execution unit delay of the source datapath adds to the execution unit delay of the destination datapath, resulting in chained execution units.

3.10 PIPELINED COMPUTATIONS VERSUS EXECUTION UNIT PIPELINING

On viewing the datapath of Fig. 3.7 and the cycle timing of Fig. 3.8, the astute reader will realize that the datapath supports an initiation period of one clock, i.e., a new input dataset of C_a, C_b, and F can be provided for every clock. For an initiation period of one clock cycle, the second input dataset

TABLE 3.3: Schedule for latency = 2, initiation period = 2

CLOCK	RESOURCES							
	INPUT (CA)	INPUT (CB)	INPUT (F)	BMULT (U2)	ONEMINUS (U1)	BMULT (U3)	SATADD (U4)	OUTPUT (CNEW)
0	ca(0)	cb(0)	f(0)	n2(0)	n1(0)	n3(0)		
1							n4(0)	
2	ca(1)	cb(1)	f(1)	n2(1)	n1(1)	n3(1)		cnew(0)
3							n4(1)	
4	ca(2)	cb(2)	f(2)	n2(2)	n1(2)	n3(2)		cnew(1)
2i	ca(i)	cb(i)	f(i)	n2(i)	n1(i)	n3(i)		cnew(i-1)
2(i+1)							n4(i)	
%utilization	50%	50%	50%	50%	50%	50%	50%	50%

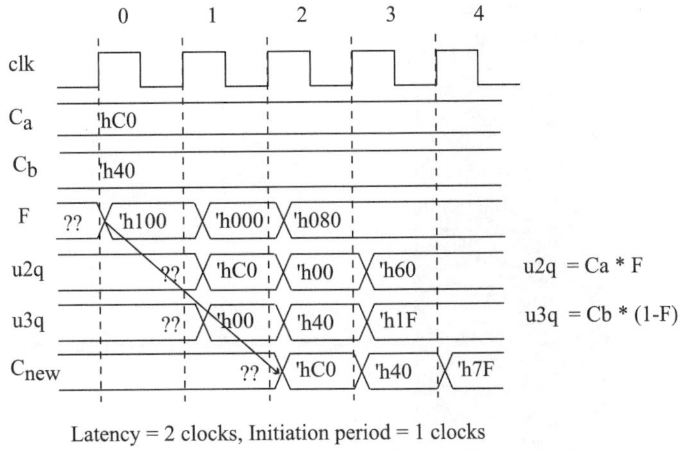

FIGURE 3.9: Cycle timing for latency = 2, initiation period = 1 clocks.

is provided before the output corresponding to first input dataset is produced. This means that the datapath has calculations on multiple datasets in progress simultaneously, with each input dataset in a different computation state. In this case, the computations for the two input datasets are said to be *pipelined*, or *overlapped*. Because the datapath resources of Fig. 3.7 are idle 50% of the time as shown by Fig. 3.8, no extra datapath resources are required to support the new initiation period of one clock cycle. Lowering the initiation period to one clock cycle doubles the throughput of the datapath, as a new result is now available with each clock instead of every two clock cycles. However, lowering the initiation period (increasing the throughput) does not affect the latency of the datapath. The cycle timing and scheduling table for an initiation period of one clock cycle are shown in Fig. 3.9 and Table 3.4, respectively. Observe that each resource is now utilized 100%, which is the best that can be achieved.

In Section 3.6, we observed that the t_{R2R} paths of Fig. 3.7 were not evenly balanced, which is undesirable as the longest t_{R2R} path determines the clock period. The excess time in the clock period for the shorter t_{R2R} paths is wasted time; distributing the delays more evenly would produce a shorter clock period. Note that the longest delay path in Fig. 3.7 contains the 1-F and multiplier units, with the multiplier having the longest delay of any execution unit. A pipeline stage inserted in the multiplier, that is, DFFs inserted within the multiplier logic, should reduce the length of this delay path. Figure 3.10 shows the blend multiplication of Fig. 3.4 modified to include a pipeline stage within the 8 × 8 multiplier. This example assumes the existence of an unsigned 8 × 8 multiplier with one pipeline stage named mult8 × 8pipe. Observe that inserting a pipeline stage in the mult8 × 8pipe component is not sufficient by itself; the other two paths through the multiplier for c[7:0] and f[8] must also have DFFs inserted so that the data streams remain synchronized when they reach the output multiplexer. If we assume that the multiplier pipeline stage perfectly divides the old combinational delay path by two, then the output and input delays of the multiplier both become

TABLE 3.4: Schedule for latency = 2, initiation period = 1

CLOCK	RESOURCES							
	INPUT (CA)	INPUT (CB)	INPUT (F)	BMULT (U2)	ONEMINUS (U1)	BMULT (U3)	SATADD (U4)	OUTPUT (CNEW)
0	ca(0)	cb(0)	f(0)	n2(0)	n1(0)	n3(0)		
1	ca(1)	cb(1)	f(1)	n2(1)	n1(1)	n3(1)	n4(0)	
2	ca(2)	cb(2)	f(2)	n2(2)	n1(2)	n3(3)	n4(1)	cnew(0)
3	ca(3)	cb(3)	f(3)	n2(3)	n1(3)	n3(3)	n4(2)	cnew(1)
4	ca(4)	cb(4)	f(4)	n2(4)	n1(4)	n3(4)	n4(3)	cnew(2)
i	ca(i)	cb(i)	f(i)	n2(i)	n1(i)	n3(i)	n4(i-1)	cnew(i-2)
%utilization	100%	100%	100%	100%	100%	100%	100%	100%

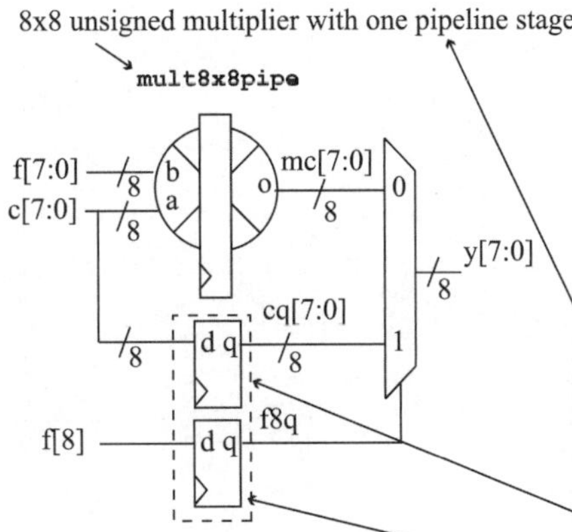


```
module bmultpipe(clk,c,f,y);
input  clk;
input  [7:0] c;
input  [8:0] f;
output [7:0] y;

wire [7:0] mc;
reg  f8q;
reg  [7:0] cq;

mult8x8pipe m1 (.clk(clk),.a(c),
                .b(f[7:0]),.o(mc));

//add DFFs to match pipeline stage
//in multiplier
always @(posedge clk) begin
 cq  <= c;
 f8q <= f[8];
end //end always

//             f8q==1     f8q==0
assign y = f8q ?  cq     :    mc;

endmodule
```

Add DFFs to c[7:0] and f[8] paths to match the one clock cycle latency that is caused by the pipeline stage in the multiplier.

multiplier output delay = T_{cq} + old delay/2
= 0.1 + 2.0/2 = 1.1
multiplier input delay = old delay/2 + T_{su}
= 2.0/2 + 0.1 = 1.1

FIGURE 3.10: Multiplication of an eight-bit color operand by nine-bit blend operand with pipeline stage.

equal to 1.1 time units as seen in Fig. 3.10. This decreased delay path comes at the cost of a clock cycle of latency through the blend multiplication unit.

Figure 3.11 shows the blend implementation of Fig. 3.7 modified to use the pipelined multiplier of Fig. 3.10. The longest t_{R2R} path has been reduced from 2.6 to 1.6 time units, at the cost of an extra clock cycle of latency.

The cycle timing for the blend implementation with the pipelined multiplier is shown in Fig. 3.12. The only difference between this timing and the timing in Fig. 3.9 is the extra clock cycle of latency. Table 3.5 shows the scheduling table for the blend implementation with the pipelined multiplier. The table entries for the bmultpipe units show two calculations, one for each pipeline stage of the bmultpipe unit. The extra clock cycle of latency in the bmultipipe units causes the satadd unit to remain idle until clock cycle two, as opposed to clock cycle one in the Table 3.4 schedule.

In comparing the cycle timings and schedules for the two clock cycle latency versus the three clock cycle latency solutions, a good question to ask is "When is it not advantageous to pipeline execution units?" Each clock cycle of latency is one more clock cycle that it takes for the pipeline to become full and for all execution units to become active. A pipelined datapath with a large latency is efficient as long as it has a continuous stream of input data. If the application using the datapath does

TABLE 3.5: Schedule for latency = 3, initiation period = 1

CLOCK	RESOURCES							
	INPUT (CA)	INPUT (CB)	INPUT (F)	*PIPE (U2)	ONEMINUS (U1)	*PIPE (U3)	SATADD (U4)	OUTPUT (CNEW)
0	ca(0)	cb(0)	f(0)	n2(0)	n1(0)	n3(0)		
1	ca(1)	cb(1)	f(1)	n2(1), n2(0)	n1(1)	n3(1), n2(0)		
2	ca(2)	cb(2)	f(2)	n2(2), n2(1)	n1(2)	n3(2), n2(1)	n4(0)	
3	ca(3)	cb(3)	f(3)	n2(3), n2(2)	n1(3)	n3(3), n2(2)	n4(1)	cnew(0)
4	ca(4)	cb(4)	f(4)	n2(4), n2(3)	n1(4)	n3(4), n2(3)	n4(2)	cnew(1)
i	ca(i)	cb(i)	f(i)	n2(i), n2(i-1)	n1(i)	n3(i), n2(i-1)	n4(i-2)	cnew(i-3)
%utilization	100%	100%	100%	100%	100%	100%	100%	100%

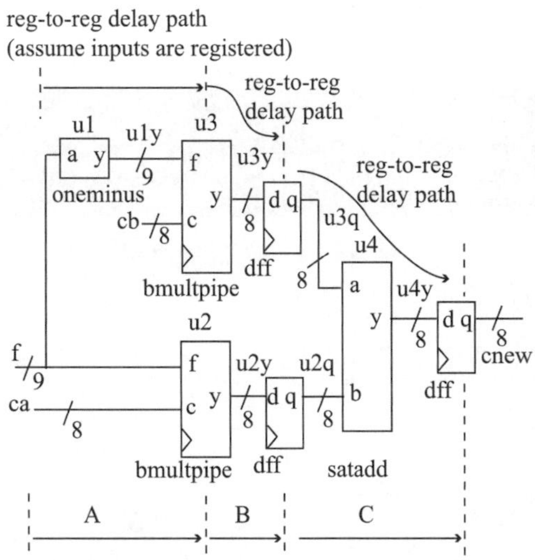

$A = T_{cq}$+oneminus + bmultpipe (input delay)
= 0.1 + 0.4 + 1.1
= 1.6 time units

$C = T_{cq}$+satadd+T_{su}
= 0.1 + 1.0 + 0.1
= 1.2 time units

B = bmultipipe(output delay) + T_{su}
= 1.1 + 0.1
= 1.2 time units

```
module blendpipe(clk,ca,
                 cb,f,cnew);
input  clk;
input  [7:0] ca,cb;
input  [8:0] f;
output [7:0] cnew;

 wire [7:0] u2y,u3y,u4y;
 wire [8:0] u1y;

 reg [7:0] u3q, u2q, cnew;

 bmultpipe u2 (.clk(clk),.c(ca),
               .f(f),.y(u2y));
 oneminus  u1 (.a(f),.y(u1y));
 bmultpipe u3 (.clk(clk),.c(cb),
               .f(u1y),.y(u3y));
 satadd u4 (.a(u3q),.b(u2q),
            .y(u4y));

 // always block that adds DFFs
 // to datapath
 always @(posedge clk)
  begin
   u3q <= u3y; //dff on u3 output
   u2q <= u2y; //dff on u2 output
   cnew  <= u4y; //dff on output
  end

endmodule
```

FIGURE 3.11: Blend equation implementation with pipelined multiplier, latency = 3.

not provide continuous input data, thus allowing the pipeline to become empty or partially empty, then the datapath throughput is significantly decreased.

Table 3.6 compares the datapaths that have been discussed to this point by clock period, latency, initiation period, and throughput.

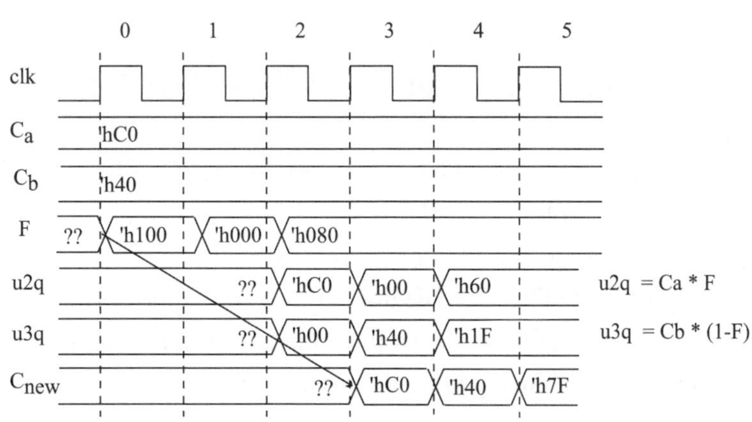

Latency = 3 clocks, Initiation period = 1 clocks

FIGURE 3.12: Cycle timing for latency = 3, initiation period = 1 clock.

TABLE 3.6: Datapath comparisons

DATAPATH	CLOCK PERIOD	LATENCY	INITIATION PERIOD	THROUGHPUT
(a) Figure 3.6	3.6	1	1	0.28
(b) Figure 3.7	2.6	2	2	0.19
(c) Figure 3.7	2.6	2	1	0.38
(d) Figure 3.11	1.6	3	1	0.63

The throughput value measures the number of input datasets processed per time unit, and is calculated by Eq. (3.2), assuming that the pipeline is filled. Decreasing either the initiation period or the clock period improves throughput, as is seen in rows (c) and (d) of Table 3.6. However, these improvements come at a cost. Decreasing the initiation period generally requires adding more datapath resources, even though this was not necessary in this simple example. Decreasing the clock period by pipelining execution units adds latency to the datapath.

$$\text{Throughput} = \frac{1}{(\text{initiation\_period} \times \text{clock\_period})} \tag{3.2}$$

3.11 A BLEND IMPLEMENTATION WITH A SINGLE MULTIPLIER

The datapaths in Sections 3.6 and 3.7 assigned each node of the DFG of Fig 3.5 to a separate execution unit. However, in more complex datapaths, resource constraints force multiple dataflow nodes to be mapped to the same execution unit. Table 3.7 gives a schedule for a blend implementation that only contains one multiplier unit. The schedule does not use overlapped computations or pipelined execution units, and has a latency of three clocks and an initiation period of three clocks. The DFG node operations n2 and n3 are both mapped to the single multiplier unit. In this case, the execution order of n2 followed by n3 is an arbitrary choice; the execution order could be reversed.

Sharing the multiplier unit brings a new set of problems to the datapath design. The first problem is that the multiplier unit's operands now change depending upon the clock cycle. In clock

TABLE 3.7: Schedule for latency = 3, initiation period = 3, single multiplier blend implementation

CLOCK	RESOURCES						
	INPUT (CA)	INPUT (CB)	INPUT (F)	BMULT (U2)	ONEMINUS (U1)	SATADD (U4)	OUTPUT (CNEW)
0	ca(0)	cb(0)	f(0)	n2(0) ca*f→rA	n1(0)		
1				n3(0) cb*u1→rB			
2						n4(0) rA+rB→rC	
3	ca(1)	cb(1)	f(1)	n2(1) ca*f→rA	n1(1)		cnew(0)
3(*i*+0)	ca(i)	cb(i)	f(i)	n2(i) ca*f→rA	n1(i)		cnew(i-1)
3(*i*+1)				n3(i) cb*u1→rB			
3(*i*+2)						n4(i) rA+rB→rC	
%utilization	33%	33%	33%	67%	33%	33%	33%

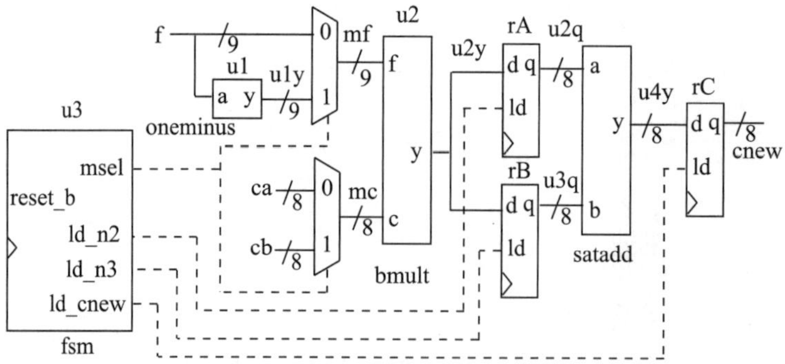


```
module blend1mult(clk,reset_b,ca,cb,f,cnew);
input  clk, reset_b;
input  [7:0] ca,cb;
input  [8:0] f;
output [7:0] cnew;

 wire [7:0] u2y,u4y,ma;
 wire [8:0] mf,u1y;
 reg [7:0] u3q, u2q, cnew;

 // muxes for the multiplier
 //             msel==1   msel==0
 assign mf = msel ? u1y    :   f;
 //             msel==1   msel==0
 assign ma = msel ?  cb    :    ca;

 bmult u2  (.c(ma),.f(mf),.y(u2y));
 oneminus  u1 (.a(f),.y(u1y));
 satadd u4  (.a(u3q),.b(u2q),.y(u4y));
 fsm u3  (.clk(clk), .reset_b(reset_b),.msel(msel),.ld_n2(ld_n2),
        .ld_n3(ld_n3), .ld_cnew(ld_cnew));

 // always block that adds registers to the datapath
 always @(posedge clk)
  begin
    if (ld_n2) u2q <= u2y;        // rA
    if (ld_n3) u3q <= u2y;        // rB
    if (ld_cnew) cnew  <= u4y;    // rC
  end //end always

endmodule
```

FIGURE 3.13: Single multiplier blend implementation.

cycle 3(i + 0), the multiplier's operands are *Ca* and *F*, while in clock cycle $3(i+1)$ the multiplier's operands are *Cb* and 1-*F*. This means that a multiplexer is needed on the multiplier's inputs to choose between the two sets of operands. The other problem is that a register is required to store the n2 result produced in clock cycle $3(i+1)$ until it is needed in clock cycle $3(i+2)$ for the n4 operation. A datapath that implements this schedule is shown in Fig. 3.13. This datapath uses *registers* instead of DFFs to break the combinational delay path and to store intermediate results. A register has a load input (LD); the register accepts a new input value only when LD is asserted and when the active clock edge occurs. By contrast, a DFF accepts a new input on each active clock edge. The three registers are named *rA*, *rB*, and *rC*. A *register transfer operation* (RTL) is added to cells in the

```
module fsm(clk,reset_b,msel,ld_n2,ld_n3,ld_cnew);
input clk,reset_b;
output msel,ld_n2,ld_n3,ld_cnew;

reg msel,ld_n2,ld_n3,ld_cnew;
reg [1:0] state, nstate;

`define s0 2'b00   //state encoding
`define s1 2'b01
`define s2 2'b11

 //dffs for finite state machine
 always @(posedge clk or negedge reset_b)
  begin
   //low-true async reset
   if (!reset_b) state <= `s0;
   else  state <= nstate;
  end

 //combinational logic for FSM
 always @(state) begin
   nstate = state;
   msel = 0; ld_n2 = 0;
   ld_n3 = 0; ld_cnew = 0;
   case (state)
    `s0 :begin
         ld_n2 = 1;nstate = `s1;
         end
    `s1 :begin
         msel = 1;ld_n3 = 1;
         nstate = `s2;
         end
    `s2 :begin
         ld_cnew = 1;
         nstate = `s0;
         end
    default :  nstate = `s0;
    endcase
 end //end always
endmodule
```

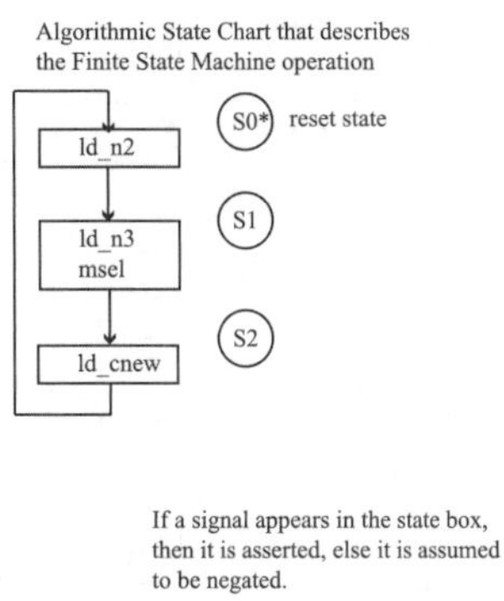

FIGURE 3.14: FSM for single multiplier blend implementation.

scheduling table for each clock cycle that register writes occurs. The RTL notation "ca*f→*rA*" for execution unit u2 in clock zero indicates that register *rA* is loaded with the result of the multiplication that has the *ca* and *f* input busses as operands. Note that the *rA* and *rB* registers controlled by the ld_n2 and ld_n3 load signals have their data inputs connected to the multiplier output u2y. The ld_n2 load signal is asserted in clock cycle $3(i + 1)$ to store the n2 result, while the ld_n3 load signal is asserted in clock cycle $3(i{+}1)$ to store the n3 result. The ld_cnew load signal is asserted in clock cycle 3(i + 2) to load the output register with the satadd n4 result. The multiplexer select signal msel is negated in clock cycle $3(i + 1)$ to pass the *Ca*, *F* operands to the multiplier, while msel is asserted in clock cycle i + 1 to select *Cb*, 1-*F* as the multiplier operands. As an optimization, register *rB* could be replaced with DFFs as its contents are only needed in the following clock cycle. The $C_{new}(i - 1)$ output value is held stable by the *rC* register for the duration of the computation; this might be useful if this value is used by a destination datapath. If this is not required, then register *rC* could also be replaced by DFFs.

A finite state machine component named FSM is responsible for driving the datapath's control lines of msel, ld_n2, ld_n3, and ld_cnew with the correct values in the appropriate clock cycles. The control signals in Fig. 3.13 are drawn with dotted lines to distinguish them from the data busses that are operated on by the execution units. The control signals and FSM component are typically not drawn in a datapath diagram; they are included here since this is the first datapath example that has required a FSM. Figure 3.14 shows the FSM implementation. Three states are required since the datapath's operation is a repeating computation covering three clock cycles.

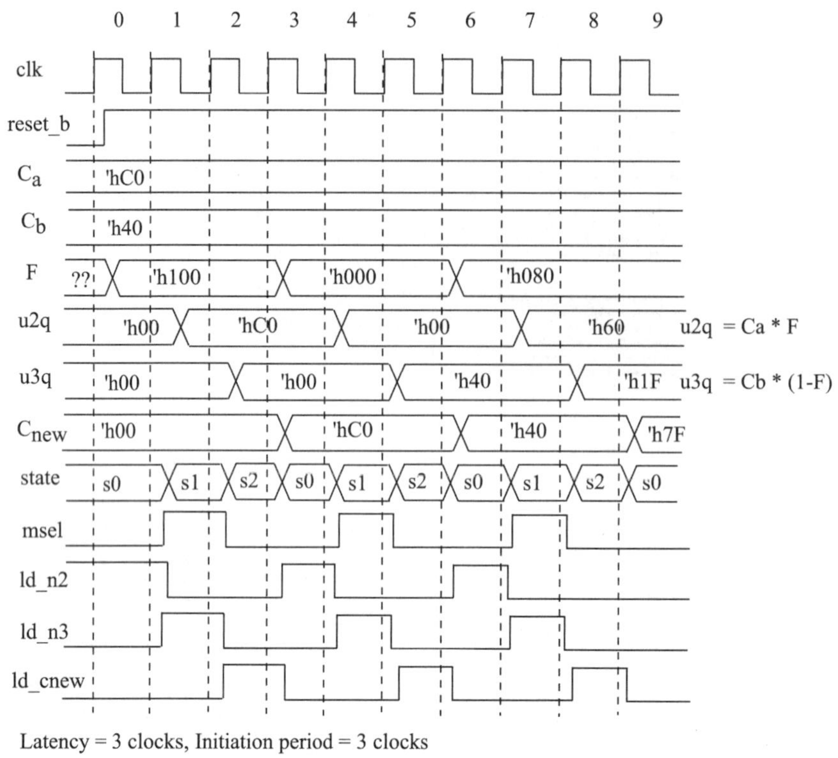

FIGURE 3.15: Cycle timing for the single multiplier blend implementation.

This FSM implementation uses two state DFFs and a grey-code encoding for the state implementation; an alternate encoding method such as one-hot encoding could have been used as well. The FSM requires an asynchronous reset input to initialize the state registers to state S0; in this example the reset signal is named reset_b and is a low-true input. The polarity choice for the reset signal, low-true or high-true, is implementation dependent.

3.12 A BLEND IMPLEMENTATION WITH HANDSHAKING

Our previous examples assumed that data is continually streaming through the datapath. However, in many cases a datapath must wait for input data to become available and must also indicate when output data is ready. Additional signals called *handshaking signals* are used by the datapath FSM for this purpose. Figure 3.16 shows the FSM of the one-multiplier blend implementation modified to add the handshaking signals irdy (input data ready) and ordy (output data ready). The differences between Fig. 3.16 and the original code in Fig. 3.14 are underlined to emphasize the changes required to support the new signals.

The FSM now remains within the S0 state until the irdy input is asserted, indicating that the input busses contain valid data, at which point the FSM transits to state S1. The ordy signal

```
module fsm(clk,reset_b,irdy,msel,ld_n2,ld_n3,ld_cnew,ordy);
input clk,reset_b,irdy;
output msel,ld_n2,ld_n3,ld_cnew,ordy;

reg msel,ld_n2,ld_n3,ld_cnew,ordy;
reg [1:0] state, nstate;

`define s0 2'b00   //state encoding
`define s1 2'b01
`define s2 2'b11

 //dffs for finite state machine
 always @(posedge clk or negedge reset_b)
  begin
   //low-true async reset
   if (!reset_b) begin
     state <= `s0; ordy <= 0;
    end
   else  begin
    state <= nstate; ordy <= ld_cnew;
    end;
  end

 //combinational logic for FSM
 always @(state or irdy) begin
   nstate = state;
   msel = 0; ld_n2 = 0;
   ld_n3 = 0; ld_cnew = 0;
   case (state)
    `s0 :begin
         ld_n2 = 1;
         if (irdy) nstate = `s1;
         end
    `s1 :begin
         msel = 1;ld_n3 = 1;
         nstate = `s2;
         end
    `s2 :begin
         ld_cnew = 1;
         nstate = `s0;
         end
    default :  nstate = `s0;
    endcase
 end //end always
endmodule
```

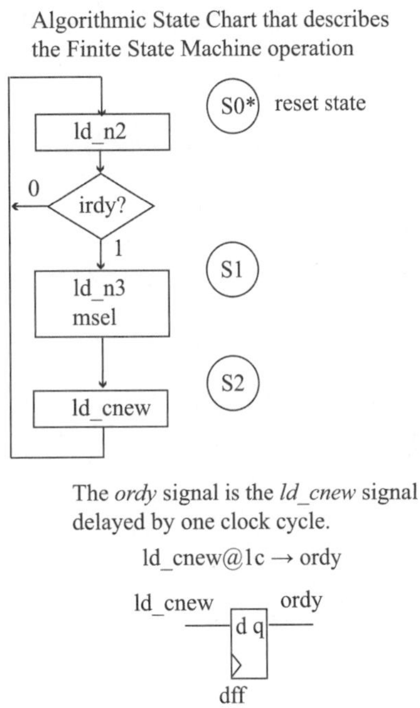

FIGURE 3.16: Handshaking added to FSM for single multiplier blend implementation.

is asserted for one clock cycle when valid data is placed on the *C*new output bus by delaying the ld_cnew signal that is asserted in state S2 for one clock cycle. This is implemented by a DFF that is synthesized via the Verilog assignment ordy < = ld_cnew within the always block used for the state registers of the FSM. In the Algorithm state chart (ASM chart), the ordy signal action is described by the annotation ld_cnew@1c→ordy, which reads, "ordy is assigned the value of ld_cnew, delayed by one clock". Figure 3.17 shows the cycle timing of the modified datapath for one computation; the assertion of irdy indicates valid input data and causes the computation to begin. The ordy signal is asserted when the *C*new output bus contains the computation result. The changes required to the blend1mult module of Fig. 3.13 to support the new handshaking signals are left as an exercise for the reader.

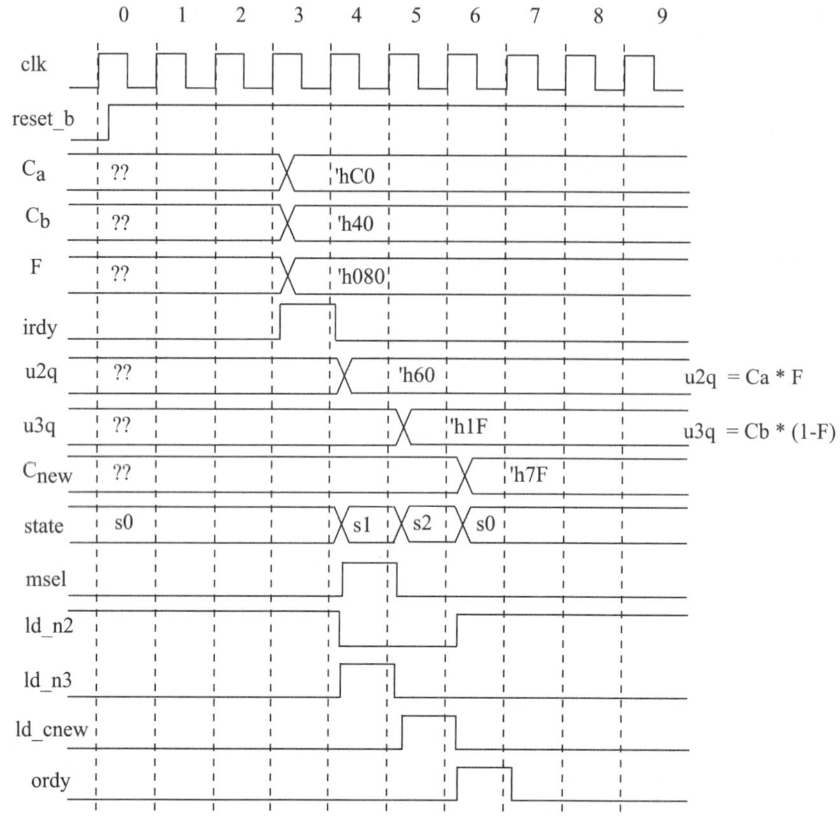

FIGURE 3.17: Cycle timing for the single multiplier blend implementation with handshaking.

3.13 A BLEND IMPLEMENTATION WITH A SHARED INPUT BUS

The previous blend implementations used separate input busses for the *F*, C_a, and C_b data values. However, input busses are resources in the same way as execution units are, and a designer may not have the luxury of using a separate input bus for each required input datum. External pins on an integrated circuit are extremely precious resources, and external pins are often time multiplexed between different functions. Table 3.8 gives the schedule for a blend implementation with latency = 4, initiation period = 4, uses a shared bus to input the *F*, C_a, and C_b data values over successive clock cycles. Only one multiplier is required; the multiplier is idle in clock cycle $i + 0$ as the C_a value is not yet available. This schedule uses a new temporary register named *rF* to hold the *F* value that is required for the n2 and n3 computations in clocks $4(i + 1)$ and $4(i + 2)$; the previous implementations assumed that the *F* value remained available on a separate input data bus for the duration of the computation.

TABLE 3.8: Schedule for latency = 4, initiation period = 4, shared input bus blend implementation

CLOCK	RESOURCES					
	INPUT (DIN)	REGISTER (RF)	BMULT (U2)	ONEMINUS (U1)	SATADD (U4)	OUTPUT (CNEW)
0	f(0)	din→rF				
1	ca(0)	f(0)	n2(0) din*rF→rA	n1(0)		
2	cb(0)		n3(0) din*u1→rB			
3					n4(0) rA+rB→rC	
4	f(1)	f(1)→rF				cnew(0)
4(i+0)	f(i)	din→rF				
4(i+1)	ca(i)	f(i)	n2(i) din*rF→rA	n1(i)		cnew(i-1)
4(i+2)	cb(i)		n3(i) din*u1→rB			
4(i+3)					n4(i) rA+rB→rC	
%utilization	75%	25%	50%	25%	25%	25%

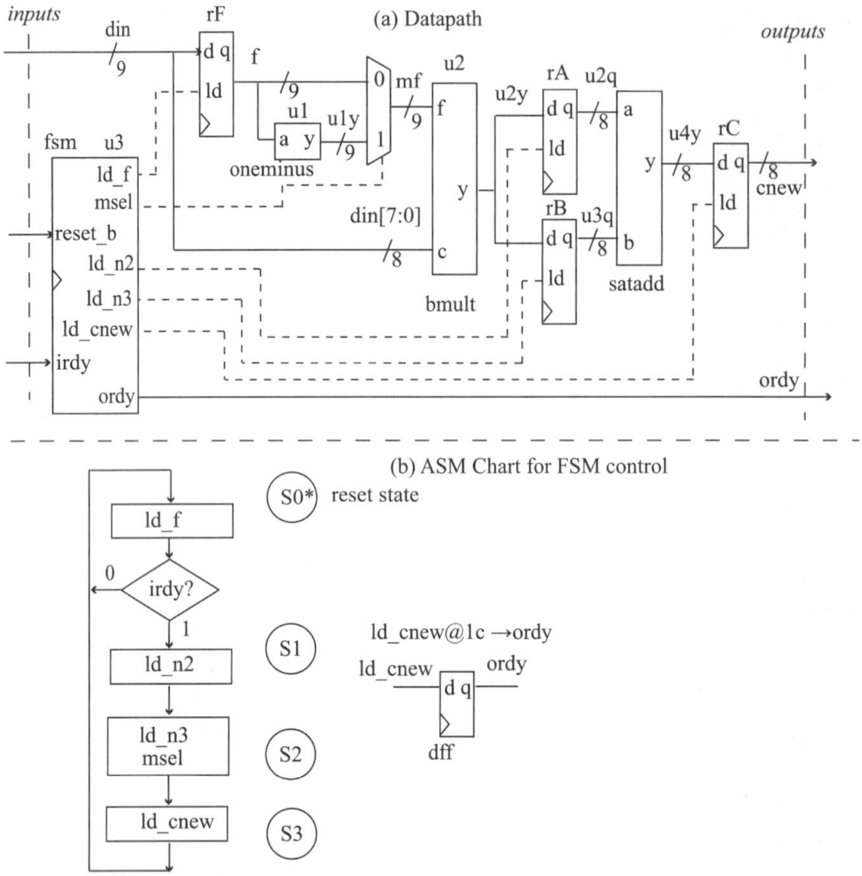

FIGURE 3.18: Shared input bus blend implementation.

Figure 3.18a shows the datapath for the blend implementation with a shared input bus. The nine-bit *din* data bus is used for the *F*, C_a, and C_b data values. The multiplexer that was used on input *c* of the *bmult* multiplier in Fig. 3.13 is no longer needed, as the C_a, C_b input values are now time-multiplexed over the *din* databus.

Figure 3.18b shows the ASM chart for the datapath's FSM control; the FSM uses handshaking in the same manner as used in Fig 3.16. The Verilog code for this implementation is left as an exercise for the reader.

3.14 RECURSIVE CALCULATIONS, INITIALIZATION VERSUS COMPUTATION

The blend equation in Eq. (3.1) is a *non-recursive* equation; its output value is not dependent upon previous output values. Eq. (3.3) gives an example of a recursive equation; the Y output is dependent upon the current input value X and a previous output value Y@1. Please note that the value Y@1

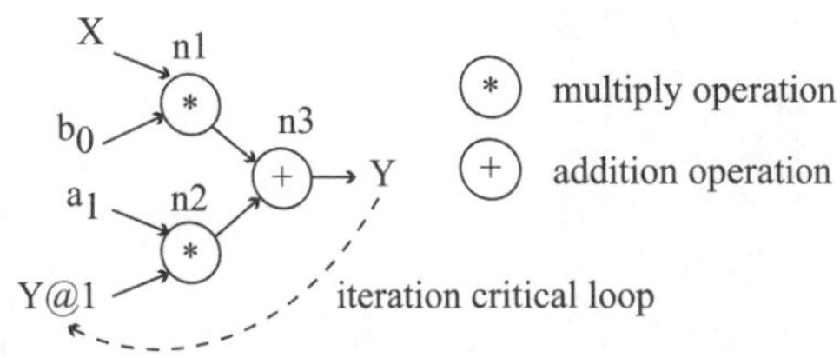

FIGURE 3.19: Dataflow graph of equation 3.3.

is the output computed from the previous input dataset, and is not the output of Y delayed by one clock cycle. A special class of digital filters known as *infinite impulse response* (IIR) filters have the general structure of (Eq. 3.3), except that multiple previous output values (Y@1, Y@2, . . .Y@n) and multiple previous input values (X, X@1, X@2, . . . X@k) are typically used as shown in (Eq. 3.4). The values a_i (a_1, a_2, a_3, . . . a_n) and b_i (b_0, b_1,..b_k) that are multiplied by the previous output and previous input values are called the *filter coefficients*, and are determined by the filter's specifications (cutoff frequencies for low pass, band pass, high pass; roll-off constraints, etc.). Each multiplication operation is called a *filter tap*, and increasing the number of filter taps improves the filter quality.

$$\mathrm{Y} = Y@1 \times a1 + X \times b0 \tag{3.3}$$

$$\begin{aligned} \mathrm{Y} = &(Y@1 \times a1 + Y@2 \times a2 ... + Y@n \times an) \\ &+ (X \times b0 + X@1 \times b1 ... + X@k \times bk) \end{aligned} \tag{3.4}$$

One of the features of a non-recursive equation is that a datapath implementation can always achieve an initiation period of one clock cycle by overlapping computations and adding the required extra resources such as input data busses, execution units, and registers. However, assuming that execution units cannot be chained, the minimum initiation period of a recursive calculation depends upon the *iteration critical loop*, which is the shortest path through the data flowgraph involving a previous output. Figure 3.19 gives the DFG of Eq. (3.3), with the iteration critical loop containing nodes n2 and n3. Each node requires one clock cycle assuming that execution unit chaining is not allowed, thus resulting in a minimum initiation period for this DFG of two clock cycles.

Table 3.9 shows a schedule for Eq. (3.3) that meets the minimum initiation period of two clock cycles. This schedule assumes that the filter coefficients are loaded into the datapath over the shared input data bus during an initialization phase, which is done before the datapath computation loop is entered.

Figure 3.20 shows the datapath and FSM control for the schedule of Table 3.17, with eight-bit data used for all calculations and 0.8 fixed-point encoding assumed. The ASM chart shows the states divided into two groups: *initialization* and *computation*. The S0 and S1 states are used to initialize the a_1, b_0 coefficient registers of the datapath with the a_1, b_0 values input over the *din* input bus in consecutive clock cycles once the *irdy* handshaking signal is asserted. States S2 and S3 form the

TABLE 3.9: Schedule for latency = 2, initiation period = 2, Eq. (3.3) implementation

CLOCK	RESOURCES				
	INPUT	MULT (U1)	MULT (U2)	SATADD (U3)	OUTPUT
0	x(0)	n1(0) b0*din→rA	n2(0) a1*rY→rB		
1				n3(0) rA+rB→rY	
2	x(1)	n1(1) b0*din→rA	n2(1) a1*rY→rB		y(0)
2(*i*+0)	x(i)	n1(i) b0*din→rA	n2(i) a1*rY→rA		y(i-1)
2(*i*+1)				n3(i) rA+rB→rY	
%utilization	50%	50%	50%	50%	50%

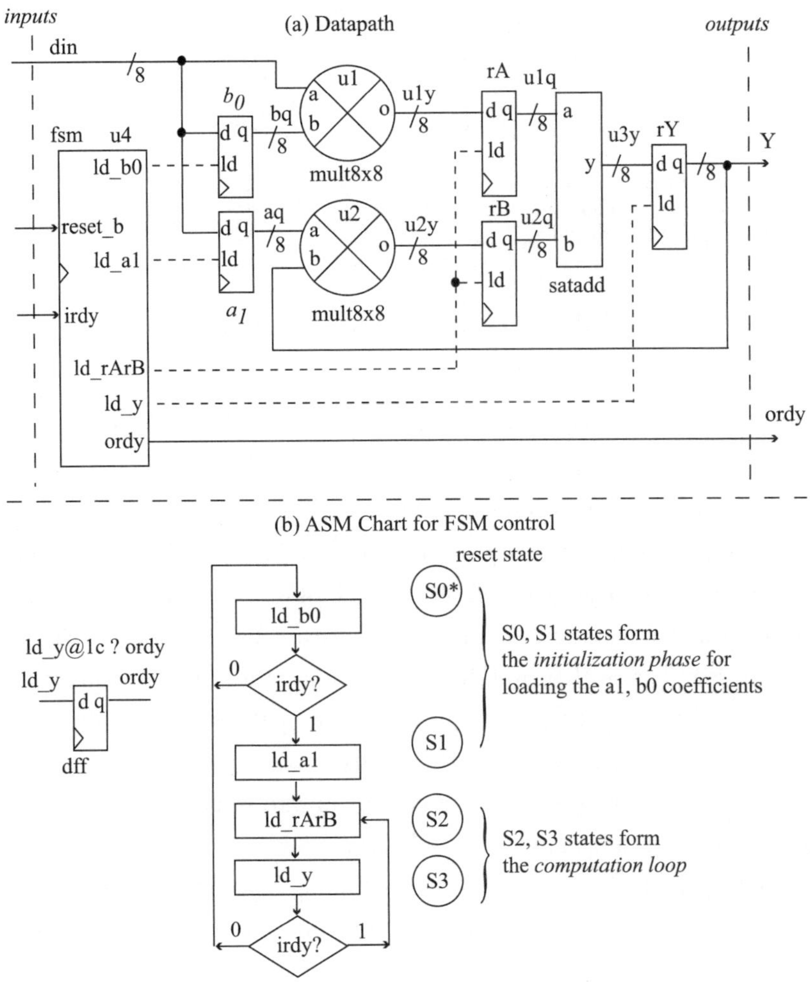

FIGURE 3.20: Datapath, FSM for equation 3.3 implementation.

computation loop, with new X values available over the *din* input bus as long as the *irdy* handshaking signal is asserted. The computation loop is exited when the *irdy* handshaking signal is negated. The *ordy* output handshaking signal is produced by delaying the *ld_y* signal of the FSM by one clock cycle. The Verilog code for this implementation is left as an exercise for the reader.

3.15 A DESIGN METHODOLOGY FOR HIGHER COMPLEXITY DATAPATHS

The previous datapath examples contained a relatively low number of operations, and scheduling the DFG operations on execution units and storing temporary results within registers was relatively straightforward. However, scheduling becomes more difficult as the target equation complexity increases, i.e., the number of operations in the target equation increases. In this section, a scheduling

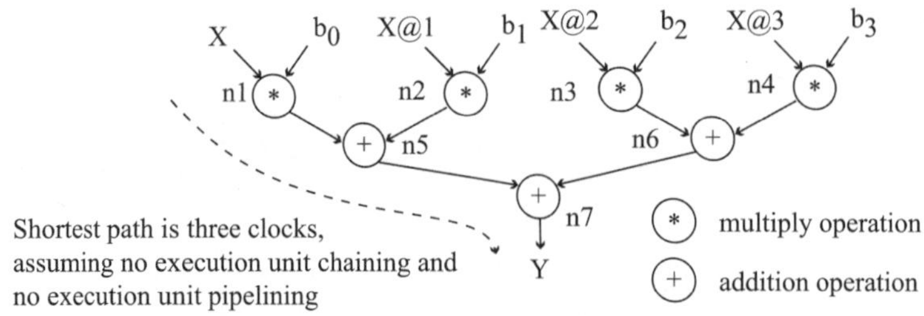

FIGURE 3.21: Dataflow graph of equation 3.5.

methodology appropriate for higher complexity datapaths is developed. This methodology does not attempt to include all of the optimizations found in behavioral synthesis methodologies [3], but rather serves to illustrate the key problems in datapath scheduling.

Equation (3.5) is a four-tap *finite impulse response* (FIR) filter, and is used as the target equation for the datapath implementations that follow. A FIR filter differs from an IIR filter (Eq. 3.4) in that it is a non-recursive equation — the filter does not use past output values. A FIR filter generally requires more filter taps than an IIR filter to achieve the same filter quality. As with the IIR equation, X@1 means the X input from the previous input dataset, and is not the X input delayed by a clock cycle. Please note that because of the regular structure of the FIR equation, an efficient datapath implementation can be done for the case of initiation period = 1, where each addition and multiplication operation are mapped to individual execution units. This equation is used in this section to illustrate the more difficult problem of mapping multiple flowgraph operations onto the same execution unit, when resource constraints prevent one-to-one mappings of operations to execution units.

$$Y = X \times b0 + X@1 \times b1 + X@2 \times b2 + X@3 \times b3 \qquad (3.5)$$

Figure 3.21 shows the DFG for Eq. (3.5). The shortest path through this DFG is three clock cycles, assuming no execution unit chaining and non-pipelined execution units. This shortest path of three clock cycles is the minimum achievable latency for this equation.

Table 3.10 shows the steps in the datapath design methodology that is followed in this section. This methodology's goal is a datapath that uses the minimum number of execution units to meet a set of target constraints.

The target constraints in this methodology is initiation period and latency, both measured in clock cycles. Step #2 computes a lower bound for each type of resource required to meet the target constraints, using Eq. (3.6). The result of Eq. (3.6) is a *lower bound* for the resource, which means that it cannot be done with any fewer resources than this value, and may actually require more than

TABLE 3.10: Datapath design methodology

STEP	ACTION
1.	Set target constraints (initiation period, latency)
2.	Compute a lower bound on the resources needed for the target constraints
3.	Attempt scheduling using the number of resources computed in step 2. If scheduling fails, go to step 4; if scheduling succeeds then go to step 5
4.	Either increase the resource(s) that has caused scheduling to fail, and loop back to step 3, or relax the constraints, and go back to step 2.
5.	Execution unit scheduling has succeeded; do register scheduling
6.	Implement the datapath

this number of resources.

$$\#\text{ of resources} = \left\lceil \frac{\#\text{operations}}{\#\text{InitiationPeriod}} \right\rceil \tag{3.6}$$

For example, assume that the target constraint for a datapath implementation of Fig 3.21 is an initiation period of three clocks, and a latency of three clocks. The number of operations for a particular resource is determined by simply counting the addition or multiplication nodes in Fig. 3.21. A lower bound on the number of adders, multipliers, and input busses needed for these target constraints are given in Eqs (3.7) –(3.9). The input bus calculation of Eq. (3.8) is somewhat superfluous as the FIR equation only requires one new X input value each clock assuming that coefficient values are loading during an initialization phase, but this calculation is included to emphasize that input busses are also resources.

$$\#\text{ of multipliers} = \left\lceil \frac{4}{3} \right\rceil = 2 \tag{3.7}$$

$$\#\text{ of adders} = \left\lceil \frac{3}{3} \right\rceil = 1 \tag{3.8}$$

$$\#\text{ of input busses} = \left\lceil \frac{1}{3} \right\rceil = 1 \tag{3.9}$$

Table 3.11 shows a scheduling attempt of Fig. 3.21 using two multipliers and one adder to meet the target constraints of latency = 3 clocks and initiation period = 3 clocks. The scheduling fails as the n7 node computation is not scheduled. In order to perform the n7 computation in clock #2, the n5 and n6 computations must both be performed in clock #1, which requires that the number

TABLE 3.11: Schedule for Figure 3.40 using two multipliers, one adder for target latency = 3, target initiation period =3

CLOCK	RESOURCES				
	INPUT	MULT (U1)	MULT (U2)	SATADD (U3)	OUTPUT
0	x(0)	n3(0)	n4(0)		
1		n1(0)	n2(0)	n6(0)	
2				n5(0)	
	Scheduling Fails! Operation n7 is not scheduled within target latency.				

of adders must be increased from one to two. However, performing the n5 and n6 computations in clock #1, requires that the n3, n4 multiply operations be performed by clock #0, which requires that the number of multipliers be increased from two to four.

Table 3.12 shows that the scheduling now succeeds with the increased resources of four multipliers and two adders for the target latency of three clocks. However, meeting this target required a doubling of the resources from their lower bound computations, which may not be acceptable if resources are limited. Relaxing the target constraints must be done if the resource requirements are too high.

If the target constraints are relaxed to initiation period = 4 clocks and latency = 4 clocks, then the new lower bound computations are shown in Eqs (3.10) and (3.11) (the input bus resource is omitted for brevity as it clearly does not affect the scheduling).

$$\text{\# of multipliers} = \left\lceil \frac{4}{4} \right\rceil = 1 \tag{3.10}$$

$$\text{\# of adders} = \left\lceil \frac{3}{4} \right\rceil = 1 \tag{3.11}$$

Table 3.13 shows that the scheduling attempt fails for these resource lower bounds, because the addition operations n5 and n7 cannot be scheduled within the target latency of four clocks. The three addition operations must begin in clock #1 if they are to be completed within the four clock latency using only one adder. If the n6 addition operation is scheduled in clock#1, then the n3 and n4 multiply operations must be scheduled in clock#0, which requires two multipliers.

TABLE 3.12: Schedule for Figure 3.40 using four multipliers, two adders for target latency = 3, target initiation period =3

CLOCK	RESOURCES							
	INPUT	MULT (U1)	MULT (U2)	MULT (U3)	MULT (U4)	SATADD (U6)	SATADD (U7)	OUTPUT
0	x(0)	n3(0)	n4(0)	n1(0)	n2(0)			
1						n6(0)	n5(0)	
2						n7(0)		
$3(i+0)$	x(i)	n3(i)	n4(i)	n1(i)	n2(i)			
$3(i+1)$						n6(i)	n5(i)	
$3(i+2)$						n7(i)		y(i-1)
%utilization	33%	33%	33%	33%	33%	67%	33%	33%

TABLE 3.13: Schedule for Figure 3.40 using one multiplier, one adder for target latency = 4, target initiation period = 4

CLOCK	RESOURCES			
	INPUT	MULT (U1)	SATADD (U2)	OUTPUT
0	x(0)	n4(0)		
1		n3(0)		
2		n2(0)	n6(0)	
3		n1(0)		
	Scheduling fails, operations n5, n7 are not scheduled within target latency.			

Table 3.14 shows that scheduling is successful for the target latency of four clocks after the number of multipliers is increased from one to two. Assuming that this resource increase is acceptable, the datapath design can continue with register scheduling.

3.16 REGISTER SCHEDULING

Register scheduling determines how temporary results are stored in registers. This can be a complex problem if a minimum number of registers are desired as the execution unit schedule also affects the register count; fortunately registers are relatively inexpensive in terms of gate count. There may be good reasons for not using the minimum number of registers; for example, it may be desirable for the register containing a previous output result to keep this value stable throughout the computation of the new result in case it is being used by a downstream datapath. Also, using the minimum number of registers may increase the multiplexer depth in front of registers, thus creating longer t_{R2R} paths. Our register scheduling methodology only determines the registers needed for a particular execution unit schedule, and does not attempt to modify the execution unit schedule to reduce the register count.

Our register scheduling methodology begins by examining the register storage requirements of each clock as shown in Table 3.15. The *Initial* column lists the data values that are present within the datapath at the beginning of the clock cycle. The *Produced* column lists the data values that are

TABLE 3.14: Schedule for Figure 3.40 using Two Multipliers, One Adder for Target latency = 4, target initiation period = 4

CLOCK	RESOURCES				
	INPUT	MULT (U1)	MULT (U2)	SATADD (U3)	OUTPUT
0	x(0)	n3(0)	n4(0)		
1		n1(0)	n2(0)	n6(0)	
2				n5(0)	
3				n7(0)	
4(i+0)	x(i)	n3(i)	n4(i)		y(i-1)
4(i+1)		n1(i)	n2(i)	n6(i)	
4(i+2)				n5(i)	
4(i+3)				n7(i)	
%utilization	25%	50%	50%	75%	25%

either produced by computations or input to the datapath during the cycle and saved for a future clock cycle. For example, in clock cycle $i + 0$, the x value in the *Produced* column is input by the datapath during that cycle and must be saved as it becomes the *x@1* value in the next dataset computation. The *Consumed* column lists items from the *Initial* column that are no longer needed after this clock cycle. The *Total Registers* column is the total number of registers needed during that clock cycle, and is computed as *Initial* + *Produced* – *Consumed*, as registers whose values are consumed can now be used to store new values. The maximum register count in the *Total Registers* column is the number of registers required by the datapath for this schedule; in this case it is seven registers. This does not include the registers required for coefficients b_0, b_1, b_2, and b_3 as they are loaded during the initialization phase and do not change during the computation loop. The total number of datapath registers is 11 (7 + 4) once the coefficient registers are included. Observe that the scheduling of node operations in Table 3.14 affects the number of registers required for a particular clock cycle. For example, if node operations *n1*, n2 were scheduled in clock $i + 0$ instead of nodes *n3*, *n4*, then the *x*@3 value would not be consumed in clock cycle $i + 0$, and the register count for that clock cycle would be seven. This does not increase the maximum number of registers for this datapath, but this may not be true for other datapaths.

TABLE 3.15: Number of required registers by clock cycle

CLOCK	REGISTER REQUIREMENTS			
	(1) INITIAL	(2) PRODUCED	(3) CONSUMED	TOTAL REGISTERS COLUMNS(1 +2 -3)
4(i+0)	x@1, x@2, x@3, y(i-1)	x, n3, n4	x@3	6
4(i+1)	x, x@1, x@2, n3, n4, y(i-1)	n1, n2, n6	n3, n4	7 (max value)
4(i+2)	x, x@1, x@2, n1, n2, n6, y(i-1)	n5	n1, n2	6
4(i+3)	x, x@1, x@2, n5, n6, y(i-1)	y(i)	n5, n6, y(i-1)	4

The registering requirements of Table 3.15 can be mapped to specific registers on a clock-by-clock basis as shown in Table 3.16. The seven registers identified in Table 3.15 are named *rA*, *rB*, *rC*, *rD*, *rE*, *rF*, and *rY*, with the register contents corresponding to the *Initial* and *Produced* columns of Table 3.15. If a register's content is changed during a clock cycle, then this is indicated by a register write operation such as "n3→rD" (the result of operation n3 is written to register *rD*) or "*rE*→*rA*" (the contents of register *rE* is written to register *rA*). This write operation is shown because this translates into a load line assertion for this register in the finite state machine control of the datapath. If a register's contents is no longer required after a clock cycle, then that table cell is shown as blank even though the register's contents has not physically changed (i.e., the n6 computation result in register *rF* is consumed in clock $i + 3$ and no new value is written to register *rF*, so the table cell entry for *rF* is blank in clock $i + 3$ even though the n6 computation result is still physically present). The *initial* row shows the assumed register contents at the beginning of the $i + 0$ clock cycle; the assignments of *x@*1, *x@*2, *x@*3 to registers *rA*, *rB*, *rC* is an arbitrary choice. Observe that register transfers in clock $i + 3$ such as "*rE*→*rA*" that writes the current *x* value to *rA* is done to get ready for the next set of computations, as *x* becomes *x@*1, *x@*1 becomes *x@*2, and *x@*2 becomes *x@*3. The register choices made in Table 3.16 affects the multiplexing requirements of the datapath; in this methodology we do not attempt to optimize the register assignments in order to reduce the multiplexing.

The execution unit scheduling of Table 3.14 and the register content scheduling in Table 3.16 is now combined into one table that completely specifies the datapath operation, as shown in Table 3.17. The execution unit operations are now specified asRTL, such as "*rC* * *b*3→*rC*" for the n4 computation done in clock $i + 0$. The table also contains a column that contains register to register transfers such as "*rE*→*rA*". Observe that the choice of a particular unused register for storing a result affects the multiplexing needed for a register input. For example, the n1 and n3 computations are both written to register *rC*, while n2 and n4 are written to register *rD*. From Table 3.17, it is seen that register *rD* receives results only from multiplier unit *u2*, and thus does not require a multiplexer on its input. However, in clock cycle $i + 0$ if register *rD* had been chosen for computation n3, and *rC* for computation n4, then register *rD* would receive results from both the *u1* and *u2* multiplier units, requiring a multiplexer on the *rD* register input. After creating initial versions of the execution unit scheduling, register scheduling, and combined execution unit/register scheduling tables, the multiplexing requirements become visible and changes can be made to register assignments to reduce the number of multiplexors in the datapaths. It should be noted that high-level synthesis tools exist that perform these optimizations automatically.

The datapath and FSM implementation of the scheduling in Table 3.17 is shown in Fig. 3.22. The FSM control such as register load signals and multiplexer select signals are not shown in the datapath; the presence of these signals is assumed. Datapath diagrams such as Fig. 3.22a quickly become unwieldy as the datapath complexity increases and are also not strictly necessary, as the

TABLE 3.16: Register contents by clock cycle

CLOCK	REGISTER CONTENTS						
	RA	RB	RC	RD	RE	RF	RY
initial	x@1	x@2	x@3				y(i-1)
4(i+0)	x@1	x@2	n3 (n3→rC)	n4 (n4→rD)	x (x→rE)		y(i-1)
4(i+1)	x@1	x@2	n1 (n1→rC)	n2 (n2→rD)	x	n6 (n6→rF)	y(i-1)
4(i+2)	x@1	x@2	n5 (n5→rC)		x	n6	y(i-1)
4(i+3)	x (rE→rA)	x@1 (rA→rB)	x@2 (rB→rC)				y (n7→rY)

TABLE 3.17: Combined execution unit and register scheduling

CLOCK	DATAPATH OPERATIONS					
	INPUT	MULT (U1)	MULT (U2)	SATADD (U3)	OUTPUT	REGISTER TRANSFERS
$4(i+0)$	x(i)	n3(i) rB * b2→rC	n4(i) rC * b3→rD		y(i-1)	x →rE
$4(i+1)$		n1(i) rE * b0 →rC	n2(i) rA * b1→rD	n6(i) rD + rC →rF		
$4(i+2)$				n5(i) rD + rC →rC		
$4(i+3)$				n7(i) rF + rC →rY		rE→rA rA→rB rB→rC

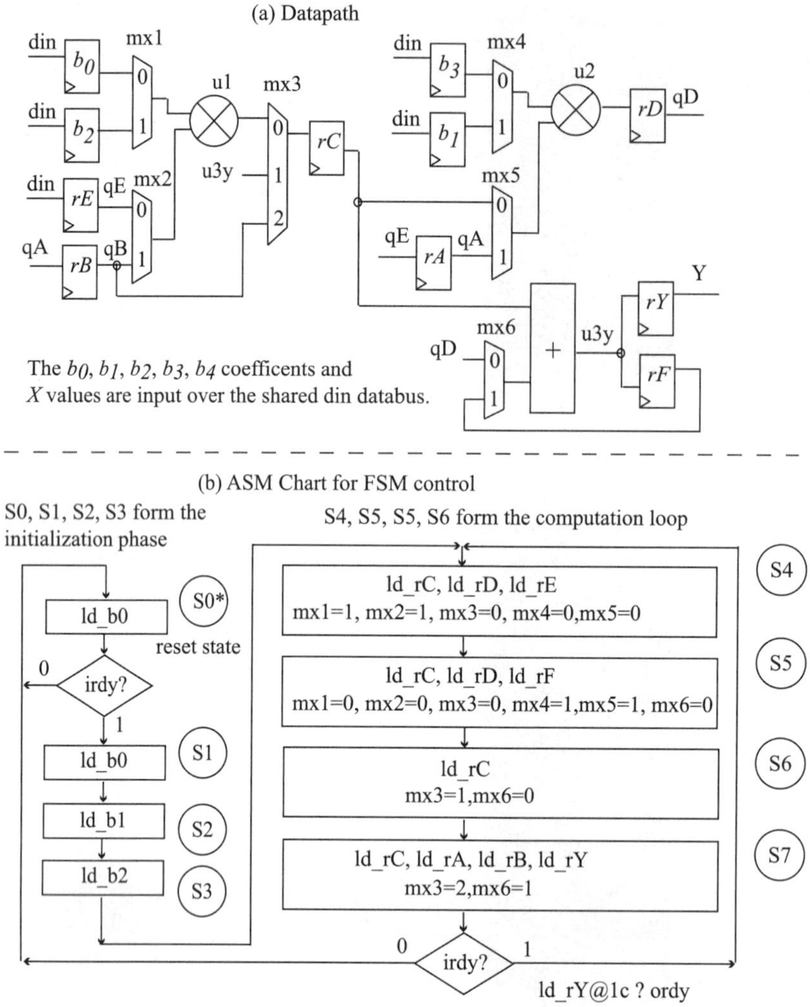

FIGURE 3.22: Datapath, FSM for implementation using Table 3.7 scheduling.

scheduling operations in Table 3.17 specify datapath operations. The Verilog code that implements the datapath is the final representation of the datapath operation, with datapath diagrams only used as an aid for visualizing the components and their interconnection that comprise the datapath. The FSM control is comprised of eight states; four states for the initialization of the coefficient registers, and four states for the compute loop. The assignments of registers to the mx1, mx2 and mx4, mx5 multiplexer inputs were done so that the select lines of these two pairs of multiplexors can be connected together. Thus, the number of multiplexer select signals in the ASM chart can be reduced from what is shown, as the mx1, mx2 and mx4, mx5 signals have the same values in each state and thus each pair can be driven by one signal. The assignments of inputs to the mx3 and mx6 multiplexer were arbitrarily chosen.

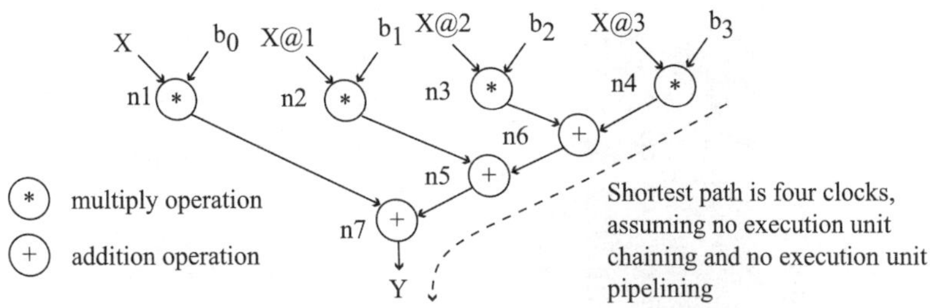

FIGURE 3.23: Restructured flowgraph for equation 3.5.

3.17 FLOWGRAPH TRANSFORMATIONS, OVERLAPPED COMPUTATIONS REVISITED

In the previous section, two multipliers were required for a latency = 4, initiation period = 4 solution to the flowgraph of Fig. 3.21.

Table 3.18 shows an attempt to remove one of the multipliers by increasing the target latency from four clocks to five clocks. However, the schedule fails because the last addition operation, n7, is not scheduled within the target latency.

For the scheduling to succeed with a latency of five clocks and only one multiplier, the three multiplication operations have to begin in clock cycle #2, with one multiplication done per clock cycle. Fortunately, the multiply-accumulate operations in Eq. (3.5) are associative, allowing the flowgraph to be *restructured* as shown in Fig. 3.23.

This illustrates the dependency of scheduling on flowgraph structure; automated high-level synthesis tools will restructure a flowgraph when searching for a scheduling solution that meets target latency and target initiation period constraints.

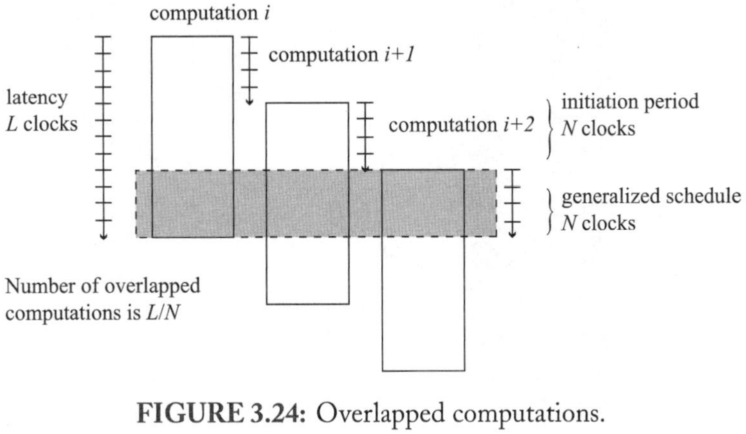

FIGURE 3.24: Overlapped computations.

TABLE 3.18: Schedule for Figure 3.40 using one multiplier, one adder for target latency = 5, target initiation period =5

CLOCK	RESOURCES			
	INPUT	MULT (U1)	SATADD (U2)	OUTPUT
$5(i+0)$	x(i)	n4(i)		
$5(i+1)$		n3(i)		
$5(i+2)$		n2(i)	n6(i)	
$5(i+3)$		n1(i)		
$5(i+4)$			n5(i)	
	Scheduling fails, operation n7 is not scheduled within target latency.			

TABLE 3.19: Schedule for Figure 3.44 using one multiplier, one adder for target latency = 5, target initiation period = 5

CLOCK	RESOURCES			
	INPUT	MULT (U1)	SATADD (U2)	OUTPUT
5(i+0)	x(i)	n4(i)		y(i-1)
5(i+1)		n3(i)		
5(i+2)		n2(i)	n6(i)	
5(i+3)		n1(i)	n5(i)	
5(i+4)			n7(i)	
%utilization	20%	80%	60%	20%

3.18 OVERLAPPED COMPUTATIONS REVISITED

As stated earlier, overlapping computations for input datasets increases throughput usually at the cost of additional resources. The methodology of Table 3.10 can also be used for overlapped computations. When determining the initiation period (N) and latency (L) constraints for overlapped computations, the initiation period should be evenly divisible into the latency. The latency divided by the initiation period (L/N) is the number of *overlapped computations* in the design, and the generalized schedule is equal to the initiation period of N clocks.

As an example, choose a target latency of four clocks, and a target initiation period of two clocks for the flowgraph of Fig. 3.21. Using Eq. (3.6), the lower bounds on the multiplier and adder execution units are given in Eqs (3.12) and (3.13).

$$\text{\# of multipliers} = \left\lceil \frac{4}{2} \right\rceil = 2 \tag{3.12}$$

$$\text{\# of adders} = \left\lceil \frac{3}{2} \right\rceil = 2 \tag{3.13}$$

The number of overlapped computations is 4/2 = 2, so the generalized schedule contains computations for datasets i and $i-1$, and the output value for computation $i-2$. Table 3.20 shows that scheduling succeeds for latency = 4 and initiation period = 2 using the lower bound estimates for execution units. Observe that the operations mapped to execution units are chosen such as to repeat the same operations on the execution unit for the initiation period's two clocks in order to reach a

TABLE 3.20: Schedule for Fig. 3.21 using two multipliers, two adders for target latency = 4, target initiation period =2

CLOCKS	RESOURCES					
	INPUT	MULT (U1)	MULT (U2)	SATADD (U3)	SATADD (U4)	OUTPUT
0	x(0)	n3(0)	n4(0)			
1		n1(0)	n2(0)	n6(0)		
2	x(1)	n3(1)	n4(1)	n5(0)		
3		n1(1)	n2(1)	n6(1)	n7(0)	
4	x(2)	n3(2)	n4(2)	n5(1)		y(0)
5		n1(2)	n2(2)	n6(2)	n7(1)	
2(*i*+0)	x(i)	n3(i)	n4(i)	n5(i-1)		y(i-2)
2(*i*+1)		n1(i)	n2(i)	n6(i)	n7(i-1)	
%utilization	50%	100%	100%	100%	50%	50%

generalized schedule. For example, it would not work to schedule the n6, n5, and n7 operations all on the u3 adder as this cannot be repeated within the two clocks of the initiation period.

Table 3.21 show that the temporary registers required by this schedule is eight, so the total number of registers needed for the datapath, including the four coefficient registers, is 12. Assuming the clock period remains the same, doubling the throughput has only cost one additional register and one extra adder. The reason for this small increase in resources is because of the low *%utilization* of the resources in the latency = 4, initiation period = 4 solution of Table 3.14.

The remaining detailed register scheduling and datapath design is left as an exercise for the reader.

3.19 SUMMARY

DFGs are useful tools for visualizing the data dependencies of a computation. Latency and initiation period constraints determine the number of registers and execution units required to implement a particular computation. A scheduling table is used to map computations to available execution units and registers. Overlapped computations and pipelined executions are both useful techniques for increasing the throughput of a datapath.

3.20 SAMPLE EXERCISES

1. Create a Verilog implementation of the datapath in Fig. 3.18.
2. Create a Verilog implementation of the datapath in Fig. 3.20.
3. Design a datapath with latency = 3, and initiation period = 3 for the DFG of Fig. 3.19 using multiplier units with one pipeline stage; use the minimum number of adder and multiplier units that meets these constraints.
4. Modify the schedule of Table 3.8 for latency = 4, initiation period = 2 and do a Verilog implementation of the datapath.
5. Do a Verilog implementation of the datapath in Fig. 3.22.
6. Use Table 3.20 and Table 3.21 to complete a schedule that contains all of register transfer operations for this datapath, create the ASM chart for the required FSM control, and implement the datapath in Verilog.

Equation 3.14 implements an operation known as *bilinear filtering* in which a new color C_{new} is produced from four colors C_{00}, C_{01}, C_{10}, C_{11} using two blend factors, u and v. As an example, if $v = 0.5$ and $u = 0.5$, then C_{new} is an equal blend of each color ($C_{new} = 0.25^*C_{00} + 0.25^*C_{01} + 0.25^*C_{10} + 0.25^*C_{11}$). The data types and operations in Eq. 3.14 are the same as in the blend equation. The colors are 0.8 fixed-point values, while u, v are nine-bit values encoded in the same manner as F in

TABLE 3.21: Number of required registers by clock cycle

CLOCK	REGISTER REQUIREMENTS			
	(1) INITIAL	(2) PRODUCED	(3) CONSUMED	TOTAL REGISTERS (1 +2 -3)
2(i+0)	x@1, x@2, x@3, n1(i-1), n2(i-1), n6(i-1), y(i-2)	x, n3(i), n4(i),n5(i-1)	x@3,n1(i-1),n2(i-1)	7+4-3 = 8 max value
2(i+1)	x, x@1, x@2, n3(i), n4(i), n5(i-1), n6(i-1),y(i-2)	n1(i), n2(i), n6(i),y(i-1)	n3(i),n4(i),n5(i-1), n6(i-1),y(i-2)	8+4-5 = 7

the blend equation.

$$\begin{aligned} C_{new} &= C00 \times (1 - v) \times (1 - u) + C01 \times (1 - v) \times u \\ &+ C10 \times v \times (1 - u) + C11 \times v \times u \end{aligned} \tag{3.14}$$

Figure 3.25 shows a DFG for Eq. (3.14) that assumes a single nine-bit input databus, with the u, v blend factors input during the datapath initialization phase and multiple four-tuples of C_{00}, C_{01}, C_{10}, and C_{11} input during the computation loop for use with these blend factors. The square boxes around C_{01}, C_{10}, and C_{11} and the arrows linking C_{00}, C_{01}, C_{10}, and C_{11} indicate that these are input operations over a shared input bus.

The following questions reference Eq. (3.14) and Fig. 3.25. Use the minimum number of execution units in all implementations.

7. Using the methodology of Table 3.10, design a datapath that has latency = 6 clocks and initiation period = 6 clocks. Assume that C_{00}, C_{01}, C_{10}, and C_{11} are available in successive clock cycles in the first four clocks of the initiation period.
8. Using the methodology of Table 3.10, design a datapath that has latency = 8 clocks and initiation period = 4 clocks. Assume that C_{00}, C_{01}, C_{10}, and C_{11} are available in successive clock cycles in the four clocks that comprise the initiation period.
9. If multiplier units with one pipeline stage are used in, then the shortest path becomes eight clocks. Using multiplier units with one pipeline stage, design a datapath that has latency = eight clocks and initiation period = eight clocks.

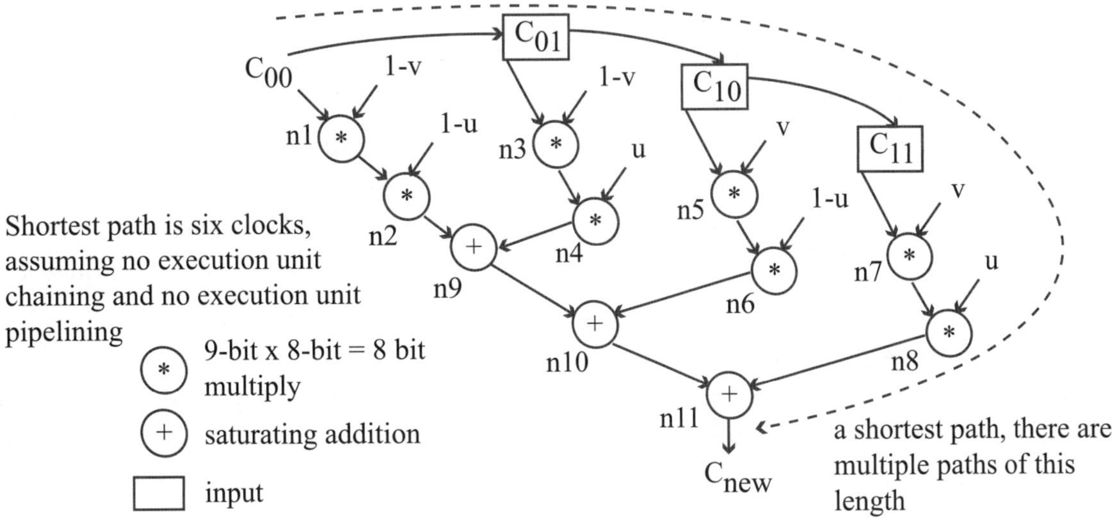

FIGURE 3.25: Dataflow Graph for Equation 3.14

10. If multiplier units with one pipeline stage are used in, then the shortest path becomes eight clocks. Using multiplier units with one pipeline stage, design a datapath that has latency = eight clocks and initiation period = four clocks.

APPENDIX: IS DATAPATH SCHEDULING A VALID TOPIC FOR MODERN DIGITAL SYSTEM DESIGN?

This chapter discusses datapath scheduling at length; it may be argued that with the high gate counts of modern FPGAs, the need for resource sharing has passed and that modern designs are mostly done in a parallel, fully pipelined manner to emphasize throughput. Another argument can be made that individual multipliers and adders are passé when tools like Xilinx Coregen can automatically generate a 1024-Point Complex Fast Fourier Transform block or the AccelDSP tool can generate a Verilog implementation for an arbitrary Matlab function.

It is the author's contention that a fundamental grounding in the concepts of latency and throughput in relation to computational intensive FSM/datapaths is important, even in the context of modern FPGAs that can contain hundreds of thousands to millions of gates. At some point, a designer will be concerned with latency/throughput of a design, and the gate count tradeoffs associated with latency/throughput. A modern designer may be using double-precision floating-point units as executions units instead of fixed-point adders and multipliers, but the finite state machine task of sequencing operations on those units and storing intermediate results will remain unchanged. Furthermore, a modern designer is usually part of a team, and may be given the task of generating a computation block to be used in a much larger design, and will probably be given latency, throughput, and clock speed constraints on that design.

Finally, even if a modern designer has a high level synthesis tool that can automatically generate an RTL design from a high-level language description in C (or some other programming language), it is important that the designer has a firm understanding of latency/throughput and clock speed because they will almost certainly be the constraints given to the high level synthesis tool when generating the design.

Datapath scheduling/RTL coding versus high-level synthesis and pre-generated IP blocks can be likened to programming in assembly language versus programming in a high level language. A modern programmer may never have the need to program in assembly language. However, it can be assured that the programmer has training in assembly language in order to understand the linkage between a high level language (HLL) and the target assembly language, and to understand the role that a compiler plays in the transformation of HLL code to assembly, and effect of HLL data types and compiler code optimizations on resulting code size and execution speed. Also, if nobody understands assembly language (datapath scheduling/RTL), who will write the compilers (write high-level synthesis tools or build embedded IP blocks)?

3.21 REFERENCES

[1] Kai Hwang, Computer Arithmetic Principles, Architecture and Design, Wiley, 1979.

[2] S. S. Bhattacharya, P.K. Murthy et al., Software Synthesis from Dataflow Graphs, Kluwer Academic Publishers, 1996.

[3] Sumit Gupta, Rajesh Gupta et al. SPARK:: A Parallelizing Approach to the High-Level Synthesis of Digital Circuits, Springer 2005, pp 262.

CHAPTER 4

Embedded Memory Usage in Finite State Machine with Datapath (FSMD) Designs

This chapter explores usage of different types of embedded memories such as read-only memories (ROMs), single-port random access memories (RAMs), first-in first-out buffers (FIFOs), and dual-port RAMs in finite state machine with datapath designs.

4.1 LEARNING OBJECTIVES

After reading this chapter, you will be able to perform the following tasks:

1. Discuss the operational differences between synchronous and asynchronous embedded memories, and between single-port, dual-port, and FIFO memories.
2. Implement FSM/datapaths that incorporate single-port synchronous RAMs.
3. Discuss application scenarios for FIFOs and dual-port memories.
4. Use two-phase and four-phase handshaking for data transfer.
5. Use a two-flop synchronizer for asynchronous input synchronization.

4.2 INTRODUCTION TO EMBEDDED MEMORIES

Modern FPGAs have various types of embedded memories available for designer usage. A simple type of embedded memory block is the asynchronous $K \times N$ ROM, as shown in Fig. 4.1a. This memory is labeled as *asynchronous* because there is no clock signal for controlling access to the memory's contents. The memory is labeled as *read-only*, because its contents are fixed at FPGA configuration time; there is no method by which the application can modify the memory's contents. The $K \times N$ parameters give the memory's *organization*; the memory has K locations with each location containing N bits, thus providing a total data storage of $K \times N$ bits. An *address* bus, labeled as `addr`, is used to access the memory's contents; the width of the address bus is $\log_2(K)$. The output data bus, labeled as `dout`, carries the data of the memory location specified by the address bus. An asynchronous ROM is a combinational logic device; the output (`dout`) changes after some delay from an input (`addr`) change. This propagation delay from a change in address value to a stable data output value is the memory's *access time* (T_{ACCESS}). In general, larger embedded memories have

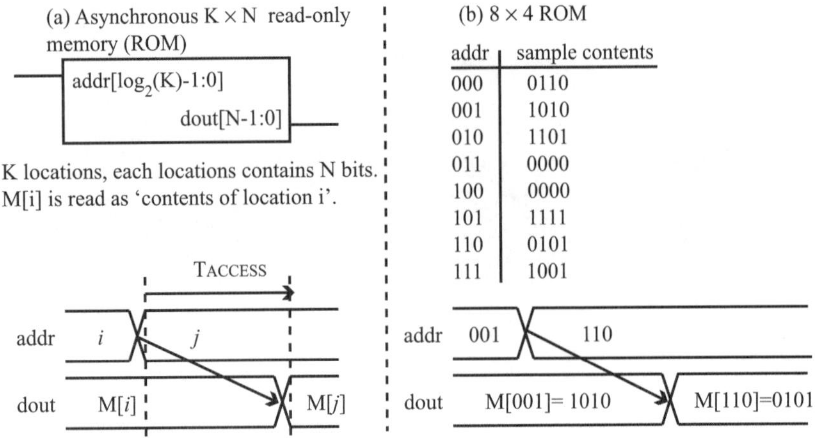

FIGURE 4.1: Asynchronous $K \times N$ read-only memory (ROM).

longer access times. Figure 4.1b shows sample contents for an 8×4 ROM; this memory requires a three-bit address bus ($\log_2(K)$) and a four-bit data output bus.

Figure 4.2 shows a synchronous version of a $K \times N$ ROM. DFFs are placed on the address inputs (i.e., these inputs are *registered*), thus latching the address inputs on a rising clock edge. The data output bus is available in both registered and unregistered versions. A designer might use the registered `dout` version if the ROM's access time is large and the designer did not want the ROM's access time summed with the datapath delay that follows the ROM's output. This is similar to the methodology used in Chapter 3 in which registers are placed between execution units (adders, multipliers) to break long combinational paths, reducing critical path length and increasing system clock frequency. The tradeoff associated with using the registered `dout` bus is a clock cycle of latency for data access; the registered `dout` value in the current clock cycle corresponds to the memory contents of the address bus value latched on the rising clock edge of the previous clock cycle. By contrast, the unregistered `dout` bus contains the memory contents of the address value latched on the rising clock edge of the current clock cycle. The registered `dout` value is available at T_{cq} propagation delay after the rising clock edge; T_{cq} is less than T_{ACCESS} time. It should be noted that the availability of both registered and unregistered `dout` buses in synchronous embedded memories is a design decision made by the FPGA vendor and thus will vary by FPGA vendor and by FPGA family. In this text, the assumption is made that both registered and unregistered `dout` buses are available.

Random Access Memory (RAM) is an embedded memory block whose contents can be modified under application control. Figure 4.3 shows an asynchronous $K \times N$ RAM; the additional signals on this embedded memory block when compared to the asynchronous ROM of Fig. 4.1 are the data input bus (`din`) and write enable (`we`) input. New data on the `din` bus is written to the current address location when the `we` enable signal experiences a high-to-low transition; there

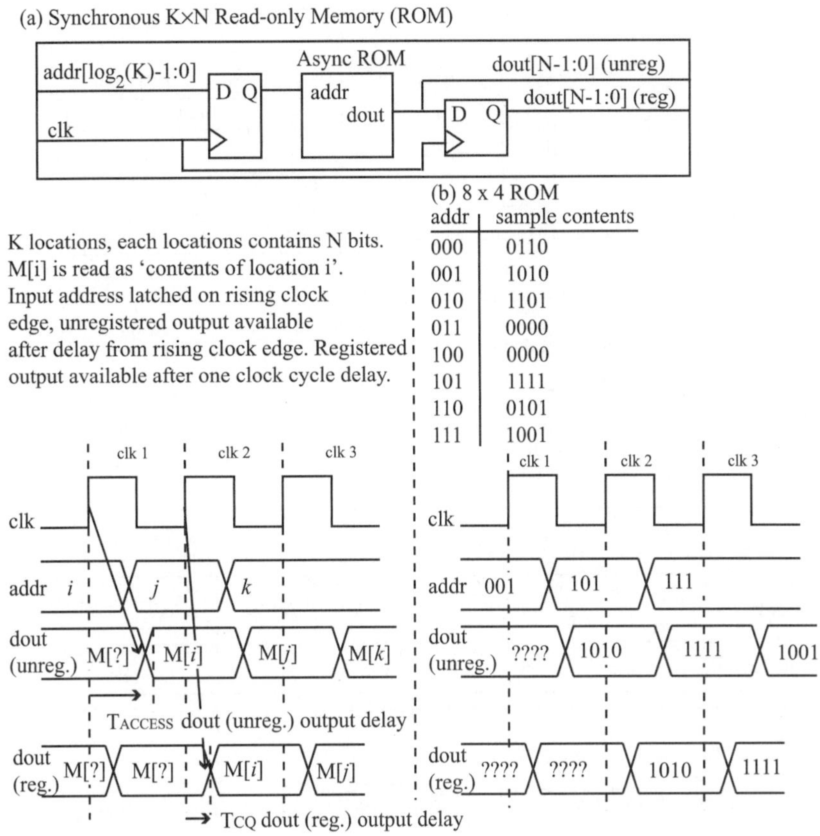

FIGURE 4.2: Synchronous $K \times N$ read-only memory (ROM).

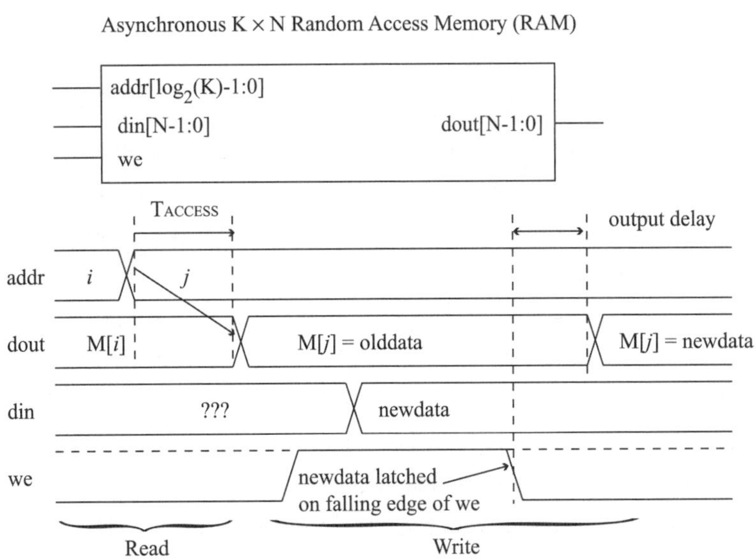

FIGURE 4.3: Asynchronous $K \times N$ random access memory (RAM).

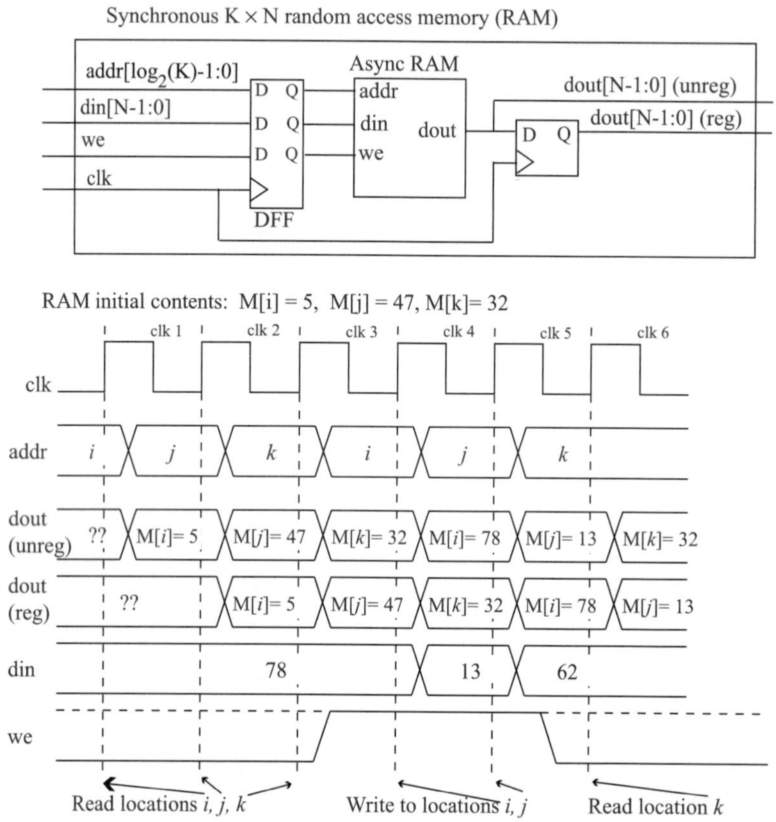

FIGURE 4.4: Synchronous $K \times N$ random access memory (RAM).

is also a minimum high pulse-width requirement on the we signal with setup (tsu) and hold (thd) constraints for `din` on the falling we edge.

Figure 4.4 shows read and write operations for a synchronous $K \times N$ RAM. The read operation for a synchronous RAM is the same as for a synchronous ROM, the address input is latched on the rising clock edge and output data is available either as an access time later (unregistered `dout`) or a T_{cq} time after the next rising clock edge (registered `dout`). In Fig. 4.4, clock cycles four and five demonstrate write operations to the RAM. The `addr`, `din`, and `we` inputs are latched on the rising clock edge; a logic one value on the `we` signal indicates a write operation. Location *i* is written with the value 78 (`din` bus value) in clock cycle four; observe that the unregistered `dout` bus reflects this new value as t_{pd} (at least T_{ACCESS} and it may be longer depending on the memory) after the rising clock edge of clock cycle four. Location *j* is written in clock five with the value 13. The `din` bus value does not affect memory operation when `we` is negated.

Synchronous RAMs are almost always preferred over asynchronous RAMs in designs in order to avoid problems with timing uncertainty during write operations. Figure 4.5 shows a finite state machine (FSM) connected to an asynchronous RAM, with the timing diagram illustrating a write

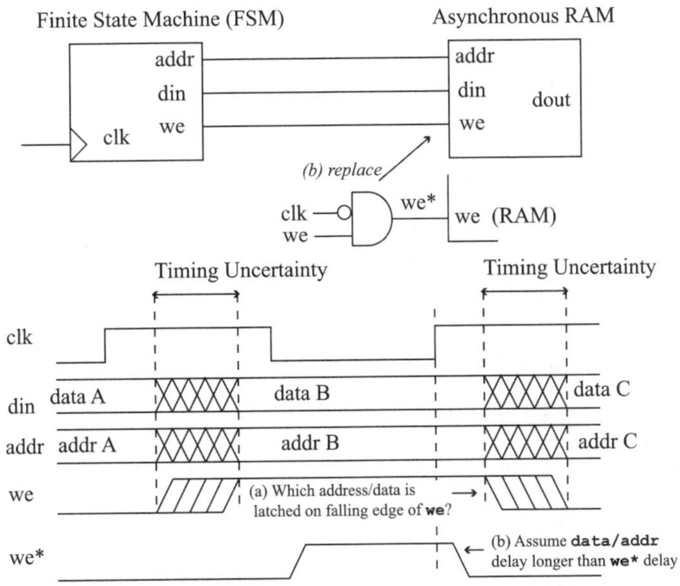

FIGURE 4.5: A problem with using an asynchronous RAM with a FSM.

operation. New `addr`, `din`, and `we` values are provided by the FSM some delay after the rising clock edge. This delay is dependent on how the signals are generated by the FSM (registered only, or registered plus combinational encoding) as well as wiring delays between the FSM and the RAM. Wire delay in FPGAs can be significant and can also vary significantly depending on the number of programmable switches that a signal passes through between the blocks. This timing uncertainty is problematic during a write operation as the input data and address values that are latched on the falling edge of the write enable signal are unknown. This problem is sometimes attacked by AND'ing the `we` signal from the FSM with the inverse of the clock signal and using this new signal (`we*`) as the RAM `we`. However, this approach relies on the assumption that the address and data input bus values have a longer delay than `we*`, which is an assumption that may not be true and whose timing may be violated if routing delays change between the FSM block and RAM.

The timing problems in Fig. 4.5 can be avoided by using a synchronous RAM, as shown in Fig. 4.6. The `data/addr/we` signals to the synchronous RAM only have to satisfy t_{su} and t_{hd} relative to the rising clock edge. The timing uncertainty for these signals can be an issue for t_{hd} , but t_{pd} of the `data/addr/we` signals after the rising clock edge is typically much larger than the RAM t_{hd},which is either zero or very small. The astute reader may observe that because a synchronous RAM is an asynchronous RAM with registered inputs, the race condition between the `addr/din` signals and the `we` signal is simply moved inside the synchronous RAM block. This is true, but it is the responsibility of the synchronous RAM designer to solve this timing problem, and it is not an issue for a designer who wishes to use synchronous RAM blocks since correct operation is guaranteed as long as the input t_{su} and t_{hd} are met.

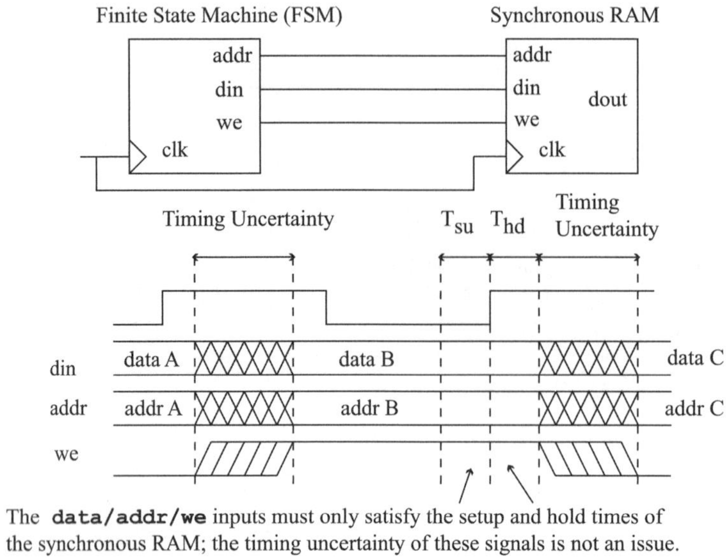

FIGURE 4.6: Using a synchronous RAM with a FSM.

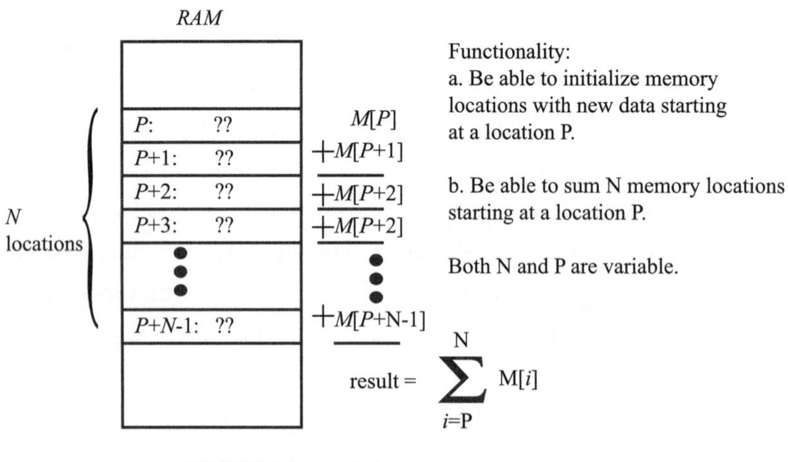

FIGURE 4.7: Memory sum overview.

4.3 SAMPLE APPLICATION: MEMORY SUM

Figure 4.7 gives an overview of a simple application used to illustrate a datapath design that contains an embedded synchronous RAM. The datapath's functionality consists of two operation modes:

- *Initialization*: the datapath initializes the RAM's content's starting at a location P. Both P and initialization data are provided from an external input data bus.
- *Computation:* the datapath sums the contents of N locations, starting at location P. Both N and P are specified by an external input data bus, and with the result given on an external output data bus.

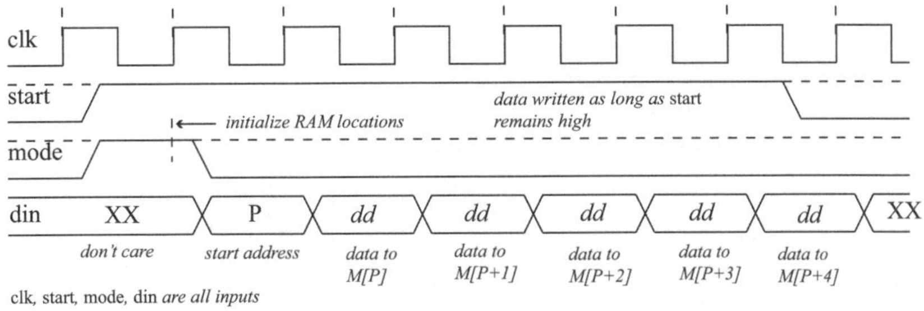

FIGURE 4.8: Initialization mode timing specification.

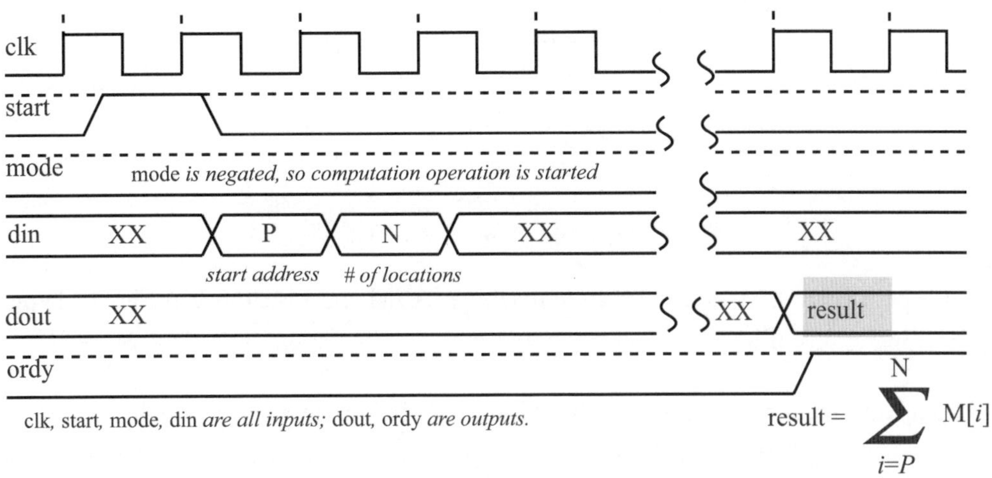

FIGURE 4.9: Computation mode timing specification.

Datapath operation is controlled by assertion of a `start` input, with a `mode` input determining if initialization or computation is performed.

The cycle timing specification for the initialization operation is shown in Fig. 4.8. The combination of `start` = 1 and `mode` = 1 causes the initialization operation to begin. The starting address *P* for the initialization operation is provided on the `din` input bus in the clock cycle following `start` assertion. Memory locations *M*[*P*], *M*[*P*+1], *M*[*P*+2], etc., are written in successive clock cycles with data provided on `din`; locations are written as long as `start` is asserted (Figure 4.8 shows writes to only four locations; more locations could have been written). The negation of `start` signals the end of the initialization operation.

Figure 4.9 gives the computation mode timing specification. The start address (*P*) and number of locations to sum (*N*) are provided in the first two clock cycles after `start` assertion with `mode` = 0. At some later time, the output ready (`ordy`) output is asserted by the datapath when the result is available on the `dout` data bus. The number of clock cycles required for the computation is implementation dependent.

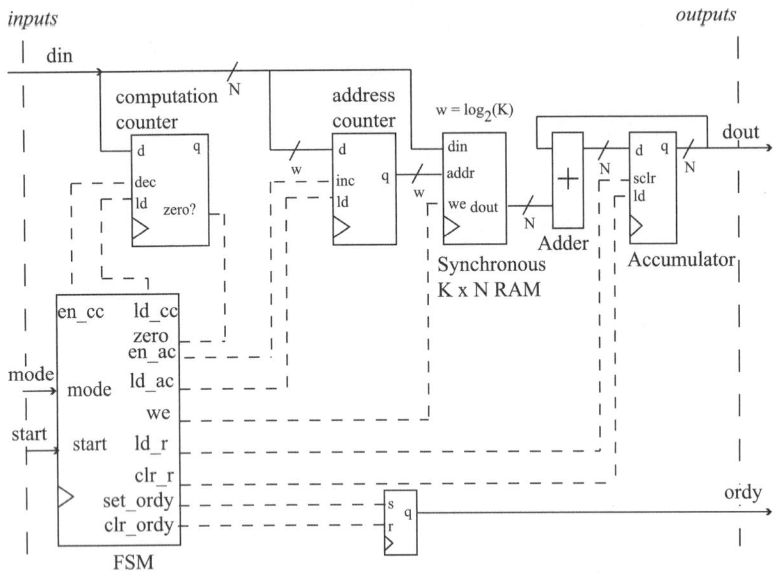

FIGURE 4.10: Memory sum datapath.

A datapath (Fig. 4.10) and finite state machine (ASM chart is shown in Fig. 4.11) performs the required operations of initialization and computation. The datapath particulars are:

- The *address* counter provides the RAM address; it is used to sequentially access memory locations during both initialization and computation operations. The counter is loaded with (P) at the start of both operation modes, and has an increment by-one functionality.
- The *computation* counter tracks the number of locations remaining to be summed during computation operation and is loaded with (N) to be summed at the beginning of this operation. The computation operation is halted when the count value reaches zero. The counter has a decrement by one functionality.
- The adder coupled with an output register provides an *accumulator* functionality, that is, successive additions add the register value with the contents of the currently accessed memory location. The register has a synchronous clear function since the register value must be zero for the first addition. The `dout` bus is the accumulator output.
- A synchronous $K \times N$ RAM is used as the embedded memory block.
- A set/reset flip-flop (SRFF) is used to implement the output ready (`ordy`) signal; an SRFF is useful when a signal must be asserted for several clock cycles.

The FSM sequences the actions on the datapath according to the ASM chart given in Fig. 4.11. State S0 waits for `start` assertion, and then branches to the first states of the initialization operation or computation operation based on the `mode` input.

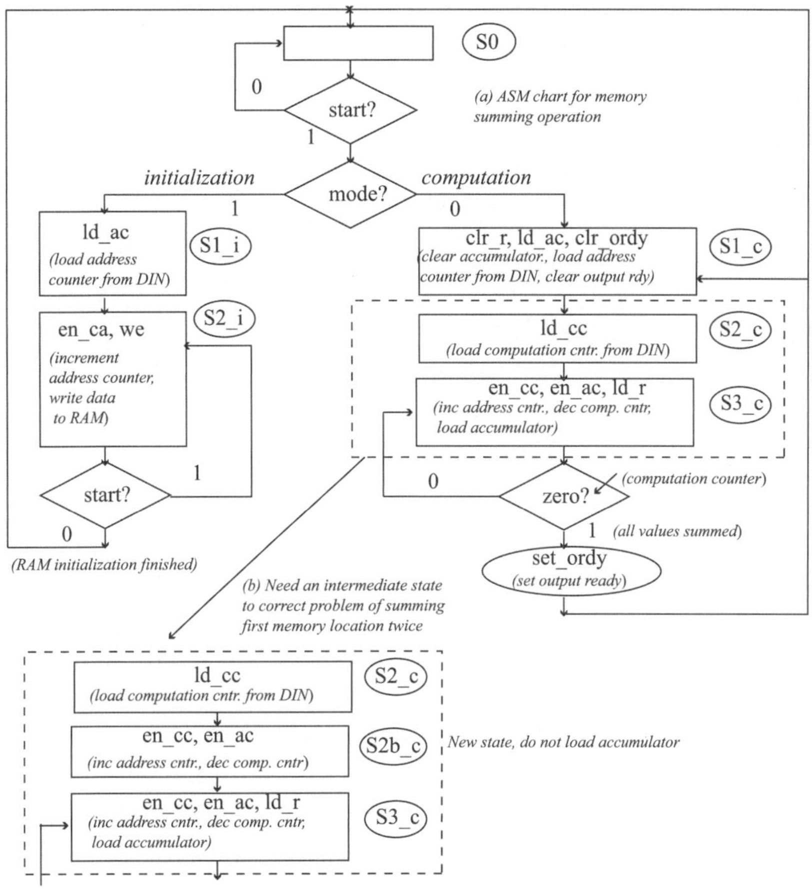

FIGURE 4.11: Memory sum ASM chart.

The initialization operation is straightforward. The first state S1_i loads the starting address into the address counter by asserting the address counter's `ld` input. The second state S2_i writes data values in the RAM by asserting the RAM's write enable; the input data is provided on the `din` data bus. The address counter is incremented in `S2_i` by assertion of the address counter's `inc` input. State S2_i returns to state S0 when `start` is negated. Fig. 4.12 is a timing diagram for the initialization operation with example data, and contains both external and internal signals. Data is written to locations four through eight on the leading rising edges of clocks four through eight. Observe that even though `start` is negated in clock cycle seven, the data in this clock cycle is written to RAM as specified in Fig. 4.8.

Two versions of the computation operation are provided— an incorrect version of three states (S1_c, S2_c and S3_c) and a correct version of four states (S1_c, S2_c, S2b_c, and S3_c). The incorrect version appears to be a straightforward implementation of the computation operation of Fig. 4.9 in that the starting address and locations to be summed are captured in states S1_*c* and S2_*c*, with state

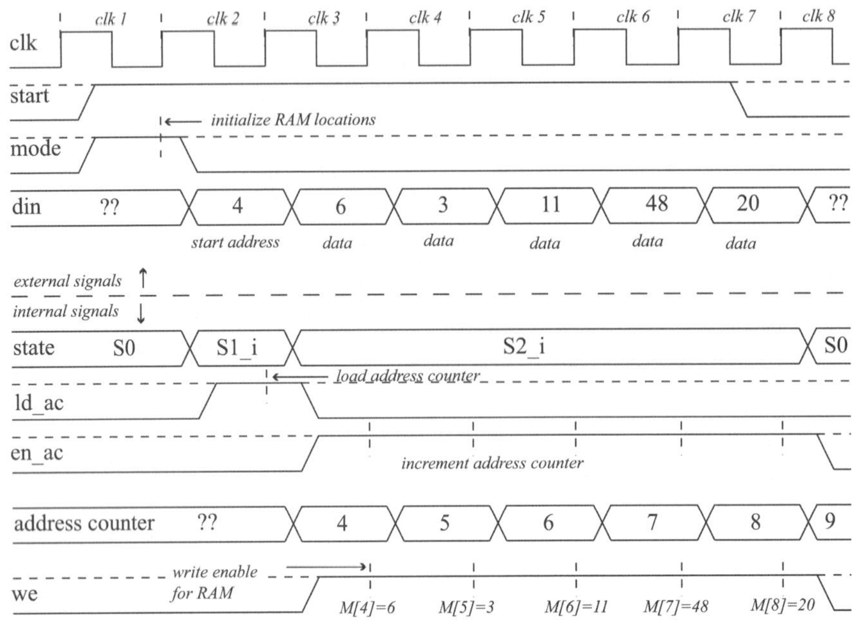

FIGURE 4.12: Initialization operation showing both external and internal signals for sample data.

S3_*c* is used to sum the memory contents. However, Fig. 4.13 illustrates the reason for the incorrect behavior by attempting to sum two locations, starting at location five. In the first clock cycle of state S3_*c* (clock 4), the memory dout bus contains $M[5] = 3$, the accumulator value is zero, and the adder output is $3 + 0 = 3$. The accumulator load signal is asserted in S3_*c*, so in clock five the accumulator becomes three, and the address counter is incremented to location six. However, even though the address counter value is now six, this value is not latched into the RAM until the next clock cycle, and thus the RAM `dout` remains at $M[5] = 3$ for clock five. This means that at the end of clock five, the new value loaded into accumulator is $3 + 3 = 6$, causing the first location to be included twice in the accumulated sum. The next clock produces $M[6] = 11 + 6$ for a final result of 17, which is incorrect. The correct result should be $M[5] + M[6] = 3 + 11 = 12$.

There are multiple ways to correct the errant behavior of Fig. 4.13; one solution is to not assert the accumulator load line in the first clock cycle after state S2_*c*. This is done by inserting a new state named S2*b*_*c* between states S2_*c* and S3_*c*; state S2*b*_*c* increments the address counter and decrements the computation in the same way as state S3_*c*, but it does not load the accumulator register. Fig. 4.14 shows the datapath/FSM operation with the new S2*b*_*c* state producing the correct sum of $M[5] + M[6] = 3 + 11 = 14$.

4.4 FIRST-IN,FIRST-OUT BUFFER

Another type of embedded memory block is a first-in, first-out (FIFO) buffer, which is a synchronous RAM block that has additional logic to give it a specialized behavior. Figure 4.15 shows the concep-

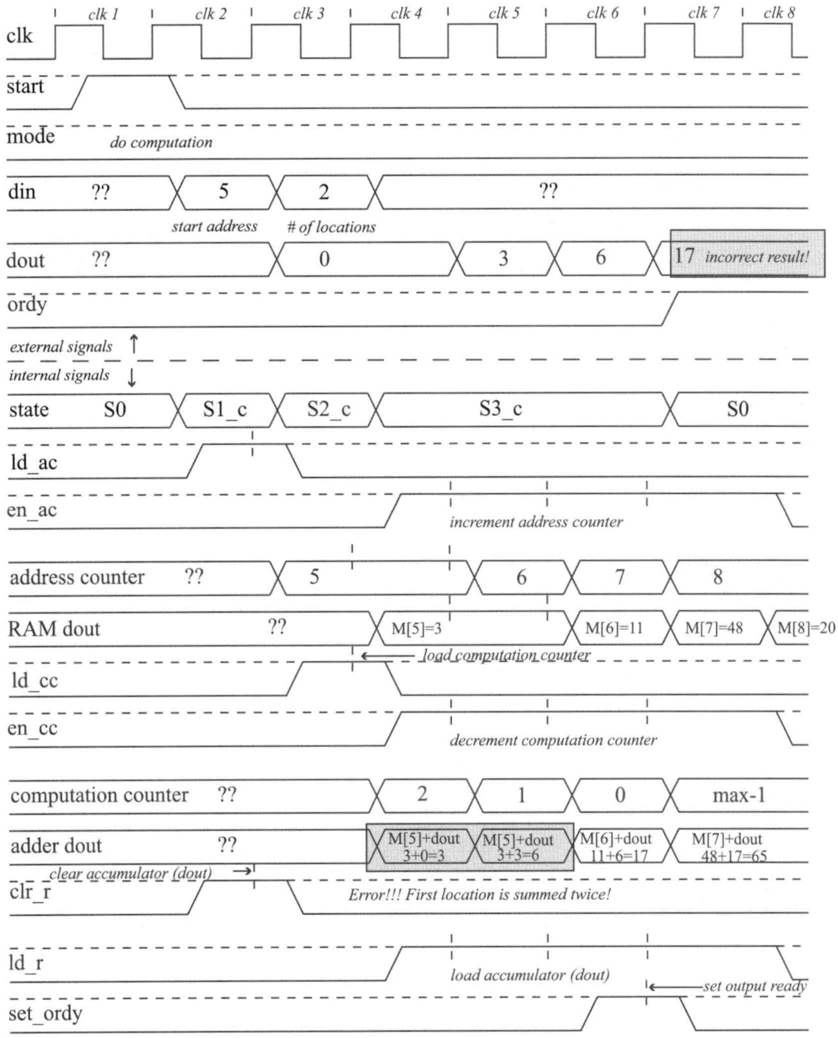

FIGURE 4.13: Sum operation (incorrect version).

tual operation for an eight-entry FIFO. A FIFO has a write port for placing data into the FIFO, and a read port for removing data from the FIFO. Figure 4.15a shows an empty eight-element FIFO. A write operation in Fig. 4.15b places *dataA* into the buffer, followed by a second write of *dataB* in Fig. 4.15c. Read operations in Fig. 4.15d and Fig. 4.15e first removes *dataA* and then *dataB*, thus illustrating FIFO nature of the buffer. Figure 4.15f shows a full FIFO after eight successive write operations.

Figure 4.16 provides two sample uses of a FIFO in a digital system. One common use is for buffering data from an external input channel as shown in Fig. 4.16a. Many input channels have the characteristic that data arrives in irregular bursts, and the individual data elements cannot always be processed by the digital system as they arrive, since the system may be busy with other tasks. The

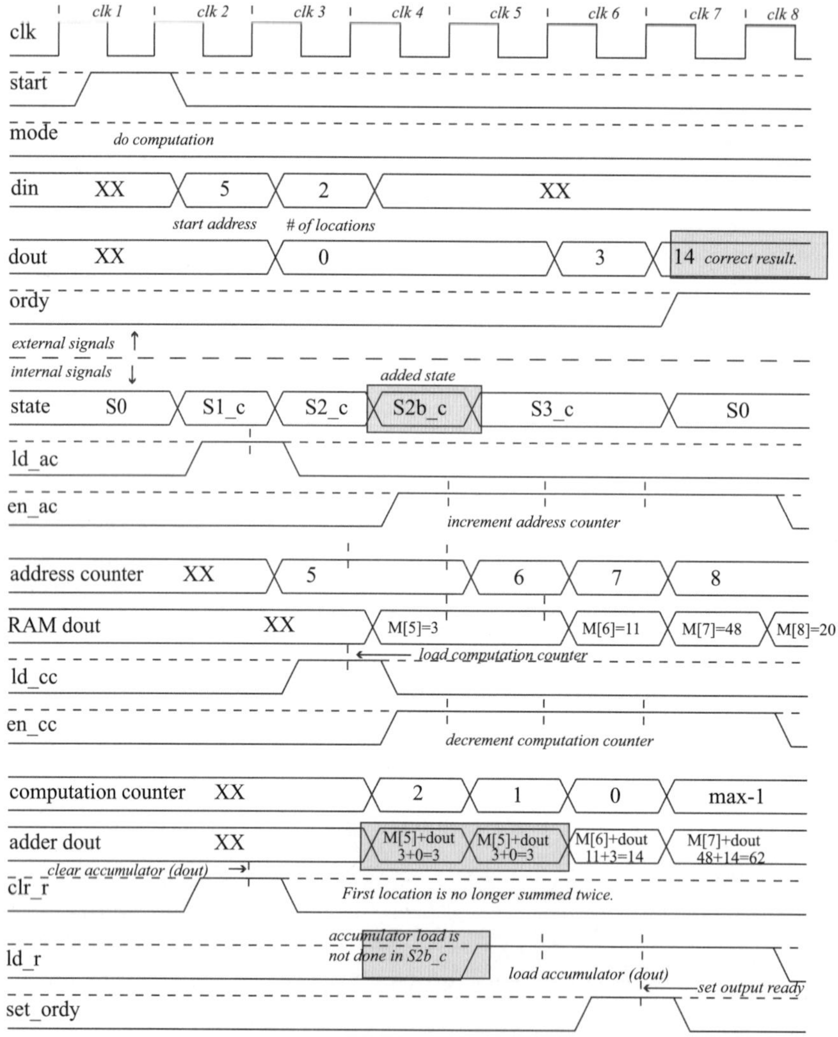

FIGURE 4.14: Sum operation (correct version).

FIFO holds the data until the system is ready for input processing. If handshaking signals are not used to regulate the data flow of the input channel, then the FIFO size is chosen to accommodate the maximum expected number of data elements to arrive between input processing tasks by the digital system.

Another typical FIFO usage is for data transfer between cooperating FSM/datapaths operating in different clock domains as shown in Fig. 4.16b. Data is written to the FIFO synchronized by clock domain A, and removed from the FIFO synchronized by clock domain B. Data transfer between two independent clock domains is an *asynchronous* transfer, that is, data can arrive at any time and is not synchronized to the receiver's active clock edge. This uncertainty in data arrival can cause

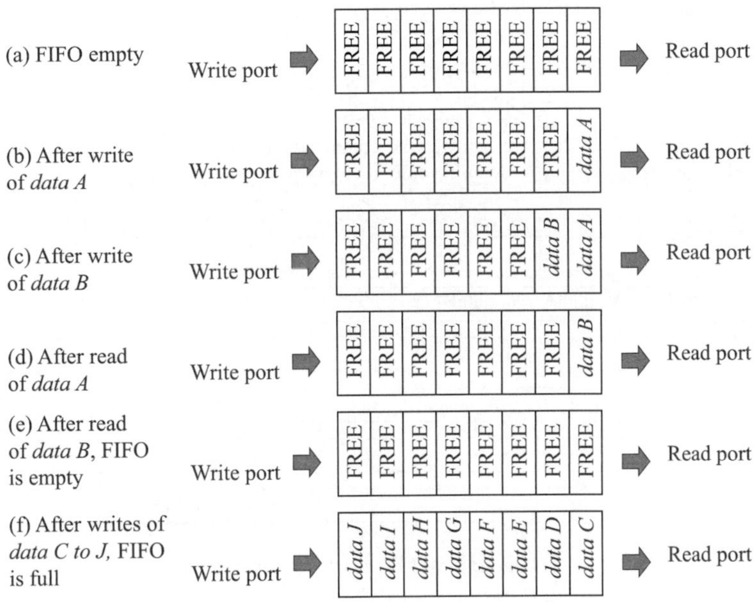

FIGURE 4.15: FIFO conceptual operation.

t_{su} and t_{hd} violations in the receiver's input register, resulting in a corrupted data transfer. A FIFO that supports independent read and write clocks is one method for solving this asynchronous data transfer problem.

The design of a FIFO with independent read/write clocks is challenging from a timing perspective, and is beyond the scope of this text, but FPGA vendors provide these as ready-to-use

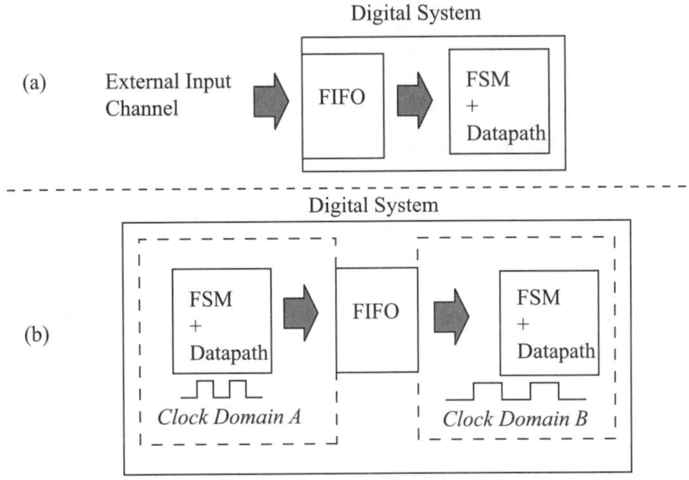

FIGURE 4.16: FIFO usage.

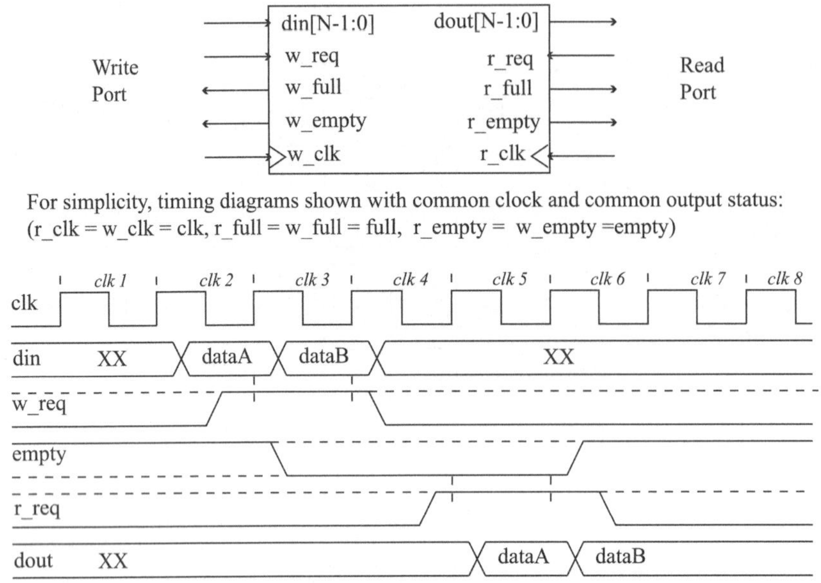

FIGURE 4.17: FIFO interface.

embedded memory blocks. Figure 4.17 shows a sample interface for a FIFO with independent read/write clocks. The write port consists of the write clock (w_clk), input data bus (din), write request input (w_req), empty status output (w_empty), and full status output (w_full). Data is written to the FIFO on the active edge of w_clk when the w_req input is asserted. The w_empty output is asserted when the FIFO is empty, and w_full is asserted when the FIFO is full, with transitions synchronized to the write clock. The read port consists of the read clock (r_clk), output data bus (dout), read request input (r_req), empty status output (r_empty), and full status output (r_full). Data is read from the FIFO on the active edge of r_clk when the r_req input is asserted. The timing diagram in Fig. 4.17 shows *dataA*, *dataB* written to an empty FIFO in clocks three and four, and data read from the FIFO in clocks five and six. Observe that the empty status output is negated after the write of *dataA* to the FIFO, and is asserted after the read of *dataB* from the FIFO. For simplicity, the timing diagrams assumes common clocks for the read and write ports. It must be noted that the timing details of FIFOs with independent read/write clocks can vary significantly from one FPGA vendor to another, and even between FPGA families of the same FPGA vendor. Thus, Fig. 4.17 is provided for example purposes only; the reader must consult the data sheets for FIFO blocks offered by a particular FPGA vendor when incorporating a FIFO into a digital system.

Some FIFO blocks have additional status signals named almost_empty and almost_full with configurable thresholds for these conditions. These signals are useful for assisting with controlling the data flow between the writing and reading digital systems. Two error conditions associated with FIFOs are:

- Writing to a full FIFO (input data is typically discarded). This condition is avoided by writing to the FIFO only when the `full` signal is negated.
- Reading from an empty FIFO (output data is unknown). This condition is avoided by reading from the FIFO only when the `empty` signal is negated.

In some FIFO implementations, the triggering of these error conditions may corrupt the internal FIFO status and produce erratic subsequent behavior, and error status signals (`read_error`, `write_error`) may be provided for system monitoring.

4.5 DUAL-PORT MEMORY

A dual-port memory has two ports, A and B, which support independent memory operations on each port. Figure 4.18 shows a typical interface for a dual-port memory. A dual-port memory that allows independent clocks for each port is sometimes referred to as a *true* dual-port memory.

Simultaneous operations to different memory locations have no timing constraints in relationship with each other. However, simultaneous operations to the same memory location will have timing constraints that vary by FPGA vendor. A typical specification for simultaneous access to the same memory location for a true dual-port memory is as follows:

- Simultaneous read access to the same location has no timing constraints.
- Simultaneous write operations to the same location produces unreliable data in that location.
- Simultaneous write and read operation to the same location produces correct data written to the location, but the read operation returns unreliable data.

The digital system designer using a dual-port memory is responsible for creating a system that avoids forbidden simultaneous operations. This usually involves external handshaking signals that coordinate access to the memory (the FIFO's empty/full signals fulfills this purpose in a FIFO design). Figure 4.19 shows two datapaths using a true dual-port memory and two handshaking signals, request (`req`) and acknowledge (`ack`), to send data from datapath A to datapath B. Figure 4.19a uses a two-phase protocol for accomplishing the data transfer; a change in the `req` signal indicates data availability from datapath A, with a corresponding change in the `ack` signal acknowledging receipt of the data by datapath B. In a two-phase protocol, data is transferred on each low-to-high transition of

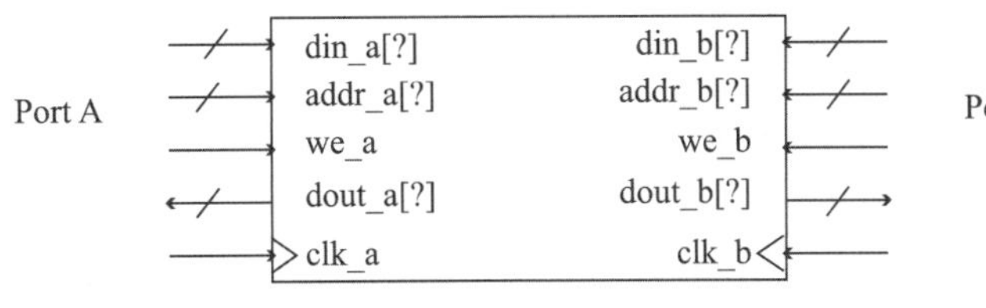

FIGURE 4.18: Dual-port memory.

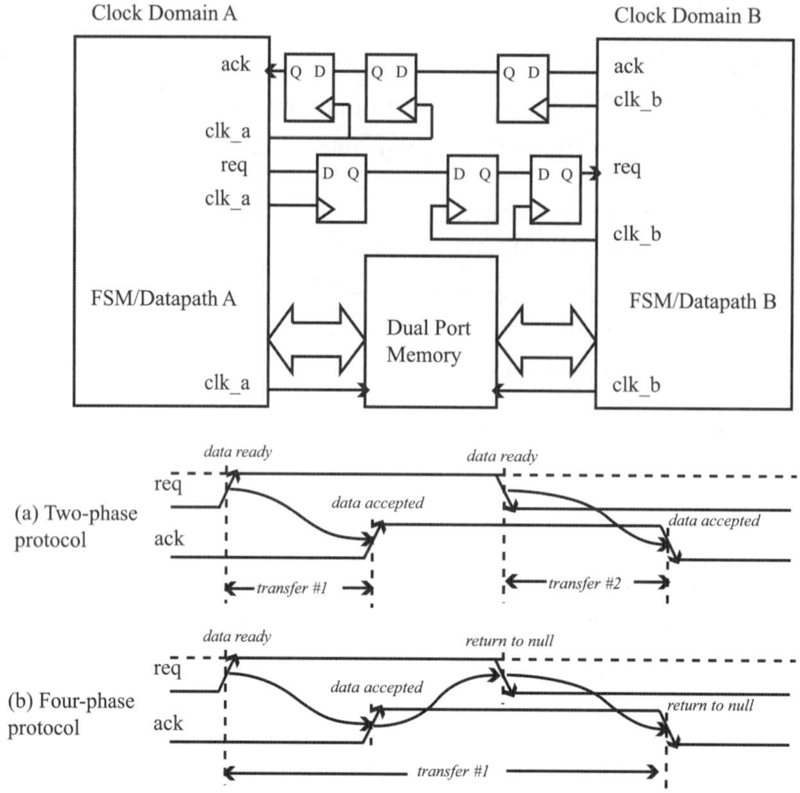

FIGURE 4.19: Dual-port memory use with handshaking.

the `req` line. A two-phase protocol requires changes in the `req` line to be detected, and is sometimes referred to as an edge-triggered or transition-sensitive protocol.

A four-phase protocol is used in Fig. 4.19b for accomplishing the data transfer; a logic one for `req` indicates data availability while a logic one for `ack` indicates data acceptance. Both the `ack` and `req` signals are negated (logic zero) before beginning a new data transfer. A four-phase protocol is referred to as a *level sensitive* protocol because the logic state of the handshaking signals indicate data availability and data acceptance.

Both four-phase and two-phase protocols can be readily expressed in modern HDLs. Some of the conventional pros/cons of two-phase versus four-phase protocols are as follows:

- A two-phase protocol requires more complex logic.
- A four-phase protocol maximizes signal transitions and thus energy consumed by those transitions.
- The return-to-null waiting period for the four-phase protocol may slow data transfers if the communication channel delay is long.

However, all of these pros/cons are technology and design dependent, with designer experience determining the protocol choice for a particular design.

The reader may question the necessity for using `req/ack` signals and instead want to indicate data availability by having datapath A write a nonzero value to a specified memory location being monitored by datapath B. This works only if the dual-port memory supports a simultaneous read during write operation to the same location, which is not the case for most true dual-port memories. It should be noted that if the two datapaths and the dual-port all share the same clock, then a simultaneous read during write operation to the same location is typically supported.

The advantages of a dual-port memory over a FIFO are that the dual-port allows bi-directional transfers between two datapaths and provides greater flexibility in data access. The disadvantage is that handshaking signals for avoiding forbidden simultaneous accesses may need to be provided by the designer.

4.6 ASIDE: SYNCHRONIZATION

In Fig. 4.19, the two DFFs clocked by `clk_a` on the `ack` input to datapath A and the two DFFs clocked by `clk_b` on the `req` input to datapath B are known as *two-flop synchronizers*. This is an accepted method for reducing the risk of an asynchronous input to a datapath input entering a *metastable* condition, in which the signal's voltage is stuck between a logic zero and logic one for an indeterminate period of time. A metastable condition can be triggered by a DFF's input failing to meet t_{su} and t_{hd} of the flip-flop. The probability of entering a metastable condition depends on many factors, some of which are:

- the internal design of the flip-flip
- the frequency at which the input signal changes
- the clock speed of the receiving system

A synchronizer is needed for any asynchronous input to a synchronous system. The reader is referred to [1] for a more complete discussion of metastability and synchronizer design.

In Fig. 4.19, the DFF clocked by `clk_b` on the `ack` output of datapath B and the DFF clocked by `clk_a` on the `req` output of datapath A are included to ensure that the `ack` and `req` outputs are glitch-free, that is, they only experience a single high-to-low or low-to-high transition during any clock period. These DFFs can be removed if these signals are already registered within the datapath. An FSM output signal that is generated by combinational gating using an FSM's state registers may experience glitches due to different delay paths through the logic gates. Because the `req` and `ack` outputs are asynchronous inputs to the receiving datapaths, these glitches could be treated as valid inputs, causing incorrect operation. If the two datapaths shared a common clock, then glitch-free outputs would not be needed because it is assumed that the outputs would be stable (satisfy t_{su} /t_{hd}) by the time the active clock edge occurred.

4.7 SUMMARY

This chapter has introduced the reader to commonly available embedded memory blocks found in modern FPGAs. Synchronous RAM blocks are preferred over asynchronous RAMs blocks because timing constraints for the designer are simplified when using synchronous RAM. Typical usage of RAM blocks requires counters to drive address lines, adding an extra clock cycle of latency from assertion of counter input to RAM output. FIFOs and dual-ports are useful for data exchange between datapaths that use different clock domains.

4.8 SAMPLE EXERCISES

1. Implement the datapath of Fig. 4.10 and ASM of Fig. 4.11 in the FPGA/HDL of your choice.
2. Modify the ASM of Fig. 4.11 to operate correctly if the registered `dout` output of the synchronous RAM of Fig. 4.10 is used instead of the unregistered `dout` output.
3. Compare the unregistered clock-to-dout time to the registered clock-to-dout time for an embedded memory block in an FPGA of your choice.
4. Using an FPGA of your choice, explore the timing characteristics for a FIFO that supports independent read and write clocks. Set the read clock to have 2/3 of the period of the write clock.
   a. How many read clock cycles does it take for the empty flag (read port side) to be negated when a write is performed?
   b. How many write clock cycles does it take for the empty flag to be asserted (write port side) when a read is performed that empties the FIFO?

   Repeat 4a, 4b with the read clock having 1/3 longer clock period than the write clock.
5. Using an FPGA of your choice, use an N-element FIFO with independent read/write clocks to create a design with the following characteristics:
   a. Set the FIFO size to be N-elements (your choice). Set the write clock to be 1/3 the period of the read clock.
   b. Create a write-side FSM that writes 2*N elements (use dummy data) to the FIFO at one write clock cycle per datum when a `start` input is asserted. Monitor the `full` signal to ensure that a write is not done to a full FIFO. Suspend writing if `full` is asserted; resume writing when `full` is negated. Halt operation when 2*N elements have been written to the FIFO.

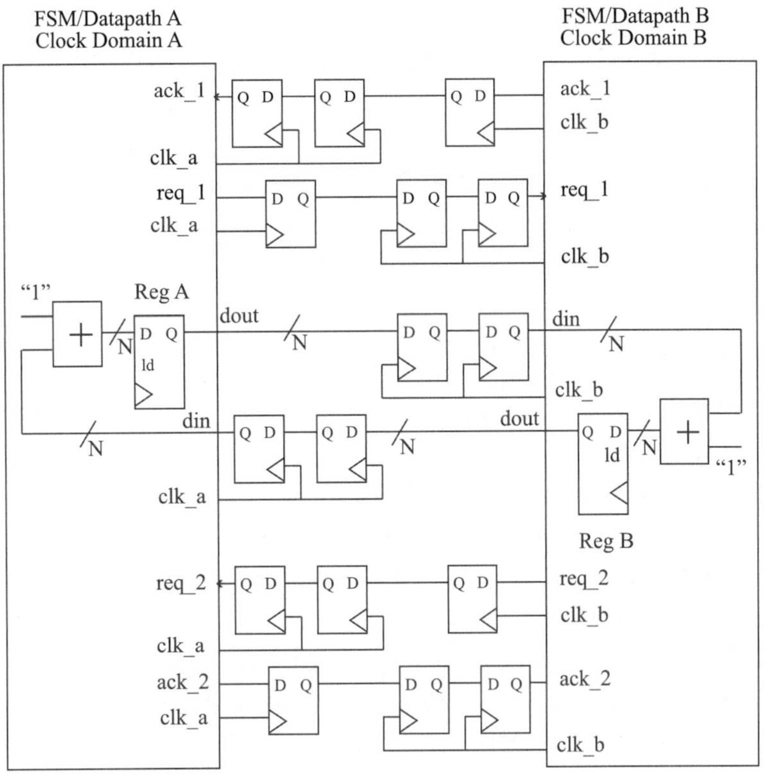

FIGURE 4.20: Asynchronous transfer.

c. Create a read-side FSM that removes elements from the FIFO whenever the `empty` signal is negated; remove data as fast as possible from the FIFO (one clock per datum). Ensure that your FSM does not attempt to read from an empty FIFO.

d. Change the read/write clocks such that the write clock has a 1/3 longer period than the read clock. Verify that your design performs as expected.

6. This problem refers to Fig. 4.20. Using four-phase handshaking and with datapath A clock 2/3 the period of datapath B, create FSMs for dapathpaths A/B that accomplish the following (steps *a* through *c* are FSM A operation, steps *d* through *f* are FSM B operation).
   a. After reset, FSM A initializes Register A to zero.
   b. FSM A then transmits the Reg A value to FSM B using the handshaking pair `req_1/ack_1` and its `dout` bus.
   c. FSM A then waits for a value to be transmitted back from FSM B on its din bus and using the handshaking pair `req_2/ack_2`. This new value is incremented by 'one' via the adder, and loaded into Reg A(at this point, FSM A loops through steps *b* and *c*, resulting in a continuously incrementing value being transmitted between FSM A and FSM B.)

d. After reset, FSM B initializes Register B to zero.

e. FSM B then waits for a value on its din bus to be transmitted from FSM A using the handshaking pair `req_1/ack_1`. This value is then incremented by 'one' via the adder, and loaded into Reg B.

f. FSM B then transmits the Reg B value to FSM A using the handshaking pair `req_2/ack_2` and its `dout` bus (at this point, FSM B loops through steps *e* and *f*, resulting in a continuously incrementing value being transmitted between FSM A and FSM B.).

7. Repeat problem #6 using two-phase handshaking.
8. Using the FPGA of your choice, create a dual-port memory design similar to Fig. 4.19 that has the following characteristics:

   a. Set the datapath A clock to be 1/3 the period of the datapath B clock. Use a four-phase handshake protocol to coordinate access to the dual-port.

   b. Using the initialization mode of Fig. 4.8 as a guide, have datapath A write the value N to location zero of the dual-port and then the data to be summed into locations one through $N + 1$. Once the dual-port has been initialized, have datapath A inform datapath B that data is ready to be summed through the handshaking protocol.

   c. Have datapath B read location zero to determine the N value, then sum the values in locations 1 through $N + 1$. Once datapath B is finished, use the handshaking protocol to inform datapath A that the data in the dual-port has been consumed, and then resume waiting for another data packet to be placed in the dual-port by datapath A.

9. Repeat problem #7 using the two-phase handshaking protocol.

4.9 PROJECT SUGGESTION

The latter part of Chapter 3 used a FIR digital filter to explore issues in datapath scheduling. The general form of an N-order FIR digital filter is:

$$y{=}x \times a0{+}x@1 \times a1{+}x@2 \times a2{+}\ldots\ldots{+}x@N \times aN \tag{4.1}$$

The x value represents the current input sample value, $x@1$ the input sample value from the previous sample period, $x@2$ the input sample value from two sample periods previously, etc. The filter coefficients $a0$, $a1$, ... aN determine the filter's performance characteristics such as low pass, high pass, band pass, etc. A JAVA applet that produces FIR filter coefficients given a filter specification is available at [2]. Typical results from the applet are given in Table 4.1.

This project's task is to build a fixed-point, *programmable* FIR filter that allows the filter order and coefficients to be dynamically loaded. As with the memory sum example of Section 4.3, the filter has an initialization mode in which the filter order and coefficients are loaded, and a

TABLE 4.1: FIR Filter Example

Rectangular window FIR filter, Filter type: Low Pass (LP), Order: 20	
Passband: 0 – 1000 Hz, Transition band: 368 Hz, Stopband attenuation: 21 dB	
Coefficients:	
a[0] = 0.00360104 (0x007)	a[11] = 0.230304 (0 x 01D7)
a[1] = 0.027779866 (0x038)	a[12] = 0.13769989 (0 x 011A)
a[2] = 0.032870565 (0x043)	a[13] = 0.03300727 (0 x 043)
a[3] = 0.009205259 (0x012)	a[14] = -0.03924712 (0 x FAF)
a[4] = −0.030985044 (0x0FC0)	a[15] = −0.057350047 (0 x F8A)
a[5] = −0.057350047 (0xF8A)	a[16] = −0.030985044 (0 x 0FC0)
a[6] = −0.03924712 (0xFAF)	a[17] = 0.009205259 (0 x 012)
a[7] = 0.03300727 (0x043)	a[18] = 0.032870565 (0 x 043)
a[8] = 0.13769989 (0x011A)	a[19] = 0.027779866 (0 x 038)
a[9] = 0.230304 (0x01D7)	a[20] = 0.00360104 (0 x 007)
a[10] = 0.26717955 (0x223)	

computation mode that accepts new input samples and produces a new output value for each input sample. Figure 4.21 gives the cycle specification for initialization mode, which is entered when `start` is asserted and `mode` is a logic one. The `start` input is negated when the last filter coefficient is entered.

In Fig. 4.22, computation mode is entered when `start` is asserted and `mode` is logic zero. The filter then waits for assertion of input ready (`irdy`), which indicates that a new sample value is present on the `din` input data bus. The filter asserts output ready (`ordy`) when the filter computation is finished and the `dout` data bus contains the final result. The filter then returns to waiting for the next assertion of `irdy`. Computation mode is exited when `start` is negated.

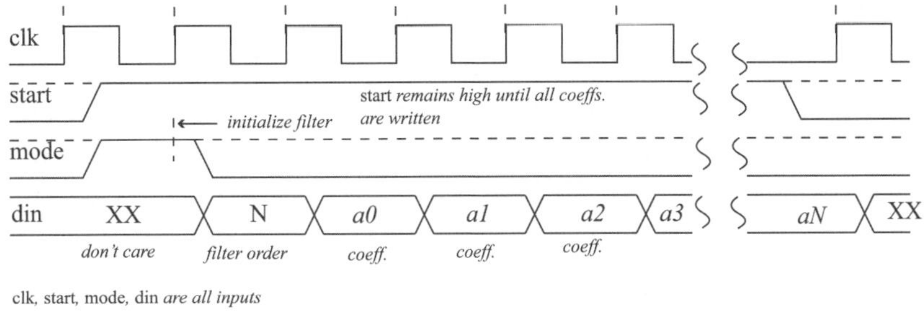

FIGURE 4.21: FIR filter initialization cycle specification.

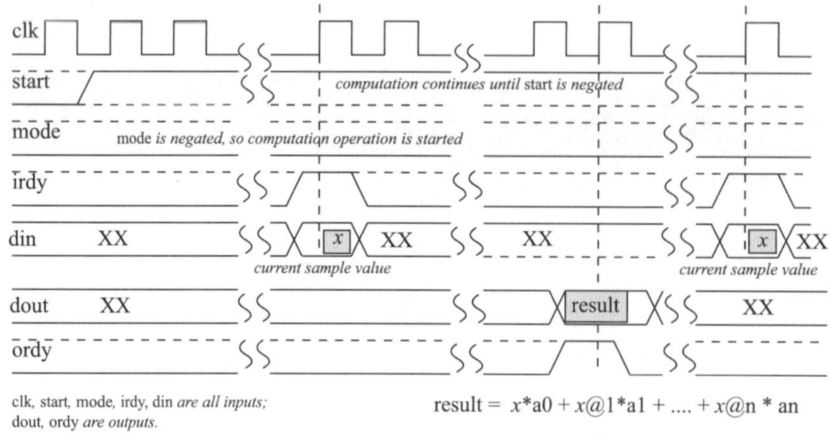

FIGURE 4.22: FIR filter computation cycle specification.

4.10 IMPLEMENTATION HINTS: SIGNED FIXED-POINT, EXAMPLE DATAPATH

The coefficients of Table 4.1 include negative values, so one choice for number representation is two's complement fixed-point representation (unsigned fixed-point number representation was explored in Chapter 3). Given N bits, two's complement represents the integer range $2^{N\text{-}1} - 1$ to $-2^{N\text{-}1}$. For example, 12-bit 2's complement represents the integer range +2047 to −2048. This range can be mapped to the number range (+1.0 to −1.0] by dividing each integer by $+2^{N\text{-}1}$. A fractional value in the range (+1.0 to −1.0] can be mapped to its binary value by multiplying it by $2^{N\text{-}1}$. The range (+1.0 to −1.0] is a good choice for a fixed-point digital filter implementation because the output of an unsigned N-bit analog-to-digital converter (ADC) that samples an analog input is easily converted to this range by subtracting $2^{N\text{-}1}$ from the ADC output code. The hex values given for the coefficients of Table 4.1 are the 12-bit two's complement representations calculated by multiplying each coefficient by 2048.

Fig. 4.23 shows an example datapath for implementing the programmable filter. Input samples are assumed to be two's complement 12-bit, mapped to the range (+1.0 to −1.0]. Two single-port RAMs are used for storing the coefficients and previous input samples.

The movement of the counters that address the sample and coefficient RAM during the calculation for a single input sample *x*0 is shown in Fig. 4.24. The coefficients are stored in the first $N + 1$ locations of the coefficient RAM, in order from *a*0 to *aN*. The $N + 1$ sample values used in a calculation (*x*0 through *xN*) are stored in the first $N + 1$ locations of the coefficient RAM, but the samples values are stored in decreasing memory locations from wherever the current sample *x*0 is stored (this is because arriving samples are stored in increasing memory addresses, so decreasing memory addresses contain *past* input samples).

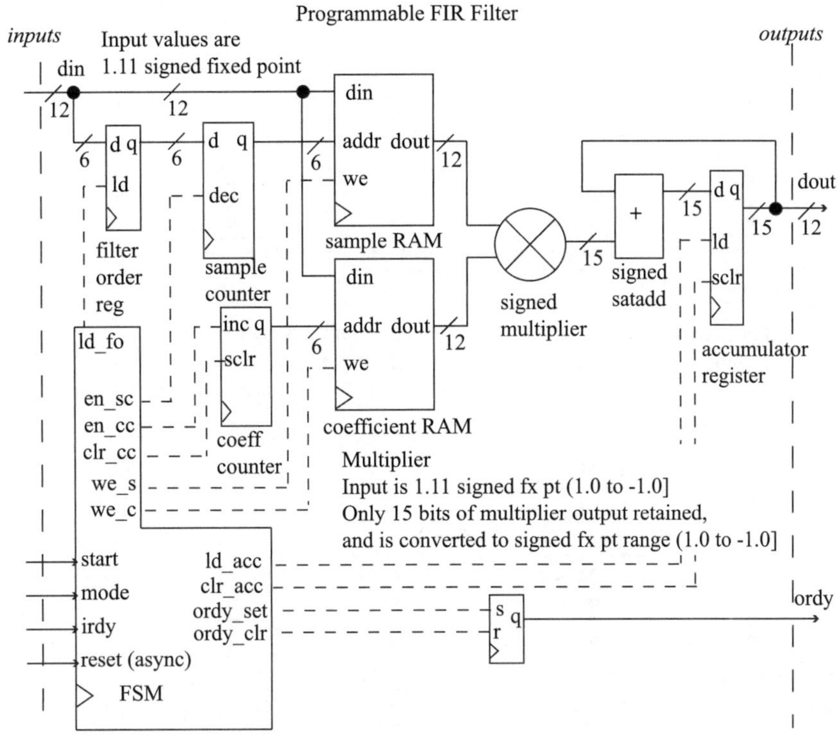

FIGURE 4.23: Sample datapath for FIR programmable filter.

Because the datapath contains only one multiplier and one adder, an FIR calculation for a new input sample requires at least $N + 1$ clocks. The multiplier is a *signed* multiplier, which is generally available as a building block from FPGA vendors. It was mentioned in Chapter 3 that a K-bit $\times K$-bit multiplier produces a $2K$-bit result. For unsigned fixed-point numbers mapped to the range (1.0 – 0.0], it was noted that the lower K-bits of the $2K$-bit product could be discarded, since these represented the K least significant bits, and the datapath size could be kept at K-bits.

However, what bits should be discarded for a signed K-bit $\times K$-bit multiplier using numbers in the range $(+1.0$ to $-1.0]$? One may intuit that it would also be the least significant K-bits, but the true answer is somewhat more complex. To illustrate, examine Eq. 4.2 that shows the multiplication of $+0.5 * -0.5$:

$$y = (+0.5) \times (-0.5) = -0.25 \qquad (4.2)$$

The numbers + 0.5, –0.5 mapped to 12-bit two's complement are + 0.5 * 2048 = 1024 = 0×400 and – 0.5*2048 = -1024 = 0×C00. The signed binary multiplication of Eq. 4.2 produces:

$$y = (0x400) \times (0xC00) = 0xF00000 \;\; (24 - \text{bit product}) \qquad (4.3)$$

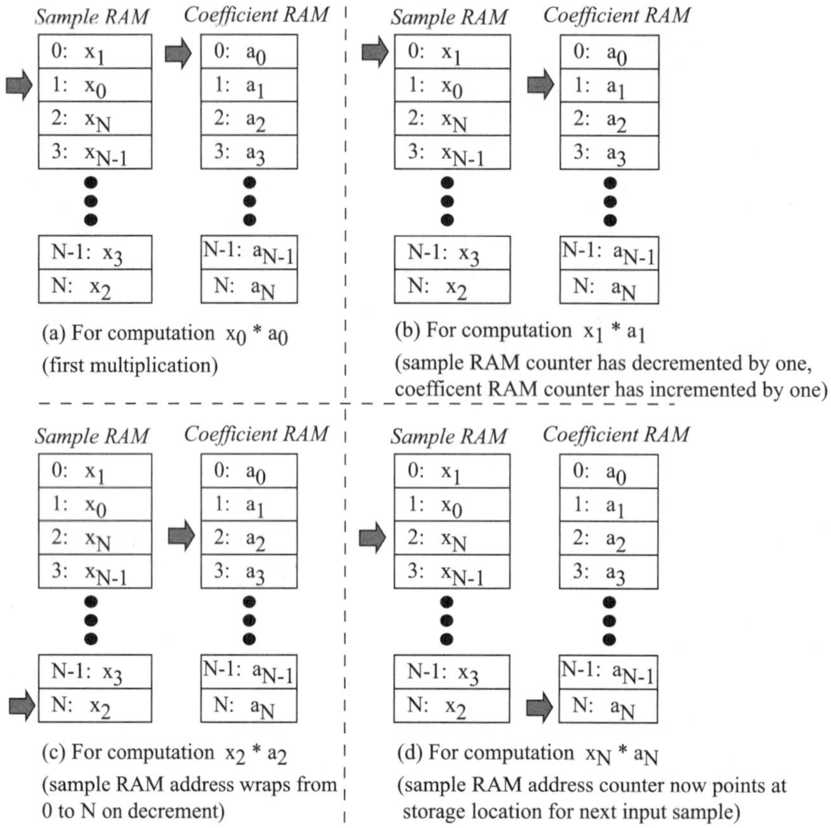

FIGURE 4.24: FIR computation.

Dropping the least significant 12-bits (last three hex-digits), the value 0 x F00 is equal to −256 as a 12-bit two's complement integer. Mapping −256 to the range (+ 1.0 to − 1.0] produces:

$$-256/2048 = -0.125 \tag{4.4}$$

which is one-half the expected value of − 0.25. Equation 4.5 shows the reason for this by examining the number range of the multiplication result:

$$(+1.0, -1.0] \times (+1.0, -1.0] = (+2.0, -2.0] \tag{4.5}$$

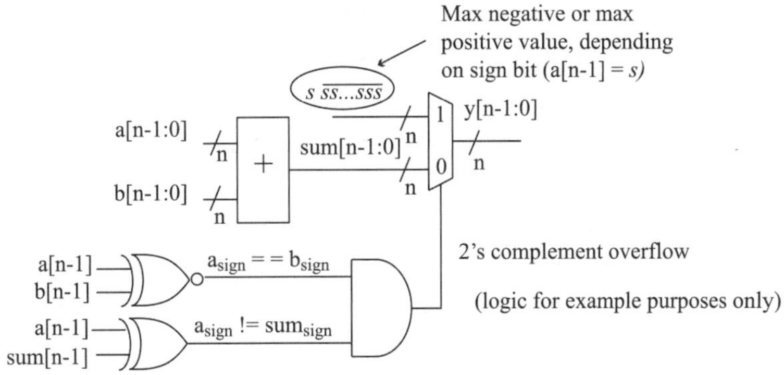

FIGURE 4.25: Two's complement saturating adder.

The multiplier output range has to be extended by an additional integer bit because the value + 1.0 is now included in the output range (because −1.0 * −1.0 = + 1.0). This means that the upper two bits of the 24-bit product are dedicated to the sign and integer portion of the result. This also has the unfortunate result that the output number range of (+2.0, − 2.0] is now different from the input number range of (+ 1.0, − 1.0]. The extra bit needed for the integer portion of the product to encode + 1.0 is wasted if the multiplier is never given the inputs of −1.0 * − 1.0. Because one of the multiplier inputs is always a coefficient, the coefficient choices can be restricted to not include −1.0. This means that actual range of values produced by the multiplier fall in the range (+1.0, − 1.0] and thus the most significant bit of the multiplier can be discarded. Note that discarding the most significant bit is the same as shifting the multiplier output to the left by one, which is multiplication by two. Multiplying the result of eq. 4.4 by two gives the expected result: −0.125 * 2 = − 0.25.

The datapath of Fig. 4.24 shows 15 bits of the 24-bit multiplier product being retained (nine bits are discarded). The bits discarded from the 24-bit product are the most significant bit, and the eight least significant bits. This gives three extra least significant bits for rounding purposes as the FIR sum is being accumulated. Only the most significant 12-bits of the accumulator register are used for the `dout` output result.

The adder shown in the datapath of Fig. 4.24 is a two's complement saturating adder, which saturates the output result to the maximum positive or maximum negative value if two's complement overflow occurs. Fig. 4.25 shows a conceptual implementation for a two's complement saturating adder (this logic works but more optimal implementations exist).

4.11 TESTING THE PROGRAMMABLE FILTER

One easy method of testing the filter is to apply an input sample of − 1.0, followed by zeros. This produces output values of − a0, − a1, − a2, − a3, . . . − aN, 0, 0, 0, etc. By implementing the FIR filter function in a programming language of choice, any arbitrary numerical input stream can be provided and the resulting output stream of the implementation is checked against expected results.

An optimum check is to provide a digitized sine wave of a particular frequency and observe the output to determine if the filter function (low-pass, high-pass, band-pass) is accomplished. The psuedo code in Listing 1 produces input values for one cycle of a sine wave for a given frequency f sampled at a frequency of S (the digital filter applet of [2] assumes a sample frequency of 8000 Hz).

Listing 4.1: PSUEDO-CODE FOR DIGITIZED SINE WAVE

```
        // f is sine wave frequency (Hz)
//S is sampling frequency of the filter (Hz)
for (t = 0, j = 0; j < (2 * π); t++, j = (t*f*2*π)/S) {
x = sin(j); //x is input sample value
}
```

Fig. 4.26 shows a sine wave input to a 20 tap LP FIR filter with a cutoff frequency of 100 Hz. The input sine wave has several cycles at 100 Hz (the edge of the pass band), followed by several cycles at 300 Hz (in the filter's transition band), followed by several cycles at 600 Hz (in the filter's stop band). The output waveform shows attenuation as the input waveform's frequency increases, which is expected for a low-pass filter.

4.12 FILTER IMPROVEMENTS

Many alternatives are possible for the example datapath shown in Fig. 4.23.

- The coefficients of N-order FIR filter are symmetric as seen in Table 4.1; a0 = aN, a1 = a($N - 1$), etc. The number of memory locations used in the coefficient RAM can be reduced from $N + 1$ to $(N/2) + 1$.
- The number of clock cycles required for producing the output given an input sample can be reduced by distributing the input samples and coefficients among multiple RAMs and including more multipliers and adders. This is the hardware resource versus computation time tradeoff examined in Chapter 3.
- The maximum clock period can be decreased at the cost of greater clock cycle latency by using the registered `dout` output of the RAM blocks and by placing a pipeline register between the multiplier and adder.
- Some FPGA vendors offer embedded RAM blocks that have built-in shift register functionality as required for digital filter implementations and could replace the counter logic that is currently used to access the RAMs.
- Some FPGA vendors offer library support for floating-point execution units; change the datapath from 12-bit fixed-point to single-precision floating-point.

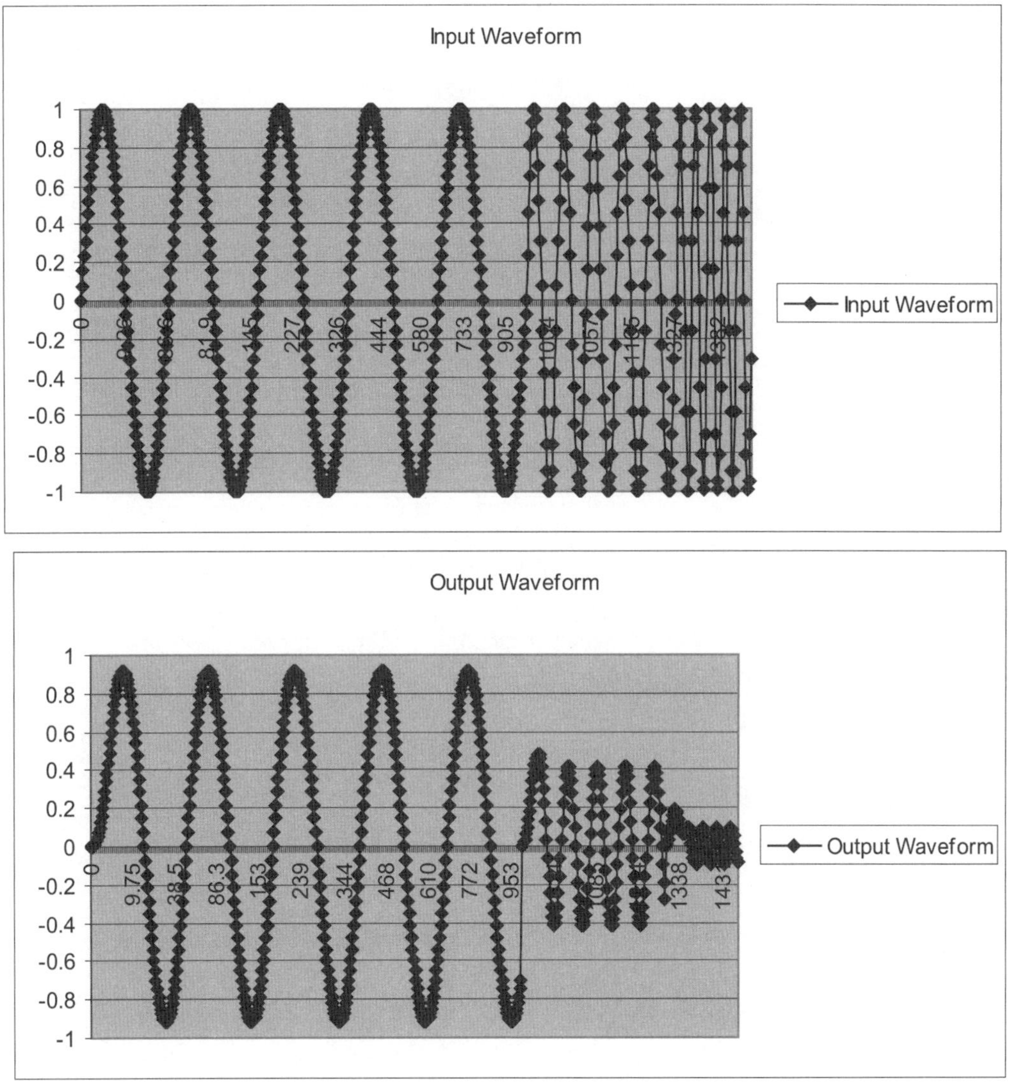

FIGURE 4.26: Filter input versus filter output.

4.13 REFERENCES

[1] R. Ginosar, "Fourteen ways to fool your synchronizer", *Proc. of the Ninth International Symposium on Asynchronous Circuits and Systems*, 12-15 May 2003, pp 89-96.

[2] FIR Digital Filter Design Applet, Online as of August 2007: http://www.dsptutor.freeuk.com/FIRFilterDesign/FIRFilterDesign.html.

Author Biography

Justin Stanford Davis received his Ph.D. in Electrical Engineering from the Georgia Institute of Technology in August 2003, as well as his M.S. and B.E.E. degrees in 1999 and 1997. During the summers of 1998 and 1999, he worked at Hewlett-Packard (now Agilent Technologies). In fall of 2003 he joined the faculty in the Department of Electrical Engineering at Mississippi State University as an Assistant Professor. In the summer of 2007 he joined Raytheon Missile Systems as a Senior Electrical Engineer. His research interests include digital design for high-speed systems, SoCs, and SoPs, as well as signal integrity and systems engineering.

Robert B. Reese received the B.S. degree from Louisiana Tech University, Ruston, in 1979 and the M.S. and Ph.D. degrees from Texas A&M University, College Station, in 1982 and 1985, respectively, all in electrical engineering. He served as a Member of the Technical Staff of the Microelectronics and Computer Technology Corporation (MCC), Austin, TX, from 1985 to 1988. Since 1988, he has been with the Department of Electrical and Computer Engineering at Mississippi State University, Mississippi State, where he is an Associate Professor. Courses that he teaches include VLSI systems and Digital System design. His research interests include self-timed digital systems and computer architecture.